Springer
Berlin
Heidelberg
New York
Barcelona
Budapest
Hong Kong
London
Milan
Paris
Santa Clara
Singapore
Tokyo

David M. Pollock   Robert F. Highsmith (Eds.)

Endothelin Receptors and Signaling Mechanisms

Springer

David M. Pollock, Ph.D.
Vascular Biology Center
Medical College of Georgia
Augusta, Georgia, U.S.A.

Robert F. Highsmith, Ph.D.
Department Molecular
and Cellular Physiology
University of Cincinnati
Medical Center Cincinnati,
Ohio, U.S.A.

ISBN: 3-540-64561-6 Springer-Verlag Berlin Heidelberg New York

Library of Congress Cataloging-in-Publication Data

Pollock, David M.; Highsmith, Robert F.
   Endothelin receptors and signaling mechanisms/ [edited by] David M. Pollock
      p. cm. — (Biotechnology intelligence unit)
   Includes bibliographical references and index.
   ISBN 1-57059-487-2 (alk. paper)
      1. Endothelins—Physiological effect. 2. Endothelins—Receptors. 3. Cellular signal
transduction I. David M. Pollock, 1955-. II. Highsmith, Rober F. III. Series.
      [DNLM: 1. Endothelins—Physiology. 2. Receptors, Endothelin—Physiology. 3. Nitric
Oxide—physiology. 4. Signal transduction—physiology. QU 68 E56545 1998]
QP55.2E538   1998
612'.015756—dc21
DNLM/DLC   98-10421
for Library of Congress   CIP

© Springer-Verlag Berlin Heidelberg and R.G. Landes Company, Georgetown, TX, U.S.A. 1998
Printed in Germany

The use of general descriptive names, registered names, trademarks, etc. in this publication does not imply, even in the absence of a specific statement, that such names are exempt from the relevant protective laws and regulations and therefore free for general use.

Product liability: The publisher cannot guarantee the accuracy of any information about dosage and application thereof contained in this book. In every individual case the user must check such information by consulting the relevant literature.

Typesetting: R.G. Landes Company, Georgetown, TX, U.S.A.

SPIN 10681662   31/3111 - 5 4 3 2 1 0 - Printed on acid-free paper

Many of us were taught that the vascular endothelium represented a cellophane lining that merely prevented the blood from making direct contact with surrounding tissues as a fundamental component of hemostasis. We now know that the powerful vasoconstrictor, endothelin, is synthesized and regulated within this thin layer of cells and plays an important role in a wide variety of functions such as the maintenance of vascular tone and growth. Along with nitric oxide, endothelin is one of the most widely investigated topics today and is the subject of great interest to a diverse group of scientists and clinicians. Within the pharmaceutical industry, scientists have worked at an unprecedented pace to identify novel receptor antagonists where they are already being clinically evaluated for the treatment of congestive heart failure, pulmonary hypertension and a variety of other vascular diseases. Despite this progress, basic scientists continue to work hard to discern the physiological role of endothelin not only in vascular but most organ systems including the kidney, lung, central and peripheral nervous system and others. Likewise, clinical scientists are still struggling to understand how this most powerful vasoconstrictor plays a role in the disease process. In only ten years since the endothelin isoforms were first purified and sequenced, our knowledge of the endothelin system has grown at such an extremely rapid rate, that staying current in the subject is difficult for the expert, much less the student, or investigator just beginning to explore the area.

This book represents an integration of our current knowledge of endothelin ligands, endothelin receptors, and the wide variety of cellular events that result from receptor activation. Topics range from the most basic biochemical interaction of endothelin ligands with their receptors to signals that reach the nucleus. It was our goal to put together a book that would provide a comprehensive overview of endothelin receptors and the biochemical and cellular events associated with endothelin action. Since the difficult task of assembling this information was too much for one individual to conquer, we had to call upon a distinguished cast of investigators, each an expert in a particular aspect of endothelin biology, to provide a detailed and expert analysis. To each of them, we give our most sincere thanks.

Finally, we would also like to acknowledge the efforts of Dr. Kristine Hickey who first discovered and initially characterized the endothelial-derived constricting factor that we now know as endothelin in her doctoral thesis of the early 1980s. Her work allowed all of us to benefit.

David M. Pollock, Ph.D.
Robert F. Highsmith, Ph.D.

CONTENTS

1. Introduction to the Endothelin System 1
   David M. Pollock

2. Endothelin Receptors and Receptor Antagonists 3
   Andrew S. Tasker and David M. Pollock
   ET Receptor Subtypes 3
   ET Receptor Antagonists 5
   Tissue Distribution of ET Receptors 10

3. Novel Endothelin Receptors 17
   Ponnal Nambi

4. "Sticky" Conundrums in the Endothelin System:
   Unique Binding Characteristics of Receptor Agonists
   and Antagonists 23
   Jinshyun R. Wu-Wong
   Introduction 23
   "Tenacity" of ET Binding 24
   Is Antagonist Binding Tenacious? 28
   Impact on Antagonist Potency 30
   Disparity Between In Vivo and In Vitro Potencies 31
   More to Consider: Interaction with Proteins 34
   Conclusions 36

5. Functional Coordination and Cooperation
   Between Endothelin and Nitric Oxide Systems
   in Vascular Regulation 41
   Michael S. Goligorsky and Joseph M. Winaver
   Introduction 41
   Functional Coupling Between ET and NO Systems 42
   ET-1 and NO in Hemodynamic Adaptations 51
   Pathophysiology of ET and NO Coordination 57
   Conclusions 61

6. Molecular and Structural Biology
   of Endothelin Receptors 67
   Maria L. Webb and Stanley R. Krystek Jr.
   Introduction 67
   Molecular Biology of ET_A and ET_B Receptors 67
   Molecular Structure 75
   Conclusion 81

7. **Cytokine Regulation of Endothelin Action** 89
   Timothy D. Warner
   Introduction 89
   Kinetics of Endothelin-1 Release 89
   Evidence for a Link Between Cytokines
      and ET-1 Production 90
   Conclusion 95

8. **Endothelin Modulation of Renal Sodium
   and Water Transport** 101
   Donald E. Kohan
   Introduction 101
   Endothelin Regulation of Water
      and Electrolyte Transport 102
   Physiologic Significance of ET-1 Regulation
      of Nephron Transport 108
   Conclusion 109

9. **Endothelin and Ion Channels** 115
   Tracy L. Keith and Robert F. Highsmith
   Ca^{2+} Mobilization: A Common Theme in ET-Signaling 115
   Actions of ET on Vascular Smooth Muscle 116
   Actions of ET in Endothelial Cells 121
   Actions of ET in the Kidney 122
   Actions of ET in Cardiac Muscle 123
   Actions of ET in Nonvascular Smooth Muscle 124
   Actions of ET in Other Cell Types 124
   Conclusions 125

10. **Endothelin and Calcium Signaling** 131
    E. Radford Decker and Tommy A. Brock
    Introduction 131
    Signal Transduction Mechanisms 131
    Endothelin Receptor-G-protein Coupling 132
    Cytosolic Calcium Signaling Mechanisms 134
    Protein Kinase C Signaling Mechanisms 137
    Calcium Sensitization of Contractile Proteins 138
    Summary 139

11. **Endothelin Regulation of Cardiac Contractility:
    Signal Transduction Pathways** 147
    Meredith Bond
    Introduction 147
    Inotropic Effects of ET in the Atrium and Ventricle 148
    $[Ca^{2+}]_c$ and the Positive Inotropic Effect of ET? 150
    PLC_{β} Dependent PI Turnover in the Positive Inotropic
       Response to ET 150

ET-dependent Intracellular Alkalinization via Activation
of the Na⁺ H⁺ Antiporter 152
ET-dependent PKC Phosphorylation
of Myofibrillar Proteins 153
Inhibitory Effects of ET Stimulation on the β-adrenergic
Pathway in Cardiac Myocytes:
Role of G$_i$ Dependent Pathways 154
ET-dependent Activation of PKC and Cross-talk
with Other Signaling Pathways 154

12. **Endothelin Signaling to the Nucleus:**
   Regulation of Gene Expression and Phenotype 163
   Michael S. Simonson
   Introduction 163
   Phenotypic Control by ET 164
   c-Src and Other PTKs in ET-1 Nuclear Signaling 167
   ET-1 Activates Nonreceptor PTKs Src
   and Focal Adhesion Kinase 169
   Src in Nuclear Signaling by ET-1 170
   Src PTKs in ET-1 Signal Transduction 171
   Conclusions 172

13. **Mechanisms of Endothelin-induced Mitogenesis**
   and Activation of Stress Response Protein Kinases 177
   Thomas L. Force
   Introduction 177
   Endothelin as a Mitogen 178
   Signal Transduction Mechanisms of Mitogenesis 181
   Tyrosine Kinase Signaling 194
   The Stress-activated MAP Kinase Signaling Cascades 199
   The Biology of SAPK and p38 Activation 202
   Conclusions 204

Color Insert 217

Index 221

EDITORS

David M. Pollock, Ph.D.
Vascular Biology Center
Medical College of Georgia
Augusta, Georgia, U.S.A.
Chapters 1,2

Robert F. Highsmith, Ph.D.
Department of Molecular and Cellular Physiology
Unversity of Cincinnati Medical Center
Cincinnati, Ohio, U.S.A.
Chapter 9

CONTRIBUTORS

Meredith Bond, Ph.D.
Department of Molecular Cardiology
The Cleveland Clinic Foundation
The Lerner Research Institute
Cleveland, Ohio, U.S.A.
Chapter 11

Tommy A. Brock, Ph.D.
Texas Biotechnology Corporation
Houston, Texas, U.S.A.
Chapter 10

E. Radford Decker, Ph.D.
Texas Biotechnology Corporation
Houston, Texas, U.S.A.
Chapter 10

Thomas L. Force, M.D.
Cardiac Unit
Massachusetts General Hospital East
Charlestown, Massachusetts, U.S.A.
Chapter 13

Tracy L. Keith, Ph.D.
Molecular and Cellular Physiology
Unversity of Cincinnati
Cincinnati, Ohio, U.S.A.
Chapter 9

Donald E. Kohan, M.D., Ph.D.
Department of Medicine,
 Division of Nephrology
University of Utah
 Health Science Center
Salt Lake City, Utah, U.S.A.
Chapter 8

Stanley R. Krystek Jr., Ph.D.
Department of Macromolecular
 Structure
Bristol-Meyers Squibb
Princeton, New Jersey, U.S.A.
Chapter 6

Michael S. Goligorsky, M.D., Ph.D.
Department of Medicine
SUNY Stony Brook
Stony Brook, New York, U.S.A.
Chapter 5

Ponnal Nambi, Ph.D.
SmithKline Beecham
Department of Renal
 Pharmacology
King of Prussia, Pennsylvania,
 U.S.A.
Chapter 3

Michael S. Simonson, M.D.
Department of Medicine,
  Division of Nephrology
Case Western Reserve University
Cleveland, Ohio, U.S.A.
Chapter 12

Andrew S. Tasker, Ph.D.
Amgen Inc.
Thousand Oaks, California, U.S.A.
Chapter 2

Timothy D. Warner, Ph.D.
The William Harvey
  Research Institute
St. Bartholomew's Hospital
  Medical College
Charterhouse Square
London, England
Chapter 7

Maria L. Webb, Ph.D.
Pharmacopeia Inc.
Cranbury, New Jersey, U.S.A.
Chapter 6

Joseph M. Winaver, M.D.
Faculty of Medcine
Haifa, Israel
Chapter 5

Jinshyun R. Wu-Wong, Ph.D.
Abbott Laboratories
Abbott Park, Illinois, U.S.A.
Chapter 4

CHAPTER 1

Introduction
to the Endothelin System

David M. Pollock

Vascular endothelial cells form a monolayer lining in all of the blood vessels of the circulation. Initially, they were thought to function as a crude filter allowing nutrients from the blood stream to diffuse through to the underlying tissues without letting proteins or blood cells escape. We now know that endothelial cells are important regulators of circulatory function, due in large measure to their recognized ability to synthesize and release many factors that regulate vascular smooth muscle tone. Endothelial-derived factors, including relaxing and contracting substances such as prostacyclin, nitric oxide (NO), and endothelin (ET), have been identified as important contributors in the regulation of vascular tone. Endothelial cells have a highly active metabolic function and are involved in clearing a number of agents from circulating blood. In addition, they have the enzyme that inactivates bradykinin and converts angiotensin I into the very potent pressor agent, angiotensin II (Ang II). Endothelial cells also generate various proteins like von Willebrand's factor, tissue plasminogen activator, growth promoting factors and lipids such as platelet activating factor. It is now clear that in addition to the regulation of vascular tone and hemodynamics, endothelial cells play a critical role in regulating growth and proliferative processes, inflammation and hemostasis.

Initially identified as the endothelium-derived constricting factor by Hickey and colleagues in 1985,[1] Yanagisawa et al set off an explosion of research when they purified and cloned a family of 21 amino acid peptides which they named endothelin.[2] Inoue and colleagues published their analysis of human genomic DNA that revealed the existence of three distinct genes encoding three ET isopeptides: ET-1, ET-2, and ET-3.[3] The only isoform thought to be constitutively released from endothelial cells is ET-1. ET-1 is synthesized as the result of a series of proteolytic cleavages of the initial gene product. A unique endothelin converting enzyme (ECE) is thought to remove the 18 amino acids from the C-terminus of the immediate precursor referred to as big ET-1.[4] Our knowledge is extremely limited in terms of tissue-specific localization of ECEs. In fact, the physiological relevance of several identified ECEs has yet to be established.

Endothelin production has been shown to be increased in a wide variety of vascular diseases and recently developed receptor antagonists are now being tested in clinical trials for use in the treatment of congestive heart failure, pulmonary hypertension and other disorders. Recent studies have shown that some of the

Endothelin Receptors and Signaling Mechanisms, edited by David M. Pollock and Robert F. Highsmith. © 1998 Springer-Verlag and R.G. Landes Company.

vascular effects of Ang II and inhibition of NO synthesis are actually mediated by ET-1 and that the activity of both of Ang II and ET-1 are regulated by NO and cyclooxygenase products.

ET-1 is the most active pressor substance yet discovered with a potency some ten times that of Ang II in most vascular beds. However, ET-1 is likely to be a local rather than a systemic regulating factor. Consistent with its putative role as an autocrine and paracrine factor, many believe that circulating levels of ET-1 represent spill-over from the local environment between the endothelium and vascular smooth muscle. There are several general observations that support this conclusion. ET-1 concentrations in plasma are below the levels thought to be necessary to activate endothelin receptors (< 5 pM). Plasma ET-1 is also degraded and rapidly cleared from plasma with a half-life on the order of a minute or so. Furthermore, endothelin-neutralizing antibodies have no hemodynamic effects. Therefore, plasma ET-1 concentrations at normal or low levels are difficult predictors of biological activity of ET-1 such that tissue concentrations and mRNA expression measurements of endothelin and its receptors are perhaps a more predictive means of studying endothelin activity.

The existence of multiple ET receptors was established soon after the identification of the ET peptides themselves. One line of evidence that suggested receptor subtypes were present was derived from the in vitro observations that ET-3 was less potent than ET-1 in producing vascular contraction yet these two peptides were equipotent in terms of producing vasodilation. Intravenous bolus infusion of ET peptides in vivo typically will produce a transient vasodilation followed by a prolonged hypertension. An ET-1-selective receptor, ET_A, was identified as being responsible for producing most, but not all, of the pressor effects of ET-1 while a nonisopeptide-selective receptor, ET_B, accounted for the vasodilatory response. Receptors have been cloned from mammalian cells and contain seven transmembrane domains of 20-27 hydrophobic amino acid residues, typical of the rhodopsin-type superfamily of G-protein coupled receptors.

The signaling mechanisms related to ET receptors have been widely studied, and although we have learned a great deal about these systems, there remains much to be discovered. The various chapters in this book will focus on our current state of knowledge of ET receptors, their function, and the wide variety of signaling and response mechanisms associated with the multifaceted actions of endothelin.

References

1. Hickey KA, Rubanyi GM, Paul RJ et al. Characterization of a coronary vasoconstrictor produced by cultured endothelial cells. Am J Physiol 1985; 248:C550-C556.
2. Yanagisawa M, Kurihara H, Kimura S et al. A novel potent vasoconstrictor peptide produced by vascular endothelial cells. Nature (London) 1988; 332:411-415.
3. Inoue A, Yanagisawa M, Takuwa Y et al. The human endothelin family: three structurally and pharmacologically distinct isopeptides predicted by three separate genes. Proc Natl Acad Sci USA 1989; 86:2863-2867.
4. Turner AJ, Murphy LJ. Molecular pharmacology of endothelin converting enzymes. Biochem Pharmacol 1996; 51:91-102.

CHAPTER 2

Endothelin Receptors and Receptor Antagonists

Andrew S. Tasker and David M. Pollock

ET Receptor Subtypes

ET_A and ET_B Receptor Subtypes

Multiple receptor subtypes for endothelin have been identified in pharmacological studies both in vitro and in vivo. As can be evidenced in the study by Yanagisawa et al, at least two receptor subtypes may be responsible for the biphasic nature of the hemodynamic response to intravenous injection of ET-1: a transient hypotension followed by a sustained increase in arterial pressure.[1] ET-1 and ET-2 are more active than ET-3 as pressor agents and as constrictors of many isolated vascular preparations.[2] ET-1 and ET-3 have similar potencies in terms of transient hypotensive effects that are thought to be mediated by the release of nitric oxide and/or prostaglandins.[3,4] These results led to the initial sub-classification of ET receptors as ET_A (ET-1-selective located on vascular smooth muscle) and ET_B (nonisopeptide-selective located on endothelial cells) which are generally responsible for vasoconstriction and vasodilation, respectively.

ET_A and ET_B receptors were first cloned from bovine and rat tissues, respectively.[5,6] Cells transfected with either ET_A or ET_B receptors possess binding characteristics consistent with the ET_A and ET_B classification. There is considerable homology (>90%) between human and bovine or rat ET_A receptors based on the DNA sequence while the ET_B receptor has slightly less, roughly 88%, sequence identity between human and rat. Within a given species, the degree of homology between ET_A and ET_B receptors is roughly 55%. Both ET_A and ET_B receptors are members of the G protein-coupled superfamily of receptors. Multiple effector pathways are activated and are the subject of many of the following chapters.

A variety of ligands have been widely used as pharmacological tools for characterizing endothelin receptors (see section on ET Receptor Antagonists below). Much of the receptor identification has been performed by contrasting the binding characteristics of ET-1 vs. ET-3 in membrane preparations extracted from various tissues or comparing potency of these isopeptides in vitro. The sarafotoxins are a family of snake venom peptides that have a high degree of sequence homology with the endothelins and are often used as endothelin agonists. Of particular

Endothelin Receptors and Signaling Mechanisms, edited by David M. Pollock and Robert F. Highsmith. © 1998 Springer-Verlag and R.G. Landes Company.

interest is sarafotoxin 6c which has a high affinity for the ET_B receptor and very little affinity for receptors of the ET_A subtype. Subsequent to the publishing of the DNA sequence of the ET_A and ET_B receptors, mRNA expression studies have helped in the localizing receptor subtypes (see section on Tissue Distribution of ET and its Receptors later in this chapter).

The ET_A/ET_B nomenclature has been advanced as a consensus view by the International Union of Pharmacology Committee on Receptor Nomenclature and Drug Classification.[7] However, since the development of receptor-selective and nonselective ligands, the two receptor model is no longer sufficient to explain the growing amount of biochemical and pharmacological evidence for additional receptor subtypes.

Subtypes of the ET_B Receptor: ET_{B_1} and ET_{B_2}

By virtue of their location on vascular smooth muscle cells, ET_A receptors are primarily responsible for producing vasoconstriction while ET-1-induced vasodilation can be attributed to ET_B receptor activation due to their location on vascular endothelial cells. This latter effect is primarily due to nitric oxide production but may be the result of prostacyclin synthesis as well. It is now well-established that non-ET_A receptors mediate at least some of the vasoconstrictor actions of ET-1. Some of the initial evidence for this mechanism comes from the early observations that ET_A receptor antagonists, such as BQ-123 and FR139317, are unable to prevent decreases in renal blood flow and glomerular filtration rate despite inhibiting the rise in mean arterial pressure produced by ET-1 infusion.[8] Experiments using isolated vascular preparations have also provided evidence to indicate that the renal vasculature and several other arterial and venous circulations contain ET_B receptors that produce vasoconstriction.[9-11] In addition, binding and mRNA expression studies have now determined that ET_B receptors exist in vascular smooth muscle. These receptors cannot be regarded as an insignificant component of the vascular response to ET-1 since infusion of ET_B agonists will elicit a hypertensive response in nearly every species examined.

The ET_B receptor responsible for non-ET_A-mediated vasoconstriction appears to represent a subtype of the ET_B receptor based on pharmacological evidence. The receptor present on vascular endothelium responsible for the release of nitric oxide (termed ET_{B_1}) has a unique pharmacological profile compared to the ET_B receptor that mediates vasoconstriction, termed ET_{B_2} (Fig. 2.1).[12,13] In isolated smooth muscle preparations, Warner et al observed that the "nonselective" antagonist, PD 142893, blocks ET_A-induced contractions and ET_B-induced relaxations, yet has no effect on non-ET_A contractions.[12] The ET_{B_1} and ET_{B_2} sub-classification is consistent with observations using membrane preparations from different brain regions that described "super-high" and "high" affinity ET_B receptor binding.[14] Additional evidence that the ET_{B_2} receptor on vascular smooth muscle has a lower affinity for ET-1 compared to the ET_{B_1} or ET_A receptor comes from the observation that ET_A receptor blockade completely prevents ET-1-induced decreases in renal blood flow at low doses but has no effect on higher doses of ET-1.[8,15] These results cannot be explained by insufficient blockade of ET_A receptors since the pressor response was completely prevented at both high and low doses of ET-1 and higher doses of ET_A antagonist yielded the same results.

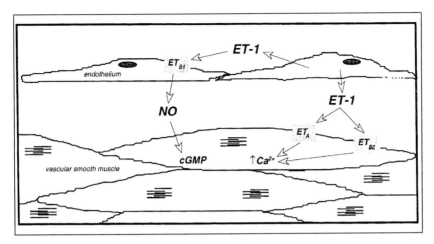

Fig. 2.1. General localization of endothelin receptor subtypes.

Despite the considerable pharmacological evidence to indicate the existence of ET_{B1} and ET_{B2} receptor subtypes, biochemical or molecular evidence has remained nonexistent for several years. In fact, RT-PCR and in situ hybridization experiments have revealed the presence of mRNA encoding both ET_A and ET_B receptors (the latter using cDNA from the endothelial ET_B receptor) in human vascular smooth muscle.[16] Therefore, the differences between the endothelium ET_B receptor, ET_{B1}, and the smooth muscle ET_B receptor, ET_{B2}, are either extremely minor or may be posttranslational in nature. There is some limited molecular evidence for variants of the ET_B receptor in human tissues. Shyamala et al recently published a study that described two distinct human ET_B receptors generated by alternative splicing from a single gene located within human brain, placenta, lung, and heart.[17] Similarly, Cheng et al discovered a novel cDNA encoding a nonselective ET receptor found in rat brain and possibly other tissues as well.[18] It has yet to be determined whether these findings could help explain some of the pharmacological characterization studies.

Evidence for subtypes of the ET_A receptor as well as an ET-3-selective receptor, ET_C, are the subject of chapter 3. The remainder of this chapter will focus on the development and characterization of ET_A and ET_B-selective antagonists and a brief review of the tissue distribution of these receptors.

ET Receptor Antagonists

Since its initial discovery there have been strong indications that an increase in the production of ET is a promoting factor in a number of diseases. This has led to an intense effort to identify inhibitors of the ET system and, in particular, to develop receptor antagonists. A plethora of agents exists. This review, however, will confine itself to a discussion of those used most widely, and to those which represent a particular advance in our understanding of receptor pharmacology. Many of these compounds have aided our investigation into the role of ET receptors in biological systems and in particular the receptor signaling mechanisms as discussed in subsequent chapters.

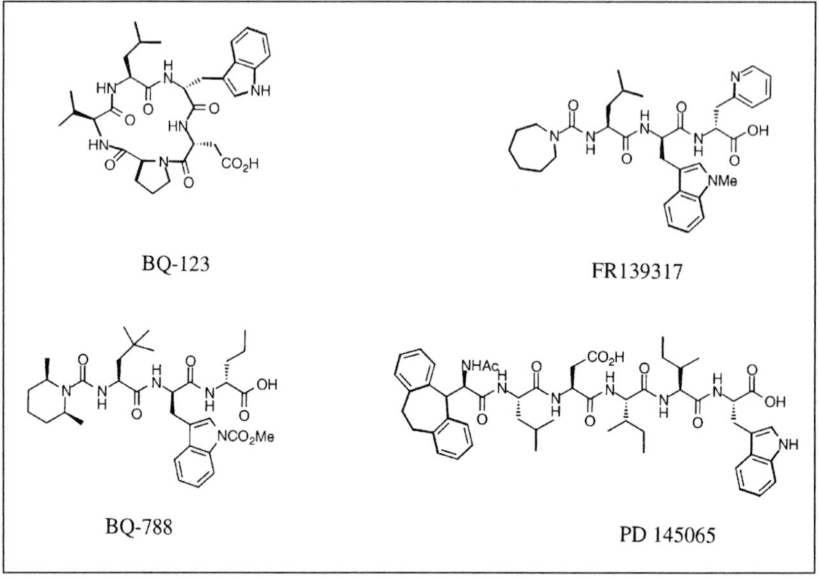

Fig. 2.2. Chemical structure of peptidic ET receptor antagonists.

Peptidic Antagonists

Discovered through screening of a streptomyces fermentation broth and subsequent ligand optimization, the most well known of the ET_A receptor selective antagonists is the cyclic pentapeptide BQ-123 shown in Figure 2.2.[19,20] The compound exhibits potency in the nanomolar range, and a thousand-fold selectivity over the ET_B receptor. Related is the linear pseudotripeptide FR139317 which is also a very potent and selective ET_A ligand (Table 2.1).[21] Numerous structural modifications of the pentapeptide resulted in the synthesis of the pseudotripeptide BQ-788.[22] The receptor selectivity is completely reversed yielding a compound with a 300-fold preference for ET_B. Extensive studies on the C-terminal tail of ET resulted in a series of hexapeptides typified by PD 145065.[23] These very potent compounds are the first reported nonselective or "mixed" antagonists.

Nonpeptidic Antagonists

The peptidic compounds all suffer limiting pharmacokinetics, but have proved extremely useful as pharmacological research tools. Driven by the need for antagonists possessing desirable pharmacokinetics, and preferably activity following oral administration, a new generation of nonpeptidic agents based around diarylsulfonamides was developed (Fig. 2.3). Ligand optimization of a lead generated by the random screening of a corporate compound library yielded Ro 46-2005.[24] This is the first reported orally bioavailable ET receptor antagonist. The compound behaves as a functional antagonist at both receptor subtypes equally, competitively blocking the contractile responses to ET-1 and sarafotoxin 6b in vitro. Tissue bath experiments with rat aortic rings yields a pA_2 of 6.50 for ET_A receptors and a pA_2 of 6.47 for ET_B receptors using rat small mesenteric arteries.[25] Further optimization of this series identified Ro 47-0203 (Bosentan). The compound dis-

Table 2.1 *Inhibition of ET-1 receptor binding by peptidic antagonists*

	Receptor Binding (μM)			
Antagonist	ET$_A$		ET$_B$	
	IC$_{50}$	Cell Source	IC$_{50}$	Cell Source
BQ-123	0.022	Porcine Aorta	18	Porcine Cerebellum
FR139317	0.0005	Porcine Aorta	4.6	Porcine Inner Medulla
BQ-788	0.280	Human Giradi Heart	0.001	Porcine Coronary Smooth Muscle
PD 145065	0.0005	Rabbit Renal Artery Vasc. Smooth Muscle	0.0007	Rat Cerebellar Tissue

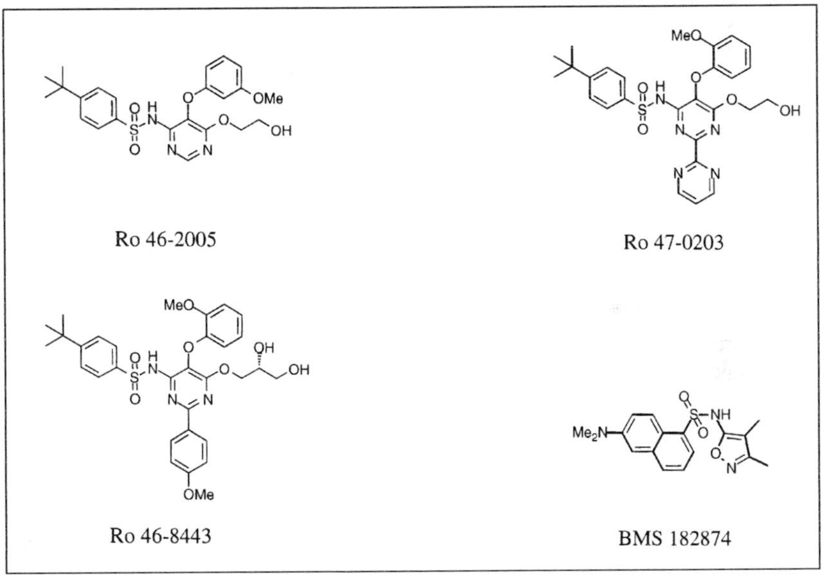

Fig. 2.3. Sulfonamide-based ET receptor antagonists.

plays both increased potency and selectivity for the ET$_A$ receptor (pA$_2$ = 7.2 in rat aorta using ET-1; pA2 = 6.0 in rat trachea using sarafotoxin 6c, an ET$_B$-selective agonist).[26]

The sulfonamidopyrimidine core of bosentan is seemingly a generic ET receptor antagonist: Altering the nature of the substituents in the 1- and 3-positions yields Ro 46-8443, an ET$_B$-selective functional antagonist (pA$_2$ = 5.7 ET$_A$ rat aorta with ET-1; pA2 = 7.1 ET$_B$ rat trachea with sarafotoxin 6c).[27]

Random screening of the Bristol-Myers Squibb compound collection identified an N-thiazolyl aminobenzenesulfonamide as a weak (69 μM) inhibitor of ET-1 binding to ET$_A$ receptors.[28] Further screening of related compounds identified a dimethylated analog that was approximately 90-fold more potent. Subsequent optimization yielded the naphthalene sulfonamide BMS 182874. This compound exhibits good binding to the ET_A receptor (K_i 55 nM) and additionally possesses high selectivity with a K_i greater than 200 μM in rat cerebellar membranes. The compound is effective as an oral agent and has been shown to reduce mean arterial pressure in the DOCA-salt rat model of hypertension.

In a modified approach, scientists at SmithKline Beecham amalgamated classical screening, molecular modeling, NMR structure studies and database methods to effect a similarity search of compounds known to be antagonists of other G-protein coupled receptors (Fig. 2.4).[29] This resulted in the identification of 1,3-diphenylindenecarboxylic acid. This compound binds selectively to ET_A receptors (K_i: ET_A 7.3 μM, ET_B > 30 μM). Overlays of the lead structure with ET-1 suggested analogy of the two pendent phenyl rings and the carboxylate with Tyr-13, Phe-14 and Asp-18 respectively. Sequential introduction of oxygen substituents on both rings to mimic that of Tyr-13, and the addition of a second carboxyl group to mimic the C-terminal Trp-21 resulted in the discovery of SB 209670 (K_i: ET_A 0.43 nM, ET_B 14.7 nM). It should be noted there follows increased affinity for the ET_B receptor on increasing the oxidation pattern and that all activity resides in the (+)-enantiomer. Intravenous administration of SB 209670 produces beneficial effects in a number of animal models of disease including acute renal failure, hypertension, radiocontrast medium-induced nephrotoxicity, and restenosis subsequent to percutaneous transluminal coronary angioplasty. The poor intestinal absorption of the diacid led to the development of an orally active analog SB 217242.

High throughput screening of a chemical library for compounds inhibiting [^{125}I]ET-1 binding in rabbit renal artery vascular smooth muscle cells (known to express only ET_A receptors) yielded the arylated hydroxybutenolide PD 012527 (IC_{50}: ET_A 2 μM; ET_B 5 μM; Fig. 2.5).[30] Guided by a Topliss decision tree protocol, ligand optimization resulted in the ring-chain tautomer sodium salt PD 156707 (Fig. 2.5). This compound is both very potent and selective for the ET_A receptor, binding to cloned human ET_A receptors expressed in LtK cells with an affinity of 0.3 nM. In cloned human ET_B receptors stably transfected into CHO-K1 cells, this agent binds with an affinity of 417 nM. Additionally the compound is reported to be 50% orally bioavailable in the rat.

Pharmacophore analysis of Ro 47-0203, BMS 182874, and a truncated SB 209670 yielded ABT-627 as shown in Figure 2.6.[31] The structural feature common to all these compounds is an acid functionality bounded by two aromatic rings. It was believed the indane core was merely a scaffold on which to append the two aromatic rings while maintaining a trans-trans relationship between the former and the seemingly indispensable carboxylic acid. In this series of compounds the pyrrolidine nitrogen served as a useful linking group to explore the receptor environment distal to the acid. This was not feasible in the indane series.

In common with the indanes, all the activity resides in the RRS (+) enantiomer. This was determined by X-ray crystallographic analysis of a phenylalanine-derived acyloxazolidinone derivative of an intermediate compound. It is a highly potent and selective functional antagonist (K_i for human ET_A receptor = 0.07 nM and 139 nM for human ET_B receptor), competitively blocking the ET-1-induced contraction in isolated rat aorta with a pA_2 of 9.2. Additionally the compound displays good pharmacokinetics with an oral bioavailability in rats of 70% and a half-life of 4.5 hours. The generic ET receptor antagonist core phenomenon observed

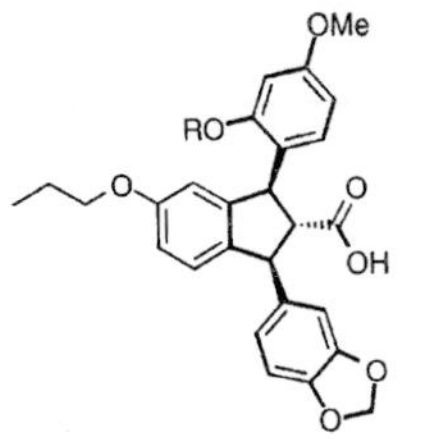

SB 209670 R = CH₂CO₂H   SB 217242 R = CH₂CH₂OH

Fig. 2.4. Indane-based ET receptor antagonists.

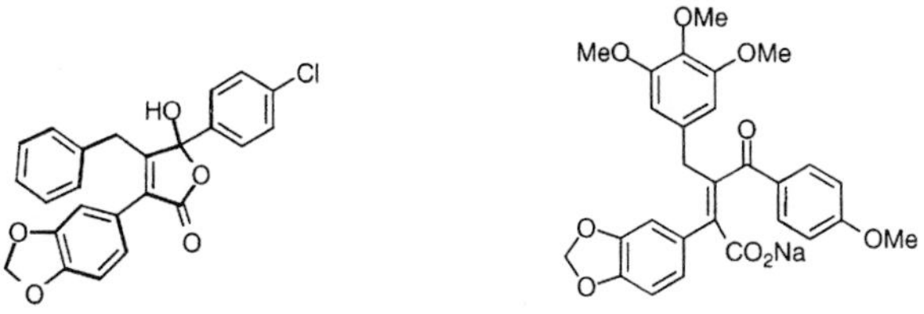

PD 012527   PD 156707

Fig. 2.5. Butenolide lead and ring-chain tautomer optimized ligand.

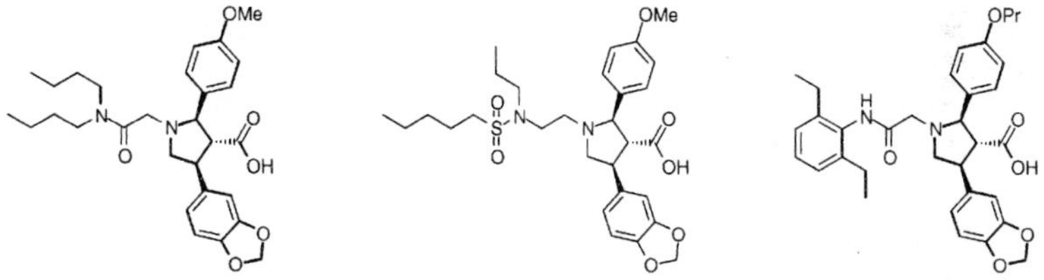

ABT-627   A-182086   A-192621

Fig. 2.6. ET$_A$-selective, nonselective and ET$_B$-selective pyrrolidine-based ET receptor antagonists.

with bosentan also holds true for the pyrrolidine core: Alkylation with subunits capable of orienting the alkyl side chains in the proper conformation yields potent agents that are either mixed antagonists or wholly ET$_B$-selective. The sulfonamide A-182086 is a potent antagonist of both receptors displaying IC$_{50}$s of 0.1 and 0.3 nM human receptors stably expressed in Chinese Hamster Ovary cells.[32] In contrast, the anilide A-192621 is 1000-fold selective functional antagonist of ET$_B$ as measured by inhibition of a downstream signaling event, phosphoinositide hydrolysis

(IC_{50} = 1100 nM for ET_A and 0.7 nM for ET_B; von Geldern et al, unpublished observations). In tissue bath studies the anilide yields a pA_2 of 5.7 in rat aorta using ET-1 (ET_A) and 7.1 in rat trachea using S6c (ET_B). Additionally both compounds display oral bioavailability.

In summary, since the discovery of the ET peptides in 1988 it has taken less than eight years to discover a wide variety of potent antagonists with desirable pharmacokinetic profiles. We are in the very fortunate position of being able to gain access to a portfolio of tools of all specifications—ET_A-selective, ET_B-selective, and nonselective.

Tissue Distribution of ET Receptors

Endothelin receptors have a widespread distribution and can be found within virtually every organ system. Considerable variability exists, however, in terms of the distribution of ET receptor subtypes between different vascular beds and among species. ET_A receptors are generally located within arterial vessels while ET_B receptors predominate on the low pressure side of the circulation.[33,34] Both ET_A and ET_B receptors can be found on a variety of other cell types aside from vascular endothelium and smooth muscle including nonvascular structures within the heart, kidney, lung, and nervous system.

Heart

Endothelin has a variety of effects on cardiac function including coronary vasoconstriction, mitogenesis, release of atrial natriuretic peptide (ANP), and inotropic effects. Endothelin can be synthesized by vascular endothelium, the conducting system, and cardiomyocytes. Plumpton et al showed that human myocardium expresses all three endothelin isoforms, ET-1, ET-2, and ET-3, based on RT-PCR, immunoreactivity, and RP-HPLC of tissue extracts.[35] Both ET_A and ET_B receptors have been identified in human atrial and ventricular myocardium, the atrioventricular conducting system, and endocardial cells.[36] ET_A receptors are found in greater numbers within the atria relative to ventricle and, in particular, the right atrium.[37-39] ET-1-induced release of ANP is mediated through ET_A receptors located in the atrial myocardium.[40,41]

In human coronary artery, several investigators have provided evidence for non-ET_A-induced vasoconstriction while others suggest that ET-1-induced constriction is predominantly mediated by the ET_A subtype.[42-45] Both ET_A and ET_B receptors have been identified on smooth muscle from human coronary vessels using RT-PCR, in situ hybridization, and membrane binding techniques.[16,46] However, the functional response attributed to non-ET_A receptors may be relatively weak compared to the degree of ET_A-mediated constrictor effects.[45,46]

Kidney

A wide variety of cell types within the kidney are capable of synthesis of ET-1 including endothelial (including glomerular and vasa recta capillaries), tubular, mesangial and interstitial cells. Tissue concentrations of ET-1 are greater in the renal medulla compared to the cortex.[47-50] Specific cellular localization studies reveal that the inner medullary collecting duct cells express the highest amount of ET-1 mRNA.[51,52]

A high density of binding sites for ET-1 have been identified in the inner medulla, inner stripe of the outer medulla, and glomeruli, with moderate to low levels of binding distributed throughout the outer cortex.[53,54] Both ET_A and ET_B receptors are found throughout the kidney although the role of each of these receptors in producing vasoconstriction has not been well established in the human renal circulation. Specific cellular localization has not been completely elucidated. In general, ET_A receptors are located on arterial smooth muscle while ET_B receptors are located within the inner medullary collecting duct.[53-55]

Lung

Pulmonary tissue probably contains the highest concentrations of ET-1 of any organ.[56] In addition to pulmonary vascular endothelial cells, ET-1 is produced by tracheal and bronchial epithelial cells.[57-59] Both ET_A and ET_B receptors have also been found throughout the pulmonary vasculature, airways and alveoli. Human pulmonary vessels contain primarily ET_A receptors while ET_B receptors predominate in the rabbit pulmonary artery.[60] Human airways contain both ET_A and ET_B receptor sites in airway smooth muscle and alveoli although the great majority within airway smooth muscle appear to be of the ET_B subtype.[61] ET-1 has been shown to be a more potent constrictor of pulmonary veins than arteries.[62-64]

Central and Peripheral Nervous System

Endothelin and its receptors have been identified throughout the central and peripheral nervous system although very little is known about the functional role of endothelin in nervous system. ET-like immunoreactivity has been observed in the paraventricular and supraoptic nuclei of the hypothalamus, glia and neurons of the cerebellum, the hippocampus, granular cerebellum, motor neurons in the spinal cord and peripheral ganglia.[65-69] The predominant endothelin isoform within the brain is thought by many to be ET-3 particularly within the hypothalamus and cerebellum which could explain why the ET_B receptor is the predominant subtype in these areas.[56,66,69] Within the cerebral circulation, ET_A receptors have also been identified on smooth muscle.[71,72] A novel finding with regard to ET receptors is the observation that ET_A receptors have been identified on human endothelial cells derived from the cerebral microcirculation.[73,74] The function of these receptors has not been elucidated. Endothelin and endothelin receptors have also been identified in the gastrointestinal tract as well as tissues responsible for reproductive and endocrine function which, because of space limitations, will not be reviewed here.

References

1. Yanagisawa M, Kurihara H, Kimura S et al. A novel potent vasoconstrictor peptide produced by vascular endothelial cells. Nature (London) 1988; 332:411-415.
2. Inoue A, Yanagisawa M, Takuwa Y et al. The human endothelin family: three structurally and pharmacologically distinct isopeptides predicted by three separate genes. Proc Natl Acad Sci USA 1989; 86:2863-2867.
3. Warner TD, Mitchell JA, de Nucci G et al. Endothelin-1 and endothelin-3 release EDRF from isolated perfused arterial vessels of the rat and rabbit. J Cardiovasc Pharmacol 1989; 13:S85-S88.
4. Gardiner SM, Compton AM, Bennett T. Effects of indomethacin on the regional haemodynamic responses to low doses of endothelins and sarafotoxin. Br J Pharmacol 1990; 100:158-162.

5. Arai H, Nori S, Aramori I et al. Cloning and expression of a cDNA encoding an endothelin receptor. Nature (London) 1990; 348:730-732.

6. Sakurai T, Yanagisawa M, Takuwa Y et al. Cloning of a cDNA encoding a nonisopeptide-selective subtype of the endothelin receptor. Nature (London) 1990; 348:732-735.

7. Masaki T, Vane JR, Vanhoutte PM. V. International Union of Pharmacology Nomenclature of Endothelin Receptors. Pharmacol Rev 1994; 46:137-142.

8. Pollock DM, Opgenorth TJ. Evidence for endothelin-induced renal vasoconstriction independent of ET_A receptor activation. Am J Physiol 1993; 264, R222-R226.

9. Bigaud M, Pelton JT. Discrimination between ET_A- and ET_B-receptor-mediated effects of endothelin-1 and [Ala^{1,3,11,15}]endothelin-1 by BQ-123 in the anesthetized rat. Br J Pharmacol 1992; 107:912-918.

10. Clozel M, Gray GA, Breu V et al. The endothelin ET_B receptor mediates both vasodilation and vasoconstriction in vivo. Biochem Biophys Res Comm 1992; 186:867-873.

11. Cristol J-P, Warner TD, Thiemermann C et al. Mediation via different receptors of the vasoconstrictor effects of endothelins and sarafotoxins in the systemic circulation and renal vasculature of the anaesthetized rat. Br J Pharmacol 1993; 108:776-779.

12. Warner TD, Allcock GH, Corder R et al. Use of the endothelin antagonists BQ-123 and PD 142893 to reveal three endothelin receptors mediating smooth muscle contraction and the release of EDRF. Br J Pharmacol 1993; 110:777-782.

13. Gray GA, Clozel M. Three endothelin receptor subtypes suggested by the differential potency of bosentan, a novel endothelin receptor antagonist, in isolated tissues. Br J Pharmacol 1994; 112:U62.

14. Sokolovsky M, Ambar I, Galron R. A novel subtype of endothelin receptors. J Biol Chem 1992; 267:20551-20554.

15. Pollock DM, Opgenorth TJ. ET_A receptor-mediated responses to endothelin-1 and big endothelin-1 in the rat kidney. Br J Pharmacol 1994; 111:729-732.

16. Davenport AP, O'Reilly G, Molenaar P et al. Human endothelin receptors characterized using reverse transcriptase-polymerase chain reaction, in situ hybridization, and subtype-selective ligands BQ-123 and BQ-3020: Evidence for expression of ET_B receptors in human vascular smooth muscle. J Cardiovasc Pharmacol 1993; 22(suppl. 8):S22-S25.

17. Shyamala V, Moulthrop TH, Stratton-Thomas J et al. Two distinct human endothelin B receptors generated by alternative splicing from a single gene. Cell Mol Biol Res 1994; 40:285-296.

18. Cheng HF, Su YM, Yeh JR et al. Alternative transcript of the nonselective-type endothelin receptor from rat brain. Mol Pharmacol 1993; 44:533-538.

19. Ihara M, Fukuroda TM, Saeki T et al. An endothelin receptor (ET_A) antagonist isolated from streptomyces misakeienis. Biochem Biophys Res Comm 1991; 178:132-137.

20. Ishikawa K, Fukami T, Nagase T et al. Cyclic pentapeptide endothelin antagonists with high selectivity. Potency- and solubility-enhancing modifications. J Med Chem 1992; 35:2139-2142.

21. Sogabe K, Nirei H, Shoubo M et al. Pharmacological profile of FR139317, a novel, potent endothelin ET_A receptor antagonist. J Pharmacol Exp Therap 1993; 264:1040-1046.

22. Ishikawa K, Ihara M, Noguchi K et al. Biochemical and pharmacological profile of a potent and selective endothelin B-receptor antagonist, BQ-788. Proc Natl Acad Sci 1994; 91:4892-4896.

23. Cody WL Doherty AM, He JX et al. The rational design of a highly potent combined ET_A and ET_B receptor antagonist (PD 145065) and related analogs. Med Chem Res 1993; 3:154-162.

24. Clozel M, Breu V, Burri K et al. Pathophysiological role of endothelin revealed by the first orally active endothelin receptor antagonist. Nature (London) 1993; 365:759-761.

25. Breu V, Loffler, B-M, Clozel M. In vitro characterization of Ro 46-2005, a novel synthetic nonpeptide endothelin antagonist of ET_A and ET_B receptors. FEBS Lett 1993; 334:210.

26. Clozel M, Breu V, Gray GA et al. Pharmacological characterization of bosentan, a new potent orally active nonpeptide endothelin receptor antagonist. J Pharmacol Exp Therap 1994; 270:228-235.

27. Breu V, Clozel M, Burri K et al. In vitro characterization of Ro 46-8443, the first nonpeptide antagonist selective for the endothelin ET_B receptor. FEBS Lett. 1996; 383:37-41.

28. Stein PD, Hunt JT, Floyd DM et al. The discovery of sulfonamide endothelin antagonists and the development of the orally active ET_A antagonist 5-(dimethylamino)-N-(3,4-dimethyl-5-isoxazolyl)-1-napthalene-sulfonamide (BMS-182874). J Med Chem 1994; 37:329-331.

29. Elliott JD, Lago AM, Cousins RD et al. 1,3-Diarylindan-2-carboxylic Acids: Potent and Selective Non-Peptide Endothelin Receptor Antagonists. J Med Chem 1994; 37:1553-1557.

30. Reynolds EE, Keiser JA, Haleen SJ et al. Pharmacological characterization of PD-156707, an orally-active ET_A receptor antagonist. J Pharmacol Exp Therap 1995; 273:1410-1417.

31. Winn M, von Geldern TW, Opgenorth TJ et al. 2,4-diarylpyrrolidine-3-carboxylic acids—Potent ET_A selective endothelin receptor antagonists. 1. Discovery of A-127722. J Med Chem. 1996; 39:1039-1048.

32. Jae HS, Winn M, Douglas B et al. Pyrrolidine-3-carboxylic acids as endothelin antagonists. 2. Sulfonamide-based ET_A/ET_B mixed antagonists. J Med Chem 1997; in press.

33. Miller VM, Komori K, Burnett Jr JC et al. Differential sensitivity to endothelin in canine arteries and veins. Am J Physiol 1989; 257:H1127-H1131.

34. Moreland S, McMullen D, Abboa-Offei B et al. Evidence for a different location of vasoconstrictor endothelin receptors in the vasculature. Br J Pharmacol 1994; 112:704-708.

35. Plumpton C, Champeney R, Ashby MJ et al. Characterization of endothelin isoforms in human heart: endothelin-2 demonstrated. J Cardiovasc Pharmacol 1993; 22(suppl. 8):S26-S28.

36. Molenaar P, O'Reilly G, Sharkey A et al. Characterization and localization of endothelin receptor subtypes in the human atrioventricular conducting system and myocardium. Circ Res 1993; 72:526-538.

37. Bax, WA, Bruinvels AT, van Suylen R-J et al. Endothelin receptors in the human coronary artery, ventricle and atrium. A quantitative autoradiographic analysis. Naunyn-Schmeid Arch Pharmacol 1993; 348:403-410.

38. Davenport AP, Morton AJ, Brown MJ. Localization of endothelin-1 (ET-1), ET-2, and ET-3, mouse VIC, and sarafotoxin S6b binding sites in mammalian heart and kidney. J Cardiovasc Pharmacol 1991; 17(suppl. 7):S152-S155.

39. Hemsen A, Franco-Cereceda A, Matran R et al. Occurrence, specific binding sites and functional effects of endothelin in human cardiopulmonary tissue. Eur J Pharmacol 1990; 191:319-328.

40. Thibault G, Doubell AF, Garcia R et al. Endothelin-stimulated secretion of natriuretic peptides by rat atrial myocytes is mediated by endothelin A receptors. Circ Res 1994; 74:460-470.

41. Williams Jr DL, Jones KL, Pettibone DJ et al. Sarafotoxin S6c: An agonist which distinguishes between endothelin receptor subtypes. Biochem Biophys Res Comm 1991; 175:556-561.

42. Bax WA, Aghai Z, van Tricht CLJ et al. Different endothelin receptors involved in endothelin-1- and sarafotoxin S6B-induced contractions of the human isolated coronary artery. Br J Pharmacol 1994; 113:1471-1479.

43. Godfraind T. Evidence for heterogeneity of endothelin receptor distribution in human coronary artery. Br J Pharmacol 1993; 110:1201-1205.

44. Opgaard OS, Adner M, Gulbenkian S et al. Localization of endothelin immunoreactivity and demonstration of constrictory endothelin-A receptors in human coronary arteries and veins. J Cardiovasc Pharmacol 1994; 23:576-583.

45. Davenport AP, Maguire JJ. Is endothelin-induced vasoconstriction mediated only by ET_A receptors in humans? Trends Pharmacol Sci 1994; 15:9-11.

46. Davenport AP, O'Reilly G, Kuc RE. Endothelin ET_A and ET_B mRNA and receptors expressed by smooth muscle in the human vasculature: majority of the ET_A subtype. Br J Pharmacol 1995; 114:1110-1116.

47. Karet FE, Kuc RE, Davenport AP. Novel ligands BQ-123 and BQ-3020 characterize endothelin receptor subtypes ET_A and ET_B in human kidney. Kidney Int 1993; 44:36-42.

48. Kitamura K, Tanaka T, Kato J et al. Immunoreactive endothelin in rat kidney inner medulla: marked decrease in spontaneously hypertensive rats. Biochem Biophys Res Comm 1989; 162:38-44.

49. Morita S, Kitamura K, Yamamoto Y et al. Immunoreactive endothelin in human kidney. Ann Clin Biochem 1991; 28:267-271.

50. Wilkes BM, Susin M, Mento PF et al. Localization of endothelin-like immunoreactivity in rat kidneys. Am J Physiol 1991; 260:F913-F920.

51. Chen M, Todd-Turla K, Wang W-H et al. Endothelin-1 mRNA in glomerular and epithelial cells of kidney. Am J Physiol 1993; 265:F542-F550.

52. Pupilla C, Brunori M, Misciglia N et al. Presence and distribution of endothelin-1 gene expression in human kidney. Am J Physiol 1994; 267:F679-F687.

53. Jones CR, Hiley CR, Pelton JT et al. Autoradiographic localization of endothelin binding sites in kidney. Eur J Pharmacol 1989; 163:379-382.

54. Kohzuki M, Johnson CI, Chai SY et al. Localization of endothelin receptors in rat kidney. Eur J Pharmacol 1989; 160:193-194.

55. Neuser D, Zaiss S, Stasch JP. Endothelin receptors in cultured renal epithelial cells. Eur J Pharmacol 1990; 176:241-243.

56. Matsumoto H, Suzuki N, Onda H et al. Abundance of endothelin-3 in rat intestine, pituitary gland, and brain. Biochem Biophys Res Comm 1989; 164:74-80.

57. Rosengurt N, Springall D, Polak J. Localization of endothelin-like immunoreactivity in airway epithelia of rats and mice. J Pathol 1990; 160:5-8.

58. Marciniak SJ, Plumpton C, Barker PJ et al. Localization of immunoreactive endothelin and proendothelin in the human lung. Pulm Pharmacol 1992; 5:175-182.

59. Giaid A, Polak JM, Gaitonde V et al. Distribution of endothelin-like immunoreactivity and mRNA in the developing and adult human lung. Am J Respir Cell Mol Biol 1991; 4:50-58.

60. Fukuroda T, Kobayashi M, Ozaki S et al. Endothelin receptor subtypes in human versus rabbit pulmonary arteries. J Appl Physiol 1994; 76:1976-1982.

61. Knott PG, D'Aprile AC, Henry PJ et al. Receptors for endothelin-1 in asthmatic human peripheral lung. Br J Pharmacol 1995; 114:1-3.

62. Horgan MJ, Pinheiro JMB, Malik AB. Mechanism of endothelin-1-induced pulmonary vasoconstriction. Circ Res 1991; 69:157-164.

63. Brink C, Gillard V, Roubert P et al. Effects of specific binding sites of endothelin in human lung preparations. Pulm Pharmacol 1991; 4:54-59.

64. Toga H, Ibe BO, Raj JU. In vitro responses of ovine intrapulmonary arteries and veins to endothelin-1. Am J Physiol 1992; 263:L15-L21.

65. Lee ME, de la Monte SM, Ng S-C et al. Expression of the potent vasoconstrictor endothelin in human central nervous system. J Clin Invest 1990; 86:141-147.

66. Yoshizawa T, Shinmi O, Giaid A et al. Endothelin: a novel peptide in posterior pituitary system. Science 1990; 247:462-464.

67. MacCumber MW, Ross CA, Snyder SH. Endothelin in brain: receptors, mitogenesis, and biosynthesis in glial cells. Proc Natl Acad Sci 1990; 87:2359-2363.

68. Giaid A, Bibson SJ, Ibrahim BN et al. Endothelin 1, an endothelium derived peptide, is expressed in neurons of the human spinal cord and dorsal root ganglia. Proc Natl Acad Sci 1989; 86:7634-7638.

69. Kobayashi M, Ihara M, Sato N et al. A novel ligand, [^{125}I]BQ-3020, reveals the localization of endothelin ET_B receptors. Eur J Pharmacol 1993; 235:95-100.

70. Elshourbagy MA, Lee JA, Korman DR et al. Molecular cloning and characterization of the major endothelin receptor subtype in porcine cerebellum. Mol Pharmacol 1992; 41:465-473.

71. De Olivera AM, Viswanathan M, Capsoni S et al. Characterization of endothelin A receptors in cerebral and peripheral arteries of the rat. Peptides 1995; 16:139-144.

72. Sagher O, Jim Y, Thai QA et al. Cerebral microvascular responses to endothelins: the role of ET_A receptors. Brain Res 1994; 658:179-184.

73. Vigne P, Breittmayer JP, Felin C. Competitive and noncompetitive interactions of BQ-123 with endothelin ET_A receptors. Eur J Pharmacol 1993; 245:229-232.

74. Stanimirovic DB, Yamamoto T, Uematsu S et al. Endothelin-1 receptor binding and cellular signal transduction in cultured human brain endothelial cells. J Neurochem 1994; 62:592-601.

CHAPTER 3

Novel Endothelin Receptors

Ponnal Nambi

Since the original discovery of endothelin (ET) by Yanagisawa et al in 1988,[1] the scientific interest in this area of research has grown exponentially. It was originally hypothesized that ET interacts with two subtypes of receptors to mediate its biological functions and subsequent cloning of two ET receptors confirmed the original hypothesis. As with any receptor, especially seven transmembrane G protein-coupled receptors, identification of receptor subtypes has always been aided by antagonists, and ET receptors are not an exception to this. According to the original definition, the ET receptors which mediated vasoconstriction were classified as ET_A and those which mediated vasodilation were classified as ET_B receptors. This classification was based on the potency of various agonists such as ET-1, ET-2, ET-3, and S6c.[2-5] In addition, radiolabeled ET-1 and ET-3 were also used to classify the two subtypes of ET receptors in various tissues. BQ-123, the very first ET receptor antagonist (selective for ET_A receptors) identified, also was very useful in this classification scheme of the subtypes of ET receptors.[6] However, just like many other seven transmembrane G protein-coupled receptors, the two receptor models appeared insufficient to explain all the biological and pharmacological data obtained for ET peptides. This would indicate that there may be additional subtypes of ET receptors. This became clear with the development of very potent nonpeptide antagonists as well as peptide agonists such as IRL-1620, 4-Ala ET-1, etc. As mentioned above, according to the original classification, ET receptors that mediated vasoconstriction were termed ET_A and those that mediated vasodilation were termed ET_B receptors.

As early as 1989, Kloog et al proposed three apparent subtypes of ET receptors which they named E-Sα, E-Sβ and E-Sγ.[7] E- Sα was shown to bind ET-1 and S6b with higher affinity compared to ET-3 and S6c. E-Sβ was shown to bind S6c with very high affinity compared to other ligands, whereas E-Sγ was shown to bind all peptides with similar affinity.[7] Using cross-linking and peptide mapping techniques, Schvartz et al suggested the presence of multiple endothelin receptors.[8] The receptor they identified in bovine cerebellum displayed similar pharmacology as the nonisopeptide-selective ET receptor that was cloned by Sakurai et al.[9] The two receptors they identified in bovine atrial membrane appear to be different from those that were cloned.[9,10] While ET-3 was as potent as ET-1 in displacing [125I] ET-1, S6c was very weak. When ET-3 was used as the radioligand, S6c was potent in displacing [125I] ET-3. These two receptors were shown to be different by peptide

Endothelin Receptors and Signaling Mechanisms, edited by David M. Pollock and Robert F. Highsmith. © 1998 Springer-Verlag and R.G. Landes Company.

mapping and cross-linking experiments.[8] The presence of subtypes of ET_B receptors were further demonstrated by Sokolovsky et al in central nervous system as well as peripheral tissues which were named ET_{B1} and ET_{B2}.[11] The authors suggested that ET_{B1}, the super high affinity site, may be involved in vasodilation and ET_{B2}, the high affinity site, may be involved in vasoconstriction.

Using BQ-123, a putative ET_A receptor antagonist, Sedo et al have demonstrated the presence of an atypical ET receptor in C6 rat glioma cells.[12] Exposure of these cells to ET-1 or ET-3 resulted in proliferation with very similar EC_{50} values, suggesting that these receptors may be of ET_B subtype. Addition of BQ-123 to these cells blocked both ET-1- and ET-3-mediated proliferation in the same concentration range. This data would suggest that either ET-3 binds to ET_A receptor with high affinity or BQ-123 interacts with ET_B receptor.

In 1993 Karne et al reported the cloning of a novel subtype of ET receptor (ET_C) specific for ET-3 from Xenopus melanophores.[13] This receptor shared 47% and 52% identity with ET_A and ET_B receptors. Displacement of $[^{125}I]$ ET-3 binding to recombinant ET_C receptors expressed in HeLa cells showed that ET-1 was 2-3 fold weaker than ET-3. At the same time, functional data demonstrated that ET-1 was 500-fold weaker than ET-3. The reason for this discrepancy is not apparent. Two additional ET receptor subtypes have been identified in Xenopus by Kumar et al and Nambi et al.[14,15] They were named ET_{AX} and ET_{BX} because of their pharmacological similarity to ET_A and ET_B receptors, respectively. While ET_{AX} receptor showed appropriate binding profile for ET-1, ET-3 and S6c, it had a very low affinity for ET_A-selective BQ-123.[14] Similarly, ET_{BX} receptor displayed appropriate pharmacology for ET-1 and ET-3 but very weak affinity for S6c.[15] In addition, ET_{AX} receptor was cloned and characterized from Xenopus heart.[16] This receptor shared 74%, 60% and 51% identities with human ET_A, human ET_B and Xenopus ET_C receptors.[16]

In addition to ET_C and ET_{AX} receptors which were cloned and characterized from Xenopus, alternately-spliced forms of mammalian ET receptors have been reported. Cheng et al reported an alternately-spliced form of ET_B receptor from rat brain.[17] There were 4 amino acid substitutions in the amino terminal region of this clone, and also the 5' and 3' noncoding regions were different between the rat brain and lung ET_B receptors. The binding profiles obtained for these two receptors were the same, although no functional data were reported.[17] An alternately-spliced form of human ET_B receptor was reported by Elshourbagy et al which showed significant differences in the last 52 amino acids at the carboxy terminal.[18] Although the binding parameters between this alternately-spliced form and the wild type ET_B receptors were the same, the alternately-spliced form of ET_B receptors was totally uncoupled.[18] Another alternately-spliced form of human ET_B receptor has been reported by Shyamala et al.[18] This novel receptor was demonstrated to be present in human brain, placenta, lung, and heart by reverse transcriptase polymerase chain reaction. Although this variant receptor has 10 additional amino acids in the second cytoplasmic domain, there were no differences in binding as well as functional parameters between this and the wild type ET_B receptors.[19]

While molecular biological approaches have led to the identification of alternately-spliced forms of ET_B receptors from different species, pharmacological as well as biochemical techniques have led to the identification of additional subtypes of ET receptors. Using BQ-123 and PD142893, Warner et al have demonstrated the presence of three ET receptors mediating smooth muscle contraction and release of EDRF.[20] They demonstrated that among the ET_B receptors identified, endothelium-dependent vasodilations were sensitive to PD142893 (ET_A/ET_B antagonist), whereas ET_B-mediated smooth muscle contractions were insensitive to PD142893.[20] In addition, ET_B receptor mediating smooth muscle contraction was insensitive to BQ-123.[20] It is important to point out in this context that in C6 glioma cells, Sedo et al demonstrated that BQ-123 blocked ET-1- as well as ET-3-mediated proliferation which would indicate the presence of ET_B receptors that are sensitive to BQ-123.[12] We have demonstrated the presence of a novel ET receptor in dog spleen and monkey spleen that is different from the cloned receptors.[21] While dog spleen displayed ET_A and ET_B receptors, there was a proportion of ET_B receptors that displayed high affinity to ET-3 but weaker affinity ($\sim$ 10-50 fold) to IRL-1620 and S6c and high affinity to BQ-123 and related ET_A antagonists.[21] The pharmacological profile of this receptor appears to be similar to that observed by Sedo et al.[12] Similar observations have been reported by Battistini et al in isolated guinea pig gall bladder.[22] In this model, S6c and other ET_B-selective agonists were weaker in causing contraction and BQ-123 as well as PD145065 inhibited ET-3-mediated contraction, suggesting the presence of non ET_A/ET_B receptor subtype.[22]

Different observations have been reported regarding the ET receptors in rabbit saphenous vein. Sudjarwo et al demonstrated the presence of ET_A and ET_B receptors, and showed that ET_A receptors were less tachyphylactic in this tissue compared to ET_B receptors.[23] In addition, they classified ET_A receptors as BQ-123-sensitive ET_{A1} and BQ-123-insensitive ET_{A2} subtypes.[23] Based on the sensitivity of ET_B antagonists, ET_B receptors also were classified as ET_{B1} and ET_{B2}.[23] Pate et al have also classified the ET_A receptors present in human saphenous vein based on the sensitivity of BQ-123.[24] In a similar study, Douglas et al have reported the presence of three pharmacologically distinct ET receptor subtypes.[25] Their classification was based on the functional characteristics of ET-3, S6c and ET-1. One of these subtypes, named ET_C, had much lower affinity for ET-1 compared to ET-3 and S6c. Of the other ET_B receptors, vasodilator receptors were termed ET_{B1} and vasoconstrictor receptors were termed ET_{B2}. They concluded that there were no ET_A receptors involved in this contraction because BQ-123 was ineffective. This is somewhat different from the observations made by Sudjarwo et al.[23] In addition, ET_B receptors present in rabbit pulmonary resistance arteries were classified based on their potency for ET-1 and sensitivity for BQ-788 (26). Using a similar approach of agonist and antagonist potencies, Miasiro et al have classified the ET_B receptors present in guinea pig ileum as ET_{B1} (sensitive to RES701 and PD145065) and ET_{B2} (less sensitive to RES701 and PD145065).[27]

In conclusion, the classification of the subtypes of ET receptors is more confusing than ever, although more and more ligands are available to answer such questions. This confusion is compounded by the facts that there are species, tissues and cellular differences for ET receptors as well as their functions. It is tempting to speculate that the ET_C receptors reported by Douglas et al[25] may be the mammalian counterpart of Xenopus ET_C receptors that were cloned by Karne et al.[13] Similarly, BQ-123-insensitive ET_{A1} receptors reported by Sudjarwo et al[23] may be the mammalian counterpart of the Xenopus ET_{AX} receptors that were cloned and characterized by Kumar et al.[16] Unless and until these novel receptors are cloned and characterized, it is difficult to come to grip with the number of ET receptor subtypes. Hopefully, future research efforts in this direction will help solve these unanswered questions.

Acknowledgments

The author would like to thank Sue Tirri for her outstanding secretarial assistance.

References

1. Yanagisawa M, Kurihara H, Kimura S et al. A novel potent vasoconstrictor peptide produced by vascular endothelial cells. Nature 1988; 332:411-415.

2. Inoue A, Yanagisawa M, Takuwa et al. The human endothelin family: three structurally and pharmacologically distinct isopeptides predicted by three separate genes. Proc Natl Acad Sci USA 1989; 86:2863-2867.

3. Warner TD, Mitchell JA, deNucci G et al. Endothelin-1 and endothelin-3 release EDRF from isolated perfused arterial vessels of the rat and rabbit. J Cardiovasc Pharmacol 1989; 13:S85-S88.

4. Gardiner SM, Compton AM, Bennett T. Effects of indomethacin on the regional hemodynamic responses to low doses of endothelins and sarafotoxin. Br J Pharmacol 1990; 100:158-162.

5. Williams DL, Jones KL, Pettibone DJ et al. Sarafotoxin 6c: an agonist which distinguishes between endothelin receptor subtypes. Biochem Biophys Res Commun 1991; 175:556-561.

6. Ihara MK, Noguchi K, Saeki T et al. Biological profiles of highly potent novel endothelin antagonists selective for the ET_A receptor. Life Sci 1992; 50:247-255.

7. Kloog Y, Bousso-Mittler D, Bdolah A et al. Three apparent receptor subtypes for the endothelin/sarafotoxin family. FEBS Lett 1989; 253:199-202.

8. Schvartz I, Ittoop O, Hazum E. Direct evidence for multiple endothelin receptors. Biochemistry 1991; 30:5325-5327.

9. Sakurai T, Yanagisawa M, Takuwa Y et al. Cloning of a cDNA encoding a nonpeptide, selective subtype of the endothelin receptor. Nature 1990; 348:732-735.

10. Arai H, Hori S, Aramori I et al. Cloning and expression of a cDNA encoding an endothelin receptor. Nature 1990; 348:730-732.

11. Sokolovsky M, Amber I, Galron R. A novel subtype of endothelin receptors. J Biol Chem 1992; 267:20551-20554.

12. Sedo A, Rovero P, Revoltella RP et al. BQ-123 inhibits both endothelin-1 and endothelin-3-mediated C6 rat glioma cell proliferation suggesting an atypical endothelin receptor. J Biol Reg Homeostatic Agents 1993; 7:95-98.

13. Karne S, Jayawickreme CK, Lerner MR. Cloning and characterization of an endothelin-3 specific receptor (ET_C receptor) from Xenopus Laevis dermal melanophores. J Biol Chem 1993; 268:19126-19133.

14. Kumar C, Nuthulaganti P, Pullen M et al. Novel endothelin receptors in the follicular membranes of Xenopus Laevis oocytes mediate calcium responses by signal transduction through gap junctions. Mol Pharmacol 1993; 44:153-157.

15. Nambi P, Pullen M, Kumar C. Identification of a novel endothelin receptor in Xenopus Laevis liver. Neuropeptides 1994; 26:181-185.

16. Kumar C, Mwangi V, Nuthulaganti P et al. Cloning and characterization of a novel endothelin receptor from Xenopus heart. J Biol Chem 1994; 269:13414-13420.

17. Cheng H-F, Su Y-M, Yeh J-R et al. Alternate transcript of the nonselective-type endothelin receptor from rat brain. Mol Pharmacol 1993; 44:533-538.

18. Elshourbagy NA, Adamou JE, Gagnon AW et al. Molecular characterization of a novel human endothelin receptor splice variant. J Biol Chem 1996; 271:25300-25307.

19. Shyamala V, Moulthrop THM, Stratton-Thomas J et al. Two distinct human endothelin B receptors generated by alternative splicing from a single gene. Cell & Mol Biol 1994; 40:285-296.

20. Warner TD, Allcock GH, Corder R et al. Use of endothelin antagonists BQ-123 and PD142893 to reveal three endothelin receptors mediating smooth muscle contraction and the release of EDRF. Br J Pharmacol 1993; 110:777-782.

21. Nambi P, Pullen M, Kincaid J et al. Identification and characterization of a novel ET receptor from dog spleen that binds both ET_A and ET_B selective ligands. Mol Pharmacol 1997; 52:582-589.

22. Battistini B, O'Donnell LJD, Warner TD et al. Characterization of endothelin receptors in the isolated gall bladder of the guinea pig: evidence for an additional ET receptor subtype. Br J Pharmacol 1994; 122:1244-1250.

23. Sudjarwo SA, Hori M, Tanaka T et al. Subtypes of endothelin ET_A and ET_B receptors mediating venous smooth muscle contraction. Biochem Biophys Res Commun 1994; 200:627-633.

24. Pate MA, Chester AH, Brown TJ et al. Atypical antagonism observed with BQ-123 in human saphenous vein. Fifth International Conf on Endothelin 1997; 74.

25. Douglas SA, Beck GR, Elliott JD et al. Pharmacological evidence for the presence of three distinct functional endothelin receptor subtypes in the rabbit lateral saphenous vein. Br J Pharmacol 1995; 114:1529-1540.

26. Docherty CC, Maclean MR. ET_B receptors in rabbit pulmonary resistance arteries. Fifth International Conf on Endothelin 1997; 48.

27. Miasiro N, Karaki H, Paiva ACM. Distinct ET_B receptors mediating the effects of sarafotoxin 6c and IRL-1620 in the ileum. Fifth International Conf on Endothelin 1997; 75.

CHAPTER 4

"Sticky" Conundrums in the Endothelin System: Unique Binding Characteristics of Receptor Agonists and Antagonists

Jinshyun R. Wu-Wong

Introduction

Endothelin (ET), originally isolated from cultured porcine aortic endothelial cells, is a highly potent vasoconstricting peptide with 21-amino acid residues.[1] Three distinct members of the ET family, namely, ET-1, ET-2 and ET-3, have been identified in humans through cloning.[2] The effects of ETs on mammalian organs and cells are initiated by their binding to high affinity G-protein linked receptors. ET receptors are found in various tissues such as brain, lung, and kidney.[3] Two major types of ET receptors in the mammalian system, ET_A and ET_B, have been characterized, isolated and their cDNAs cloned.[4-7] ET_A receptors are selective for ET-1 and ET-2, while ET_B receptors bind to ET-1, ET-2 and ET-3 with equal affinity. Pharmacologically defined subtypes of ET_A and ET_B receptors have also been reported.[8,9] Various antagonists and agonists for ET receptors have been developed. Comprehensive reviews on the pharmacological properties of these ligands are available.[10-13] For discussion purposes, some of the ET receptor ligands along with the IC_{50} values against ET_A and ET_B receptors are listed in Table 4.1.[14-27]

Since the discovery of ET in 1988,[1] intense research in this field has answered many of the fundamental questions about ET, the receptors, ET-evoked intracellular signaling, and the role of ET in the pathogenesis of various diseases, as evidenced by the abundant information presented in this book. Moreover, there is keen interest to further develop ET receptor antagonists for clinical utilization. However, quite a few questions about the ET system remain unanswered. In this report I intend to discuss a few difficult questions related to the binding characteristics of ET receptor agonists and antagonists. First, why is ET binding to the receptor so "sticky" to a point that the binding appears "irreversible"? Secondly, is the binding of antagonists as "sticky" as that of agonists? If not, does the difference in the binding tenacity between agonists and antagonists affect the potency of ET antagonists? Thirdly, why is there a disparity between the in vitro and in vivo potencies of ET antagonists?

Endothelin Receptors and Signaling Mechanisms, edited by David M. Pollock and Robert F. Highsmith. © 1998 Springer-Verlag and R.G. Landes Company.

Table 4.1 The IC$_{50}$ values of ET receptor ligands against [^{125}I]ET-1 and [^{125}I]ET-3 binding to human ET$_A$ and ET$_B$ receptors

	IC$_{50}$ values, nM[a]			
Ligand	ET$_A$	ET$_B$	Ref.	Comments
ET-1[b]	0.28	0.14	14	Nonselective
ET-3[b]	475	0.08	15	ET$_B$ selective
IRL1620[b]	4263	14	16	ET$_B$ selective
Ro46-2005	230	1101	17	Nonselective
Ro47-0203	7.13	474.8	18	Nonselective
PD 156707	0.23	2457.7	19	ET$_A$ selective
FR139317	0.99	10311.9	20	ET$_A$ selective
A-127722	0.09	128.1	21	ET$_A$ selective
L-749,329	44.59	1878.9	22	Nonselective
L-754,142	**0.35**	**26**	23	Nonselective
SB209670	0.32	351	24	Nonselective
BMS-182874	307	67,320	25	ET$_A$ selective
BQ-123	7.6	34,405	26	ET$_A$ selective
TAK-044[c]	**6.4**	**60**	27	Nonselective

a. The IC$_{50}$ values in bold are obtained from the literature. Other IC$_{50}$ values are obtained from competition studies using the following condition: Binding assays were performed in 96-well microtiter plates pretreated with 0.1% BSA. Membranes (10 µg of protein) from CHO cells stably transfected with human ET$_A$ or ET$_B$ receptors were incubated with 0.1 nM of [^{125}I]ET-1 (ET$_A$) or ET-3 (ET$_B$) in binding buffer (20 mM Tris, 100 mM NaCl, 10 mM MgCl$_2$, pH 7.4, with 0.2% BSA, 0.1 mMPMSF, 5 µg/ml Pepstatin A, 0.025% bacitracin, and 3 mM EDTA) in a final volume of 0.2 ml in the presence or absence of unlabeled ligand for 3 h at 25°C. After incubation, unbound ligand was separated from bound ligand by vacuum filtration using glass-fiber filter strips in PHD cell harvesters (Cambridge Technology, Inc., MA), followed by washing the filter strips with saline (1 ml) for three times.
b. ET-1, -3 and IRL 1620 are agonists. The rest are antagonists.
c. Rabbit ET$_A$ and ET$_B$ receptors used for TAK-044.

"Tenacity" of ET Binding

We have previously shown in membranes prepared from a number of diverse tissues and cell types that bound ET-1 and ET-3 are difficult to dissociate from the receptor.[28-34] The observation is most striking when ET-1 is compared to angiotensin II (Ang II), another potent vasoconstrictor, in parallel in binding studies. As shown in Figure 4.1A, [^{3}H]Ang II binding to membranes prepared from rat liver was time-dependent, reaching a plateau after 30 min of incubation. Addition of unlabeled Ang II at 40 min dissociated ~11% of bound [^{3}H]Ang II after 40 min of incubation. If guanosine 5'-O-3-thiotriphosphate (GTPγS), a nonhydrolysable GTP analog which is known to interfere with ligand binding to G-protein-linked receptors, was added in addition to unlabeled Ang II, ~60% of bound [^{3}H]Ang II was dissociated within 20 min of incubation. In contrast, [^{125}I]ET-1 binding did not reach a plateau until after 120 min of incubation. Addition of ET-1 at 150 min dissociated <10% of bound [¹²⁵I]ET-1 after 240 min of incubation (Fig. 4.1B). Moreover, addition of GTPγS did not induce more dissociation (Fig. 4.1B). In similar studies we have shown that, even up to 20 h of incubation with a high concentration of unlabeled ET-1 or ET-3 plus GTPγS, very little dissociation of bound [¹²⁵I]ET-1 or ET-3 was observed.[32] The results clearly demonstrate that the binding of ET to its receptors is tenacious and may explain why ET, but not Ang II, exhibits a long-lasting vasoconstricting effect.[1] Similar results showing the irreversible binding characteristics of ET have been reported by others using membranes prepared from various tissues and cells.[35-36] Obviously the tenacious binding of ET results from the formation of an unusually stable receptor-ligand complex. The stability of the receptor-ligand complex is such that Takasuka et al[36] have shown that the complex remains intact in the SDS-PAGE analysis even though no chemical cross-linking is performed.

These studies raise the following questions. Is this observation unique for ET-1 and ET-3? How about other ET receptor agonists? The binding of BQ-3020, an ET$_B$-selective agonist, has been shown to be more reversible than ET-3.[37-38] When another ET$_B$-selective agonist IRL 1620 is examined, it is found that bound [¹²⁵I]IRL 1620 is easier to dissociate than bound [¹²⁵I]ET-3: the addition of unlabeled IRL 1620 or ET-1 at 4 h dissociates bound [¹²⁵I]IRL 1620 by ~45% after 20 h of incubation.[32] Interestingly, it seems that the reversibility of BQ-3020 and IRL 1620 binding may be dependent on the species from which the receptors are derived. For example, IRL 1620 binding to dog, pig, or human receptors is more reversible than that to rat receptors.[39-40] A similar observation is made for BQ-3020.[38]

Why is ET-1 or -3 binding so "sticky"? It is not a question with obvious answers. As mentioned above, ET binding results in a stable receptor-ligand complex. However, how is this stable complex formed? Several hypotheses have been proposed. Studies using site-directed mutagenesis and ET$_A$/ET$_B$ chimeric receptors have suggested that ET ligands may interact with the receptors at more than one site,[41-42] which may contribute to the phenomenon. Additional studies along this line reveal that, at least in the case of ET$_B$ receptor, the 29 amino acids in the N-terminal region, especially Asp-75 and Pro-93, may be responsible for forming a stable complex with the ligands.[43] Nambi et al[40] have shown that there is a good correlation between the effect of guanine nucleotide on the binding of an agonist and the reversible/irreversible binding property of the agonist, hence implying that the interaction between ET receptors and the G proteins induced by the binding of an agonist may play a role in the formation of a stable complex. Other hypotheses, such as disulfide interchange in the ET receptor-ligand complex,[44] continuous ET receptor externalization,[45] or colocalization of ET receptor with caveolin in caveolae,[46] have also been proposed to explain the stability of the ligand-receptor complex. Despite various efforts by our group and by others, it is still unclear why ET-1 binding is irreversible.

Is this unique binding characteristic of ET analogs of any pharmacological importance? Or will it remain as a curious observation in the laboratory without any impact on our understanding of the ET system, and on our efforts directed at developing ET receptor antagonists as therapeutic agents? At least two pieces of evidence suggest that this tenacious binding characteristic of ET may be of practical significance.

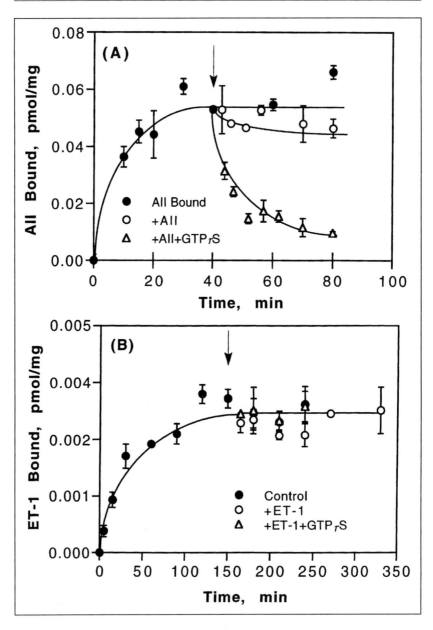

Fig. 4.1. Dissociation studies. Rat liver membranes (0.35 mg) were incubated with (A) 19 nM of [3H]Ang II or (B) 0.06 nM of [125I]ET-1 in 0.2 ml of binding buffer for different periods of time at 25°C (●). The binding study was done as described in Table 4.1. In (A), after 40 min of incubation, 1 μM unlabeled Ang II (O) or Ang II plus GTPγS (100 μM) (Δ) was added. In (B), after 150 min of incubation, 1 μM unlabeled ET-1 (O) or ET-1 plus GTPγS (100 μM) (Δ) was added. Each value represents the mean ± S.D. of 3 determinations. The maximal amounts of ligand bound were approximately 1% (A) and 10% (B) of free ligands at time 0.

First, the unique binding nature of ET may necessitate the use of special methods for the determination of binding parameters. Calculation of K_d values by Scatchard analysis may not be appropriate for the ET system. Although it is common, and probably useful to report K_d values for ET binding derived from Scatchard analysis, a more accurate way to calculate K_d values is by kinetic analysis using the association and dissociation rate constants as shown below.

$$[L] + [R] \underset{k_{-1}}{\overset{k_1}{\rightleftharpoons}} [LR] \tag{1}$$

L: ligand, R: receptor, and LR: ligand-receptor complex

$$K_d = \frac{k-1}{k_1} \tag{2}$$

k_1: association rate constant; k_{-1}: dissociation rate constant; k_d: equilibrium dissociation constant
To calculate k_{-1},

$$[LR] = [LR]_o \cdot e^{-tk_{-1}} \tag{3}$$

$[LR]_o$: concentration of occupied receptor at t_o
$[LR]$: concentration of occupied receptor at time t.
From eqn. (3),

$$t \cdot k_{-1} = \ln \frac{[LR]_o}{[LR]} \tag{4}$$

When $\ln \dfrac{[LR]_o}{[LR]}$ is plotted versus t, the slope will give k-1.

To calculate k_1,

$$[LR]_t = [LR]_{eq} \cdot (1 - e^{-tk_{obs}}) \tag{5}$$

$[LR]_t$: concentration of occupied receptor at time t
$[LR]_{eq}$: concentration of occupied receptor at equilibrium.
k_{obs} : pseudo-first order association rate constant

$$k_{obs} = k_1 L + k_{-1} \tag{6}$$

$L = L_o$ (ligand concentration at time 0), when $[L] >>> [R]$.
From eqn. (5),

$$t \, k_{obs} = \ln \frac{[LR]_{eq}}{[LR]_{eq} - [LR]_t} \tag{7}$$

When $\ln \dfrac{[LR]_{eq}}{[LR]_{eq} - [LR]_t}$ is plotted versus t, the slope is equal to k_{obs}, and k_1 can

be calculated using eqn. (6)
For more details on this subject, please see references 35 and 47. Taking Figure 4.1 as an example, the association and dissociation rate constants (K_1 and K_{-1}) for Ang II binding are calculated to be 0.0031 min^{-1} nM^{-1} and 0.036 min^{-1},

respectively. The K_d value is 12 nM since $K_d = \dfrac{k-1}{k_1}$, which is similar to values reported in the literature by Scatchard analysis.[48] From Figure 4.1, it is possible to estimate the association rate constant for ET binding as 0.27 min^{-1} nM^{-1}. The dissociation rate constant of ET binding can also be estimated as 2.0 x 10^{-5} min^{-1}. Consequently, the K_d value for ET-1 binding to rat liver membranes is calculated to be 0.075 pM, which is much lower than the sub-nanomolar values calculated from Scatchard analysis. It is worth mentioning that Waggoner et al[35] have suggested that the radioligand binding analysis program "Kinetic" (Biosoft, MO, U.S.A.) can be modified to calculate kinetic data for ET binding. They have shown that the dissociation half-life ($t_{d, 0.5}$) for bound [^{125}I]ET-1 is >30 h in membranes prepared from rat heart, lung, brain, and porcine vascular smooth muscle with K_d values in the sub-picomolar range. Using a different approach based on biological data and logic, Frelin and Guedin[49] suggest that K_d values for ET binding are probably much lower than those derived from Scatchard analysis. As a result, ET receptor densities can be higher than K_d values, and stoichiometric binding conditions may account for many unique features in the ET system.

If the calculated K_d values of ET binding are actually in the sub-picomolar range, rather than in the sub-nanomolar range, then it may not be surprising to observe that ET at 1-20 pM can evoke a biological effect such as the elevation of the intracellular calcium concentration or an increase in the vascular tone.[50-51] In addition, the sub-picomolar K_d values may provide a clue to the question of whether ET in circulation has a functional role or not. The level of ET in circulation is usually in the range of 1-5 pM. Even under pathological conditions in which ET plays a role, the level of ET in circulation is seldom increased to the sub-nanomolar range. It is generally thought that ET in circulation is a "spillover" from a local synthesis site, and the concentrations of ET at local sites are higher. Because of this, it is suggested that ET is not an endocrine factor, but mainly acts as a paracrine/autocrine factor. However, if receptor-bound ET is difficult to dissociate, and if the K_d values for ET binding are in the sub-picomolar range, then perhaps the systemic level of ET does have physiological and/or pathological significance even though it seldom rises above 50 pM.

Secondly, as mentioned above, it has been suggested that the "stickiness" of ET binding may be linked to the long-lasting biological effect of ET.[1,46,52] Although the biological significance of the long-lasting effect of ET is still not fully understood, it raises an intriguing question: how can the effects of ET be ameliorated if the binding of ET is irreversible? To answer this question, it is necessary to compare whether the binding of antagonists is less "sticky" or more reversible than that of the agonists.

Is Antagonist Binding Tenacious?

How "sticky" is antagonist binding in comparison to agonist binding? Ihara et al[53] have shown that [^{3}H]BQ-123, an ET$_A$ selective antagonist, bound to membranes prepared from human neuroblastoma cells can be readily dissociated by the addition of unlabeled BQ-123 (10 μM). The dissociation rate constant for BQ-123 is 0.51 min^{-1}. The binding of [^{3}H]SB209670, a nonpeptide ET receptor antagonist, to both human ET$_A$ and ET$_B$ receptors is shown to be reversible by the addition of unlabeled SB209670 at 1 μM.[54] The binding of [^{125}I]-PD 151242, another ET$_A$ selective antagonist, is also reversible (dissociation rate constant: 0.00144 min^{-1}), although less so than BQ-123.[55]

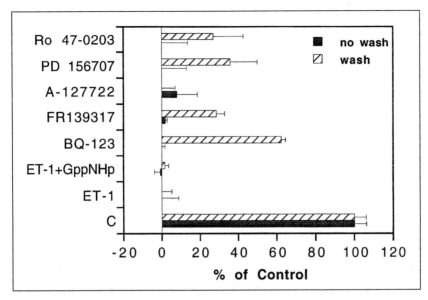

Fig. 4.2. Bind-and-wash experiments. Membranes (200 µg) prepared from rat pituitary MMQ cells expressing ET_A receptor were incubated with or without ligand at indicated concentrations in 1 ml of Wash Buffer (20 mM Tris, 100 mM NaCl, 10 mM $MgCl_2$, pH 7.4, 0.1 mM PMSF, 5 µg/ml Pepstatin A, 0.025% bacitracin, and 3 mM EDTA) for 3 h at 25°C (first incubation period). After the incubation, 10 ml of Wash Buffer was added and the mixture was centrifuged at 30,000xg for 30 min. The pellet was resuspended into 25 ml of Wash Buffer and centrifuged again. The final pellet was resuspended into 0.5 ml of Wash Buffer and a small portion was used to determine the protein content. BSA was added to a final concentration of 0.2% and then membranes (10 µg per well) were assayed for [125I]ET-1 binding (second incubation period). [125I]ET-1 binding was performed as described in Table 4.1 with a 3 h incubation period at 25°C. Hatched bar (wash): Ligands (0.1 µM ET-1, 0.1 µM ET-1 + 100 µM Gpp(NH)p (a nonhydrolysable GTP analog), 5 µM BQ-123, 0.5 µM FR139317, 0.1 µM A-127722, 0.2 µM PD 156707, 3.5 µM Ro 47-0203) were added in the first incubation period to membranes; membranes were washed extensively, and then assayed for [125I]ET-1 binding. Solid bar (no wash): No ligand was added in the first incubation period, although membranes were processed and washed as in the other group; ligands were added in the second incubation period with [125I]ET-1. Specific [125I]ET-1 binding to control membranes (also washed and processed but without adding ligands) was 27.2 ± 1.7 fmol/mg. Data shown are specific binding (% of control) that has been corrected for nonspecific binding (determined in the presence of 1 µM ET-1). Each value represents the mean ± S.D. of 4 determinations.

Is it a general phenomenon that antagonist binding is more reversible than agonist binding? How do the dissociation rate constants of different antagonists differ? To answer these questions directly, it will be necessary to prepare radiolabeled ligands for the various antagonists for dissociation studies. Before the radiolabeled forms of some antagonists become available, an indirect "bind-and-wash" procedure can be employed to compare the "stickiness" of antagonists. Figure 4.2 shows results from such a study for six ET receptor ligands. The details of the procedure have been reported previously,[31-32] and are briefly described in the legend of Figure 4.2. To best illustrate the differences among the ligands, all the data are expressed as % of control (i.e., [^{125}I]ET-1 binding determined in the absence of competing ligands). From Figure 4.2, unlabeled ET-1 at 0.1 μM completely inhibits [^{125}I]ET-1 binding when added together with [^{125}I]ET-1. When membranes are pretreated with ET-1 followed by extensive washing, specific [^{125}I]ET-1 binding is still 100% inhibited, consistent with the results in Figure 4.1 that ET-1 binding is tenacious, and cannot be reversed. On the contrary, PD 156707 at 0.2 μM also completely inhibits [^{125}I]ET-1 binding when added together with [^{125}I]ET-1. However, when membranes are pretreated with PD 156707 followed by extensive washing, >30% of specific [^{125}I]ET-1 binding is restored. Similar results are observed for other antagonists except for A-127722. Furthermore, washing restores [^{125}I]ET-1 binding to different degrees for different antagonists. It is important to note that ligands are tested at different concentrations because the differences in potencies (as shown by the IC$_{50}$ values in Table 4.1) have to be taken into consideration. The concentration of a ligand used in Figure 4.2 is approximately 500 to 1000-fold of its IC$_{50}$ so that ligands are compared at roughly equivalent potency. If the % binding restored after washing directly reflects whether the bound ligand is easy to dissociate from the receptor, then the results in Figure 4.2 suggest that the degree of reversibility of ligand binding is in the order of BQ-123 > PD 156707 > Ro 47-0203 ≥ FR139317 > A-127722 ≥ ET-1. These results suggest that, in general, antagonist binding is more reversible than ET-1 binding, and the binding of different antagonists may exhibit different degrees of "stickiness".

Impact on Antagonist Potency

Next, let us examine whether the differences in "stickiness" between agonists and antagonists impact the potencies of the antagonists. Intuitively one may reason that, since ET-1 binding is less reversible than antagonist binding, the potency of an antagonist in inhibiting ET-1 binding to the receptor should decrease when the incubation time is increased. Indeed, as shown in Figure 4.3, when ET-1 (0.1 nM), BQ-123 (5 nM) or FR139317 (1 nM) was coincubated with [^{125}I]ET-1 and MMQ cell membranes over a period of 24 h at 25°C, the level of % inhibition by ET-1 was maintained at a constant level. However, the levels of percent inhibition by BQ-123 and FR139317 decreased when the incubation time increased, suggesting that the potency is inversely affected by time. The decrease in the potency of an given antagonist as a function of incubation time is best demonstrated in Figure 4.4. Again, using MMQ cell membranes in competition binding studies, it is clear that the competition curves of ET-1 against [^{125}I]ET-1 binding did not exhibit a significant shift at 1 h, 3 h or 24 h of incubation, and that the IC$_{50}$ values of ET-1 remained in the sub-nanomolar range at the three different time points. On the contrary, the IC$_{50}$ values of Ro 47-0203 changed from 2.7 nM at 1 h to 5.8 nM at 3 h, and to 56.7 nM at 24 h of incubation. Table 4.2 summarizes the IC$_{50}$ values for various ligands determined at the three different incubation time points. These results suggest that the potencies of most ET receptor antagonists are critically dependent on the incubation time because antagonist binding is more reversible than ET binding. Furthermore, different antagonists exhibit different reversible characteristics and are affected differently by the length of incubation time.

Why are there differences in the reversibility of antagonists? It is unlikely that these differences can be explained by different modes of inhibition (e.g., competitive vs. noncompetitive) since we and others have shown in K$_i$ and pA$_2$ studies that most, if not all, of these compounds, e.g., FR139317, PD 156707, L-749329, Ro 47-0203,

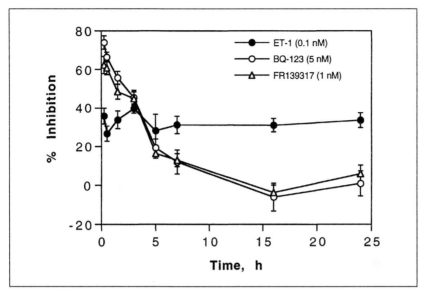

Fig. 4.3. The inhibition of [^{125}I]ET-1 binding by antagonists or unlabeled ET-1 at different incubation time points. MMQ cell membranes (0.01 mg) in binding buffer (as in Table 4.1) with 0.2% BSA were incubated with 0.1 nM [^{125}I]ET-1 for different periods of time at 25°C. At each time point, [^{125}I]ET-1 binding was determined with or without the test ligand. Specific [^{125}I]ET-1 binding with test ligands was normalized to control binding (without test ligands) of the corresponding time point to calculate percent inhibition. Each value represents the mean ± S.D. of 3 determinations. Concentrations of unlabeled test ligands are as indicated. Reprinted with permission from Wu-Wong JR et al. J Cardiovascular Pharmacol 1995; 26:S380-S384. ©1995 Lippincott-Raven Publishers.

and A-127722, are competitive inhibitors of ET binding.[56] A more plausible explanation is that each antagonist interacts with different regions of the receptor. Indeed, Breu et al[57] have shown in site-directed mutagenesis studies that binding sites for Ro 47-0203 and BQ-123 are different.

Disparity Between In Vivo and In Vitro Potencies
Although in vitro binding studies show that the potencies of ET receptor antagonists decrease when the incubation time with ET increases, numerous reports have shown that ET-induced responses in vivo (whole animal studies) and ex vivo (studies using tissues or cells) can be reversed by antagonists. Furthermore, ET receptor antagonists have been proven to be efficacious in ameliorating various pathological conditions in which ET levels are elevated. Review articles on this subject are numerous: see references 13, 21, 58-64 for reviews published in 1996.

However, one may ask: if data from binding studies suggest that the potencies of ET receptor antagonists decrease during a prolonged incubation with ET, how can ET receptor antagonists be efficacious in vivo and/or ex vivo, especially in chronic situations? Is there any link between the in vitro binding results and in vivo/ex vivo results? To answer these questions, it is necessary to consider ET-1-induced receptor internalization. Unlike binding studies using membranes in which

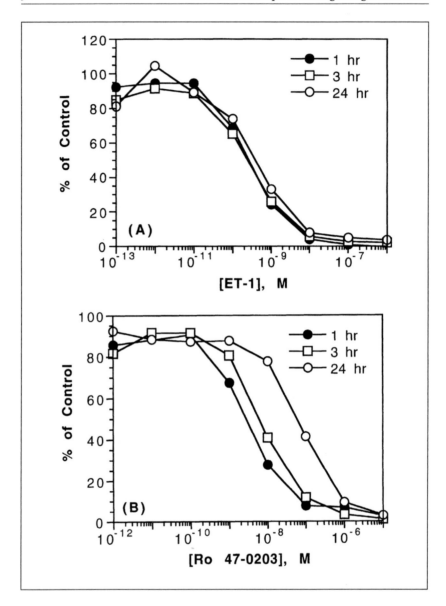

Fig. 4.4. The effect of incubation time on the potencies of ET-1 (A) and Ro 47-0203 (B) in inhibiting [^{125}I]ET-1 binding to ET$_A$ receptor. Competition studies were performed as described in Table 4.1 using membranes prepared from MMQ cell except that the incubation was allowed to continue for 1 h, 3 h, or 24 h at 25°C. Nonspecific binding, determined in the presence of 1 μM ET-1, was subtracted from total binding to give specific binding. The results are expressed as percent of control, with binding in the absence of the antagonist as 100%. Each value represents the mean (± S.D.) of 3 determinations.

Table 4.2. IC$_{50}$ values of antagonists at different incubation time

Incubation time	1 h	IC$_{50}$, nM 3 h	24 h
ET-1	0.35	0.64	0.90
PD 156707	0.24	0.76	3.2
A-127722	0.39	0.36	1.0
BQ-123	1.9	4.7	35.3
Ro 47-0203	2.7	5.8	56.9
L-749,329	39	70	312
Ro 46-2005	210	370	2,900

receptor internalization is eliminated, when ET is allowed to interact with receptors in cells, binding induces receptor internalization and the receptors may subsequently recycle back to the cell surface. Previously we and others[33,45,65-67] have shown that the binding of ET induces rapid internalization of the ligand-receptor complex. Marsault et al[45] have also shown that the reappearance of ET binding sites at the cell surface is slow, with about 40% detectable after a 60-min incubation period. Therefore, although ET binding is "irreversible", presumably free, unoccupied ET receptors will recycle back to the membrane surface and are free to bind ligands. However, since receptor recycling takes time, an antagonist may have to be present continuously and in an abundant amount in order to exert an effect. If so, then it may be expected that the potency of an antagonist in vivo or in blocking ET-1-induced chronic effects in cells will be less than that determined from the in vitro binding studies. In addition, it may be expected that the reversal of ET-induced functional effects by subsequent application of antagonists may be slow. Indeed, we have previously demonstrated that, in human smooth muscle cells, antagonists are potent in blocking ET-induced immediate responses such as PI hydrolysis and arachidonic acid release, but are less potent if the effects of ET involve prolonged incubation such as in the case of ET-stimulated DNA synthesis.[30] Furthermore, Warner et al[68] have shown that an established elevation in mean arterial pressure in rats and sustained contractions of rat aortic rings induced by ET-1 is reversed slowly by the subsequent application of BQ-123 and PD 145065.

To demonstrate that a disparity exists between the in vitro and in vivo potencies of an ET receptor antagonist, the following exercise is performed. As reported, the IC$_{50}$ value of A-127722 for the ET$_A$ receptor as determined by in vitro binding studies is 0.1 nM, and A-127722 at 100 nM completely inhibits ET-1 binding.[56] Based on the plasma elimination half-life of 3.5 hr and the peak plasma concentration of 1.1 µg/ml at 5 mg/kg in the rat,[21] we can estimate that the dose of A-127722 required to reach a plasma concentration of 100 nM is ~0.25 mg/kg. At this dosage, A-127722 should completely inhibit the increase in arterial blood pressure induced by ET-1. However, we have shown that approximately 10 mg/kg of A-127722 is required to exert a maximal inhibitory effect on an ET-1-induced increase in arterial blood pressure.[21] Is this disparity unique for an ET$_A$-selective antagonist, such as A-127722, since ET$_B$ receptor has also been shown to play a role in vasoconstriction?[12] To answer this question, the calculation can be repeated for a nonselective receptor antagonist such as SB217242. In a report by Ohlstein et al,[69] the K_i values of SB217242 from in vitro binding studies are 1.1 nM for the ET_A receptor, and 111 nM for the ET_B receptor. SB217242 at 100 nM (100-fold K_i for ET_A) is required to inhibit ET-1 binding to the ET_A receptor. When dosed i.v. at 3 mg/kg, SB217242 has a plasma elimination half-life of 3.3 hr and the peak plasma concentration is 700 ng/ml. Again, it can be estimated that the dose of SB217242 required to reach a 100 nM plasma concentration is ~0.2 mg/kg. At this dosage, SB217242 should inhibit the increase in arterial blood pressure induced by ET-1. However, 0.3 mg/kg of SB217242 hardly exhibits an effect and a dose of 30 mg/kg is required to inhibit an ET-1-induced increase in arterial blood pressure. Similar calculations can be done for other ET receptor antagonists. It is evident that disparities do exist between the in vivo and in vitro potencies for ET receptor antagonists.

More to Consider: Interaction with Proteins

Can the differences in "stickiness" in binding between agonists and antagonists fully explain the disparity between the in vitro and in vivo potencies for an ET receptor antagonist? I think at least one more factor, the interaction of ET receptor ligands with plasma proteins, has to be considered. Previously we have shown that ET-1, -3 and the ET receptor antagonists, PD 156707, L-749329, Ro 47-0203, and A-127722, exhibit a high degree of binding to human plasma proteins, especially serum albumin.[70] When ET-1 binding to the receptor was examined, 5% (v/v) human plasma inhibited ET-1 binding to both ET_A and ET_B receptor by 80-90%. Because the protein concentration in human blood is 5.5-8% and 55-60% of that is albumin, it is possible that albumin is a major protein component in plasma which interferes with ET-1 binding to the receptor. Indeed, we have shown that human serum albumin can inhibit ET-1 binding.[70] Furthermore, the addition of increasing doses of human serum albumin (HSA) can incrementally decrease the potency of an antagonist. As shown in Figure 4.5, in the absence of HSA, the IC_{50} values for A-127722 and L-749329 were 0.22 nM and 0.29 nM, respectively. When the amount of HSA increased, the IC_{50} values of both compounds increased, with L-749329 being affected more than A-127722. Similar studies were also performed for Ro 47-0203 and PD 156707. The results are summarized in Table 4.3. These studies suggest that ET and ET receptor antagonists exhibit a high degree of binding to plasma proteins, especially serum albumin. Consequently serum albumin can inhibit ET binding to its receptors and decrease the potencies of ET receptor antagonists. In addition, some antagonists are affected to a greater degree than others.

Why does HSA have such an impact on the potencies of ET receptor antagonists? First, it is not surprising that ET receptor antagonists exhibit strong binding to plasma proteins, especially serum albumin, since albumin readily binds to lipophilic acids, a common structural characteristic of ET antagonists.[70] Thus, it is likely that HSA acts as a "binding protein" for ET receptor antagonists, and antagonists which bind to HSA are no longer free to bind to ET receptors. As a result, a decrease in the potency is observed. How does the HSA factor contribute to the disparity between the in vitro and in vivo potencies of an antagonist? One possible explanation is that the in vitro potency is usually determined in a buffer system which contains at most 0.5% of serum albumin. For example, HSA or bovine serum albumin (BSA) at 0.01% was used in the binding assays by Williams et al[23] and Sogabe et al.[20] BSA at 0.1 % was used in the assays by Webb et al[25] and Reynolds et al,[19] while 0.5% BSA was used by Clozel et al.[18] Therefore, this protein-binding factor

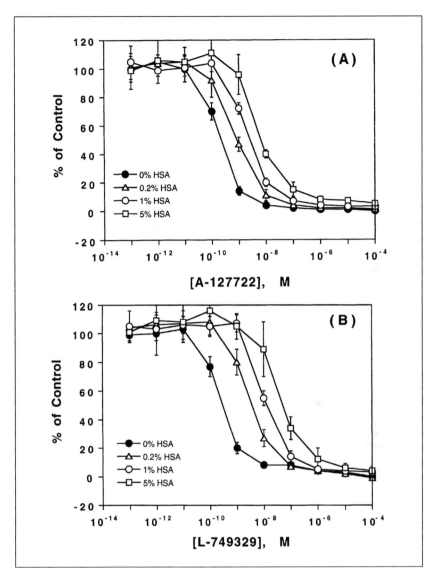

Fig. 4.5. The effect of HSA on the potencies of A-127722 (A) and L-749329 (B) in inhibiting [^{125}I]ET-1 binding to human ET$_A$ receptor. Receptor binding was performed as described above using membranes prepared from CHO cells stably transfected with human ET$_A$ receptor except that different concentrations of HSA were used to replace 0.2% BSA in the binding buffer. Nonspecific binding (7-18.5 fmol/mg), determined in the presence of 1 μM ET-1, was subtracted from total binding to give specific binding. Specific binding was 326, 173, 104, and 60 fmol/mg at 0, 0.2%, 1% and 5% of HSA, respectively. The results are expressed as % of control, with binding in the absence of the antagonist as 100%. Each value represents the mean (± S.D.) of 3 determinations. Reprinted with permission from: Wu-Wong JR et al. Life Science 1996; 58:1839-1847. ©1996 Elsevier.

Table 4.3 The effects of HSA on the potency of ET receptor antagonists in inhibiting [^{125}I]ET-1 binding to human ET_A receptor

HSA (%)	IC_{50}, nM			
	0%	0.2%	1%	5%
PD 156707	0.37	5.68	20.5	62.8
L-749329	0.29	3.64	13.1	50.2
Ro47-0203	5.7	23	47.5	122.7
A-127722	0.22	0.85	2.75	6.72

Reprinted with permission from Wu Wong JR et al. Life Science 1996; 58:1839-1847. ©1996 Elsevier.

may be minimized if the potency of an antagonist is determined in a buffer system containing a high concentration of serum albumin. However, it may be difficult to pick a single albumin concentration because the effect of HSA may not be the same for the various antagonists. One may then ask: why are the potencies of different ET receptor antagonists affected by HSA differently? Again, it is unlikely that the differences can be explained by the modality of inhibition since many of these antagonists are competitive inhibitors as discussed above. A possible explanation is that the more reversible the binding of an antagonist to the receptor, the greater the impact HSA has on its potency. In fact, the data in Table 4.2 suggest that the binding of L-749329 to ET receptors is more reversible than that of A-127722, an observation that is consistent with the results in Table 4.3 that serum albumin exerts a more profound effect on L-749329 than A-127722.

Conclusions

It is now well recognized that ET binding to the receptors is tenacious to a point that the binding appears irreversible. Although Scatchard analysis is commonly used in the literature to calculate K_d values for ET binding, kinetic analysis may be more accurate for this purpose because of the tenacious binding characteristic. K_d values of ET binding calculated by kinetic analysis are in the subpicomolar range, much lower than those reported by Scatchard analysis. If the lower K_d values are more accurate, then we may suspect that the systemic level of ET, in the range of 1-5 pM, may elicit functional effects, and may have physiological or pathological consequences. Future studies will help to define the roles of ET as a paracrine/autocrine factor, and possibly as an endocrine factor.

Although in general antagonist binding is more reversible than agonist binding, different antagonists exhibit different degrees of binding tenacity. If the binding of an antagonist is much more reversible than ET, then the potency of the antagonist in inhibiting ET-1 binding will decrease when the incubation time is increased. Consequently the potency of the antagonist in blocking ET-1-induced chronic effects in cells will be much less than that determined from in vitro binding studies. Is this a curse or a blessing for such an antagonist in drug development? Will it be more beneficial to have an antagonist which exhibits irreversible binding characteristics? Answers to questions such as these may be obtained when more data are gathered from the clinical trials of various ET receptor antagonists.

In this report I hope that I have convinced the reader that a disparity does exist between the in vitro and in vivo potencies of an ET receptor antagonist. This disparity can be partially explained by the differences in binding tenacity between ET receptor agonists and antagonists. However, the interaction between antagonists and other proteins may also play a role, as exemplified by the binding of ET receptor antagonists to serum albumin. If the concentration of serum albumin indeed affects the potencies of antagonists, then special precautions shall be taken when the potencies of ET receptor antagonists from different reports are compared, because HSA/BSA is frequently included in assays for the determination of potencies of ET receptor antagonists, and investigators do not always use same concentrations of HSA/BSA in their studies. Furthermore, I would like to suggest that, when reporting the characteristics of a new ET receptor ligand, the amount of serum albumin used be clearly defined and that one or more "reference" compounds be evaluated side-by-side with the compound of interest.

As mentioned in the beginning of the chapter, although intense research in the past nine years has answered many of the fundamental questions about the ET system, more remains to be learned about ET receptors, ET-evoked intracellular signaling, and especially the role of ET in the pathogenesis of diseases. Currently the most exciting development in the field may be the ongoing clinical trials with various ET receptor antagonists. In the near future researchers in the ET field will likely witness receptor antagonists being developed into therapeutic agents for the treatment of various diseases.

Acknowledgments

The author would like to thank Drs. Bruce Credo, Tom von Geldern, Terry Opgenorth and Jerry Wessale for their critical comments, and William Chiou for his excellent work.

References

1. Yanagisawa M, Kurihara H, Kimura S et al. A novel potent vasoconstrictor peptide produced by vascular endothelial cells. Nature (London) 1988; 332:411-415.
2. Inoue A, Yanagisawa M, Kimura S et al. The human endothelin family: Three structurally and pharmacologically distinct isopeptides predicted by three separate genes. Proc Natl Acad Sci 1989; 86:2863-2867.
3. Sokolovsky M. Structure-function relationships of endothelins, sarafotoxins, and their receptor subtypes. J Neurochem 1992; 59:809-821.
4. Kozuka M, Ito T, Hirose S et al. Purification and characterization of bovine lung endothelin receptor. J Biol Chem 1991; 266:16892-16896.
5. Wada K, Tabuchi H, Ohba R et al. Purification of an endothelin receptor from human placenta. Biochem Biophys Res Commun 1990; 167:251-257.
6. Arai H, Hori S, Arimori I et al. Cloning and expression of a cDNA encoding an endothelin receptor. Nature 1990; 348:730-732.
7. Sakurai T, Yanagisawa M, Takuwa Y et al. Cloning of a cDNA encoding a nonisopeptide selective subtype of the endothelin receptor. Nature 1990; 348:732-735.
8. Sudjarwo SA, Hori M, Tanaka T et al. Subtypes of endothelin ET_A and ET_B receptors mediating venous smooth muscle contraction. Biochem Biophys Res Commun 1994; 200:627-633.
9. Douglas SA, Beck Jr GR, Elliott JD et al. Pharmacological evidence for the presence of three distinct functional endothelin receptor subtypes in the rabbit lateral saphenous vein. Brit J Pharcol 1995; 114:1529-1540.

10. Spellmeyer DC. Small molecule endothelin receptor antagonists. Annual Reports in Medicinal Chemistry 1994; 29:65-71.
11. Warner TD. Endothelin receptor antagonists. Cardiovascular Drug Reviews 1994; 12:105-122.
12. Opgenorth TJ. Endothelin receptor antagonism. Adv Pharmacol 1995; 33:1-65.
13. Ohlstein EH, Elliott JD, Feuerstein G et al. Endothelin receptors: receptor classification novel receptor antagonists, and potential therapeutic targets. Medicinal Research Reviews 1996; 16:365-390.
14. Elshourbagy NA. Korman DR, Wu H-L et al. Molecular characterization and regulation of the human endothelin receptors. J Biol Chem 1993; 268:3873-3879.
15. Hechler U, Becker A, Haendler B et al. Stable expression of human endothelin receptors ET_A and ET_B by transfected baby hamster kidney cells. Biochem Biophys Res Commun 1993; 194:1305-1310.
16. Takai M, Umemura I, Yamasaki K et al. A potent and specific agonist, Suc-[Glu^9, Ala^{11,15}]-endothelin-1(8-12), IRL 1620, for the ET_B receptor. Biochem Biophys Res Commun 1992; 184:953-959.
17. Clozel M, Breu V, Burri K et al. Pathophysiological role of endothelin revealed by the first orally active endothelin receptor antagonist. Nature 1993; 365:759-761.
18. Clozel M, Breu V, Gray GA et al. Pharmacological characterization of bosentan, a new potent orally active nonpeptide endothelin receptor antagonist. J Pharm Exp Ther 1994; 270:228-235.
19. Reynolds EE, Keiser JA, Haleen SJ et al. Pharmacological Characterization of PD 156707, an orally active ET_A receptor antagonist. J Pharm Exp Ther 1995; 273:1410-1417.
20. Sogabe K, Nirei H, Shoubo M et al. Pharmacological profile of FR139317, a novel, potent endothelin ET_A receptor antagonist. J Pharm Exp Ther 1993; 264:1040-1046.
21. Opgenorth TJ, Adler AL, Calzadilla S et al. Pharmacological Characterization of A-127722: An orally active and highly potent ET_A-selective receptor antagonist. J Pharm Exp Ther 1996; 276:473-481.
22. Walsh TF. Progress in the development of endothelin receptor antagonists. In: Annual Reports in Medicinal Chemistry. Academic Press 1995:91-100.
23. Williams Jr DL, Murphy KL, Nolan NA et al. Pharmacology of L-754,142, a highly potent, orally active nonpeptidyl endothelin antagonist. J Pharm Exp Ther 1995; 275:1518-1526.
24. Ohlstein EH, Nambi P, Douglas SA et al. SB 209670, a rationally designed potent nonpeptide endothelin receptor antagonist. Proc Natl Acad Sci 1994; 91:8052-8056.
25. Webb ML, Bird JE, Liu ECK et al. BMS-182874 is a selective, nonpeptide endothelin ET_A receptor antagonist. J Pharm Exp Ther 1995; 272:1124-1134.
26. Ihara M, Ishikawa K, Fukuroda T et al. In vitro biological profile of a highly potent novel endothelin (ET) antagonist BQ-123 selective for the ET_A receptor. J Cardiovascular Pharmacol 1992; 12:S11-S14.
27. Watanabe T and Fujino M. TAK-044: An endothelin receptor antagonist. Cardiovascular Drug Reviews 1996; 14:36-46.
28. Wu-Wong JR, Chiou W, Magnuson SR et al. Identification and characterization of type A endothelin receptors in MMQ cells. Mol Pharmacol 1993; 44:285-291.
29. Wu-Wong JR, Chiou W and Opgenorth TJ. Phosphoramidon modulates the number of endothelin receptors in cultured Swiss 3T3 fibroblasts. Mol Pharmacol 1993; 44:422-429.
30. Wu-Wong JR, Chiou WJ, Huang Z-J et al. Endothelin receptor in human smooth muscle cells: antagonist potency differs on agonist-evoked responses. Am J Physiol 1994; 267:C1185-C1195.

31. Wu-Wong JR, Chiou W, Naugles Jr KE et al. Endothelin receptor antagonists exhibit diminishing potency following incubation with agonist. Life Sciences 1994; 54:1727-1734.

32. Wu-Wong JR, Chiou W, Magnuson SR et al. Endothelin receptor agonists and antagonists exhibit different dissociation characteristics. Biochim Biophys Acta 1994; 1224:288-294.

33. Wu-Wong JR, Chiou W, Magnuson SR et al. Endothelin receptor in human astrocytoma U373MG cells: binding, dissociation, receptor internalization. J Pharm Exp Ther 1995; 274:499-507.

34. Wu-Wong JR, Chiou W, Dixon DB et al. Dissociation characteristics of endothelin ET_A receptor agonist and antagonists. J Cardiovascular Pharmacol 1995; 26:S380-S384.

35. Waggoner WG, Genova SL and Rash VA. Kinetic analyses demonstrate that the equilibrium assumption does not apply to [^{125}I]endothelin-1 binding data. Life Science 1992; 51:1869-1876.

36. Takasuka T, Horii I, Furuichi Y et al. Detection of an endothelin-1-binding protein complex by low temperature SDS-PAGE. Biochem Biophys Res Commun 1991; 176:392-400.

37. Ihara M, Saeki T, Fukuroda T et al. A novel radioligand [^{125}I]BQ-3020 selective for endothelin (ET_B) receptors. Life Science 1992; 51:47-52.

38. Nambi P and Pullen M. [^{125}I]-BQ3020, a potent ET_B-selective agonist displays species differences in its binding characteristics. Neuropeptides 1995; 29:191-196.

39. Nambi P, Pullen M and Spielman W. Species differences in the binding characteristics of [^{125}I]IRL-1620, a potent agonist specific for endothelin-B receptors. J Pharm Exp Ther 1993; 268:202-207.

40. Nambi P, Pullen M and Aiyar N. Correlation between guanine nucleotide effect and reversible binding property of endothelin analogs. Neuropeptides 1996; 30:109-114.

41. Sakamoto A, Yanagisawa M, Sawamura T et al. Distinct subdomains of human endothelin receptors determine their selectivity to endothelin$_A$-selective antagonist and endothelin$_B$-selective agonists. J Biol Chem 1993;268:8547-8553.

42. Becker A, Haendler B, Hechler U et al. Mutational analysis of human endothelin receptors ET_A and ET_B. Identification of regions involved in the selectivity for endothelin-3 or cycle-(D-Try-D-Asp-pro-D-Val-Leu). Eur J Biochem 1994; 227:951-958.

43. Takasuka T, Sakurai T, Goto K et al. Human endothelin receptor ET_B. J Biol Chem 1994; 269:7509-7513.

44. Spinella MJ, Kottke R, Magazine HI et al. Endothelin-receptor interaction. FEBS lett 1993; 328:82-88.

45. Marsault R, Feolde E and Frelin C. Receptor externalization determines sustained contractile responses to endothelin-1 in the rat aorta. Am J Physiol 1993; 264:C687-C693.

46. Chun M, Liyanage UK, Lisanti MP et al. Signal transduction of a G protein-coupled receptor in caveolae: Colocalization of endothelin and its receptor with caveolin. Proc Natl Acad Sci 1994; 91:11728-11732.

47. Keen M and MacDermot J. Analysis of receptor by radioligand binding. In: Receptor Autoradiography. Oxford University Press 1993; 22-55.

48. Hancock AA, Surber BW, Rotert G et al. [^{3}H]A-81988, a potent, selective, competitive antagonist radioligand for angiotensin AT_1 receptors. Eur J Pharmacol 1994; 267:49-54.

49. Frelin C and Guedin D. Why are circulating concentrations of endothelin-1 so low? Cardiovascular Research 1994; 28:1613-1622.

50. Sokolovsky M, Shraga-Levine Z and Galron R. Ligand-specific stimulation/inhibition of cAMP formation by a novel endothelin receptor subtype. Biochemistry 1994; 33:11417-11419.
51. Bkaily G, Wang S, Bui M et al. ET-1 stimulates Ca^{+2} currents in cardiac cells. J Cardiovascular Pharmacol 1995; 26:S293-S296.
52. Enoki T, Miwa S, Sakamoto A et al. Long-lasting activation of cation current by low concentration of endothelin-1 in mouse fibroblasts and smooth muscle cells of rabbit aorta. Br J Pharmacol 1995; 115:479-485.
53. Ihara M, Yamanaka R, Ohwaki K et al. [³H]BQ-123, a highly specific and reversible radioligand for the endothelin ET_A receptor subtype. Eur J Pharmacol 1995; 274:1-6.
54. Nambi P, Pullen M, Wu H-L et al. Nonpeptide endothelin receptor antagonists. VII: Binding characteristics of [³H]SB 209670, a novel nonpeptide antagonist of endothelin receptors. J Pharm Exp Ther 1996; 277:1567-1571.
55. Peter MG and Davenport AP. Selectivity of [¹²⁵I]-PD151242 for human, rat and porcine endothelin ET_A receptors in the heart. Br J Pharmacol 1995; 114:297-302.
56. Wu-Wong JR, Dixon DB, Chiou W et al. Endothelin receptor antagonists: effect of serum albumin on potency and comparison of pharmacological characteristics. J Pharm Exp Ther 1997; 281:791-798.
57. Breu V, Hashido K, Broger C et al. Separable binding sites for the natural agonist endothelin-1 and the nonpeptide antagonists bosentan on human endothelin-A receptor. Eur J Biochem 1995; 231:266-270.
58. Moller S and Henriksen JH. Endothelins in chronic liver disease. Scand J Clin Lab Invest 1996; 56:481-490.
59. Noll G, Wenzel RR and Luscher TF. Endothelin and endothelin antagonists: Potential role in cardiovascular and renal disease. Mol Cell Biochem 1996; 157:259-267.
60. Love MP and McMurray JJV. Endothelin in chronic heart failure: current position and future prospects. Cardiovascular Res 1996; 31:665-674.
61. Mathew V, Hasdai D and Lerman A. The role of endothelin in coronary atherosclerosis. Mayo Clin Proc 1996; 71:769-777.
62. Ferro CJ and Webb DJ. The clinical potential of endothelin receptor antagonists in cardiovascular medicine. Drugs 1996; 51:12-27.
63. Michael JR and Markewitz BA. Endothelins and the lung. Am J Resp Crit Care 1996; 154:555-581.
64. Goldie RG, Knott PG, Carr MJ et al. The endothelins in the pulmonary system, Pulmonary Pharmacology 1996; 9:69-93.
65. Hildebrand P, Mrozinski Jr JE, Mantey SA et al. Pancreatic acini possess endothelin receptors whose internalization is regulated by PLC-activating agents. Am J Phsyiol 1993; 264:G984-G993.
66. Chun M, Lin HY, Henis YI et al. Endothelin-induced endocytosis of cell surface ET_A receptors. J Biol Chem 1995; 270:10855-10860.
67. Wu-Wong JR, Chiou W, Magnuson SR et al. Human astrocytoma U138MG cells express predominantly type-A endothelin receptor. Biochim Biophys Acta 1995; 1311:155-163.
68. Warner TD, Allcock GH and Vane JR. Reversal of established response to endothelin-1 in vivo and in vitro by the endothelin receptor antagonists, BQ-123 and PD 145065. Br J Pharmacol 1994; 112:207-213.
69. Ohlstein EH, Nambi P, Lago A et al. Nonpeptide endothelin receptor antagonists. VI: Pharmacological characterization of SB 217242, a potent and highly bioavailable endothelin receptor antagonist. J Pharm Exp Ther 1996; 276:609-615.
70. Wu-Wong JR, Chiou W, Hoffman DJ et al. Endothelins and endothelin receptor antagonists: binding to plasma proteins. Life Sciences 1996; 58:1839-1847.

Functional Coordination and Cooperation Between Endothelin and Nitric Oxide Systems in Vascular Regulation

Michael S. Goligorsky and Joseph M. Winaver

Introduction

The recent identification and characterization of two potent vasoactive substances produced by the endothelium, ET-1 and nitric oxide (NO), resulted in a substantial revision of contemporary vascular physiology. While ET-1 has gained prominence among endothelium-derived vasoconstrictors, NO has emerged as a major molecular species responsible for the endothelium-derived relaxation.[1-4] The enzyme responsible for the synthesis of NO derived from endothelial cells, endothelial NO synthase (eNOS or NOS III), is a constitutively active, calcium-dependent enzyme encoded by locus 7q35-q36 in the human genome (for review, see reference 4). A targeted disruption in eNOS gene locus results in the development of hypertension, directly demonstrating the role of NO in vascular regulation.[5] The promoter region of eNOS gene contains potential binding sites for several transcription factors (e.g., activator proteins 1 and 2 elements, shear stress response element, cAMP response element, among others) which mediate eNOS expression in response to shear stress, cyclic strain, transforming growth factor-β, basic fibroblast growth factor, or lysophosphotidylcholine. Various agonists elevating cytosolic calcium concentration in endothelial cells (e.g., acetylcholine and bradykinin) stimulate the enzyme, resulting in a burst of NO generation. The calcium-independent regulation of enzyme activity, however, is responsible for ~75% of NO produced (for review, see reference 6). Thus produced, NO acts on heme-containing soluble guanylyl cyclase in vascular smooth muscle cells causing vasorelaxation, among other actions. Hence, NO is responsible, in significant part, for the endothelium-derived relaxing activity.

Endothelin Receptors and Signaling Mechanisms, edited by David M. Pollock and Robert F. Highsmith. © 1998 Springer-Verlag and R.G. Landes Company.

The question that has emerged soon after discoveries of endothelium-derived vasoconstrictor and vasodilator activities and persisted throughout investigations of each separately was: Are there any points of interaction along the metabolic and/or signaling pathways between these two reciprocally acting regulators of vascular tone, ET-1 and NO? In this chapter we shall present evidence for multiple points of convergence between ET-1 and NO systems resulting in coordinated and cooperative regulation of vascular tone and vascular adaptation, as well as illustrate the consequences of discoordination within these systems in several pathophysiological situations.

Functional Coupling Between ET and NO Systems

Feedback Regulation of NOS By ET-1

Several investigators have provided functional evidence for expression of an ET receptor, probably the ET_B receptor, in endothelial cells (EC).[7] Using different ET isopeptides and a ligand specific for ET_B receptor, [Glu9]-Sarafotoxin 6b, the specific binding of ET-3 and Sarafotoxin analog to EC and the ability of these ligands to trigger vasodilation in rat aortic rings were demonstrated. In search for direct evidence of ET_B expression by human umbilical vein endothelial cells (HUVEC) and its functional role as a convergence point between ET-1 and NO pathways, we performed a series of experiments in genetically engineered Chinese hamster ovary (CHO) cells and in HUVEC.[7]

First, we cloned the ET_B receptor from human endothelial cells. The nonisopeptide-selective human endothelin receptor was originally isolated from a human placenta cDNA library. To determine if the identical receptor is expressed in endothelial cells, the polymerase chain reaction (PCR) was used to isolate the cDNA from human EC. Sequence analysis confirmed the identity of EC and placental-derived primary translation products, thus establishing the expression of the authentic receptor in cultured ECs.

Functional expression of ET_B receptor and eNOS in CHO cells. ET_B receptor was stably expressed in CHO cells deficient in dihydrofolate reductase, using the eukaryotic expression plasmid pMT2, as previously detailed.[7] Subsequently, transformants were selected in nucleoside-free medium and individual colonies isolated and cloned by limiting dilution. Resistant clones were initially evaluated for receptor integration and expression by genomic hybridization using the ^{32}P-radiolabelled ET_B receptor cDNA as probe. Individual clones displaying optimal integration were isolated and receptor expression was maximized by propagating resistant colonies in increasing concentrations of methotrexate. One such optimally expressing clone was further isolated, expanded and used for further functional analyses.

For double transfectants, CHO-ET_B cells were re-transfected with the cDNA for eNOS using the eukaryotic expression plasmid pMEP4 which contains an inherent hygromycin B resistance gene and an inducible metallothionein promoter upstream from the NOS cloning site. Stable transfectants (CHO-ET_B/NOS) were selected using culture medium supplemented with hygromycin B. Resistant colonies were ring cloned, propagated and functionally evaluated for NO expression using an NO-selective electrode. NOS expression was maximized by supplementation of the media with 50 M $ZnCl_2$ for 5 h prior to functional measurements.

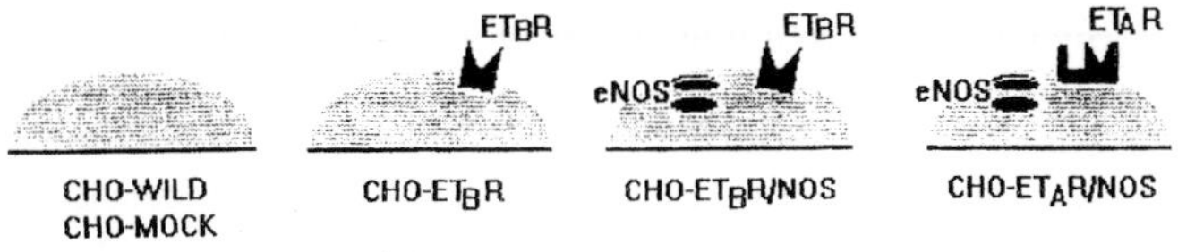

	CHO-WILD / CHO-MOCK	CHO-ET$_B$R	CHO-ET$_B$R/NOS	CHO-ET$_A$R/NOS
RESPONSE TO ET-1:				
- Cytosolic [Ca^{2+}]	-	+	+	+
- NO release	-	-	+	-
Effect of inhibitors:				
- Calmodulin			+	
- PTK			+	
RESPONSE TO IONOMYCIN:				
- Cytosolic [Ca^{2+}]	+	+	+	+
- NO release	-	-	+	+

Fig. 5.1. Summary of ET-1 effects on cytosolic calcium concentration and nitric oxide production in chimeric Chinese hamster ovary cells.

Direct coupling between the ET_B receptor and eNOS in double-transfected CHO cells: Previous attempts to assign functional coupling of ET_B receptors to different targets in endothelial cells have suggested interaction of this receptor with NO production, although these studies have left unresolved the question whether the receptor-enzyme coupling is direct or it requires the synthesis of intermediate autocrine messengers. To further ascertain the possibility of direct functional coupling between the ET_B and NOS, experiments were performed in CHO-ET_B cells stably transfected with the NOS cDNA. Application of 10^{-10}-10^{-8} M ET-1 to CHO-ET_B/NOS cells resulted in an enhanced release of NO, in sharp contrast to the CHO-ET_B cells mock-transfected with the wild-type pMEP4 vector or CHO cells transfected with NOS only, where no NO release was evident (Fig. 5.1). In addition, comparative studies were performed using CHO-NOS cells stably expressing either the ET_A (CHO-ET_A/NOS) or the ET_B receptors. Both cell types responded to administration of 10^{-8} M ET-1 with a prompt increase in cytosolic Ca^{2+} concentration, thus confirming the functional competence of the expressed receptors. Nonetheless, only the CHO-ET_B/NOS cells displayed the ability to release substantial amount of NO into the incubation medium, in contrast to ionomycin-induced release of NO by both cell types, further indicating that the coupling between the ET_B receptor and NOS is specific for this receptor subtype and not necessarily governed by a mere elevation in the concentration of cytosolic Ca^{2+}. These observations were further confirmed using specific inhibitors of L-arginine and second messenger pathways. ET-1-induced NO release by CHO-ET_B/NOS cells was blocked by pretreatment with a competitive inhibitor of NOS, N^G-nitro-L-arginine methyl ester (L-NAME), as well as by a calmodulin inhibitor, calmidazolium (CMZ), cell-permeant calcium chelator, [1,2-bis(2)aminophenoxy]ethane *N,N,N,N*-tetraacetic acid (BAPTA) and by an inhibitor of tyrosine kinase, genistein (GEN). These findings established a direct, highly specific coupling between the ET_B receptor and NOS in a simplified cell system.

ET-1- and IRL 1620-induced NO release from HUVEC: Next, similar studies were completed in HUVEC. Direct monitoring of [NO] in the incubation medium conditioned by HUVEC showed that both ET-1 and IRL 1620 resulted in the stimulation of NO production. As shown in Figure 5.2, these ligands resulted in an immediate and dose-dependent NO generation. These data are consistent with those obtained in $CHO\text{-}ET_B$/NOS cells. Certain distinctions, however, exist and they are described below.

Mapping of the ET_B receptor: Using a biotinylated ligand specific for the ET_B receptor, we next attempted to perform immunofluorescence mapping of binding sites in HUVEC. Vital staining and staining of fixed, permeabilized HUVEC resulted in a poor visualization of ET_B receptors, although receptors were conspicuously detected using antiserum to the ET_B receptor. In contrast to HUVEC, CHO-ET_B cells were readily stained with the biotinylated ligand and showed competitive displacement of this ligand by ET-1. Hence, these data confirmed the specificity of the synthetic biotinylated ligand to the ET_B receptor expressed in CHO cells.

Several possible explanations could be invoked to explain the failure of the biotinylated ligand to map receptors in HUVEC. First, translational defects or post-translational modifications could have accounted for the inability of biotinylated ligand to recognize ET_B receptors in HUVEC. Second, normally expressed ET_B receptors could be partially occupied by ET-1 which is constitutively produced by HUVEC. Third, receptors could be cryptic for other unknown reasons. Given the fact that CHO-ET_B cells, which lack endogenous ET production, were successfully mapped with the biotinylated ligand, the second possibility, namely, the ET_B receptor occupancy in HUVEC, was studied.

We argued that conditions resulting in the release of ET-1 by ECs may facilitate mapping of ET_B receptors. Both the application of thrombin and the exposure to shear stress have been implicated as potent stimuli for ECs to release ET-1, and were utilized in the next series of experiments. HUVEC were exposed to the above stimuli for ET-1 release, washed, fixed, and stained with the biotinylated ligand. Pretreatment resulted in a significant increase in cell-associated fluorescence.[7] Visualization of ET_B receptors following the release of ET-1, but not before it, suggests that the receptors in HUVEC are continuously occupied by ET-1.

Two tentative predictions can be made from these observations. First, due to the constant occupancy of this receptor by the specific ligand and the release of ET-1 upon cell stimulation, it is quite possible that ET_B receptor serves a function of a storage site for the peptide. It is well-accepted that ET-1 has no intracellular storage compartment and is released from the cell as soon as it is produced. The proposed receptor-storage would have several advantages by providing some degree of regulation to ET-1 secretion via modulation of both the binding and burst-release of the ligand. The second prediction to be made, which stems from the observed occupancy of ET_B receptor with ET-1 and the signaling from this receptor to eNOS, is concerned with the current belief in the constitutively active eNOS in endothelial cells. Since all studies supporting this idea were performed in the presence of the persisting ET-1 production, hence, in the presence of a constant stimulatory input to maintain the activity of eNOS, it remains to be clarified whether the enzyme is not just constitutively expressed, rather than being constitutively

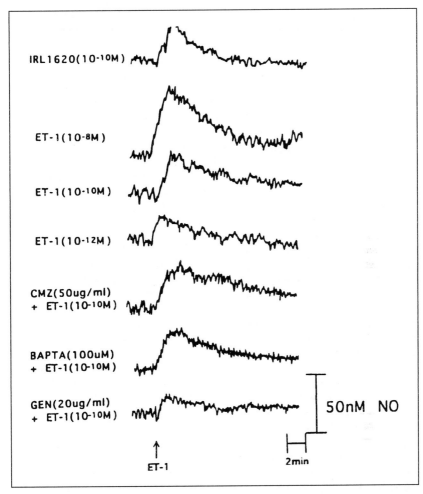

Fig. 5.2. Recordings of nitric oxide release from human umbilical vein endothelial cells. Abbreviations: CMZ-calmidazolium, BAPTA-[1,2-bis(2)aminophenoxy]ethane N,N,N,N,-tetraacetic acid, GEN-genistein. Reprinted with permission of the American Society for Biochemistry and Molecular Biology from reference 7.

active. It would be interesting, therefore, to examine eNOS activity under conditions when ET-1 production is suppressed. These peculiar properties of EC, probably, underscore certain mechanistic differences in ET_B receptor-eNOS coupling, as described below.

Tyrosine kinase inhibitors uncouple ET_B receptor and NOS in HUVEC: In our experiments, 10^{-10}- 10^{-8} M concentrations of ET-1 (which resulted in a significant release of NO) did not affect cytosolic Ca^{2+} concentration in HUVEC, although CHO-ET_B cells showed Ca^{2+} mobilization in response to 10^{-8} M ET-1 (7). Ionomycin alone resulted in a significant NO release from HUVEC, and this effect was inhibited with CMZ or BAPTA. To elucidate the role of calcium-calmodulin pathway in

ET-1-induced increase in NO production, in the next series of experiments, HUVEC were pretreated with CMZ prior to the application of 10^{-10} M ET-1. In contrast to CHO- ET_B/NOS, this pretreatment did not inhibit the ET-1-induced increase in NO production by HUVEC, suggesting that the stimulation of the calcium-calmodulin pathway is not an obligatory prerequisite for the activation of NOS following exposure of endothelial cells to ET-1 (but not to ionomycin the effect of which is blocked by CMZ or BAPTA). Similarly, BAPTA pretreatment did not affect NO release from stimulated HUVEC, while it significantly inhibited NO release in CHO-ET_B/NOS cells. In contrast to the results obtained with the inhibition of calcium/calmodulin system, an inhibitor of protein tyrosine kinase activity, GEN, virtually abolished the ET-1-induced increase in NO release from HUVEC. Using a panel of different tyrphostin derivatives, we demonstrated that ET-1-induced NO release from HUVEC was attenuated in a dose-dependent fashion by methyl-2,5-dihydroxycinnamate and lavendustin A, but not by an inactive analog, tyrphostin 1. These data suggest that the activation of ET_B receptor leads to the stimulation of eNOS via a novel alternative pathway, namely activation of protein tyrosine kinase.

The data obtained with real-time monitoring of [NO] using the NO-selective electrode imply that the ET_B receptor is coupled to NOS. The observations made with different inhibitors of signaling pathways suggest that the involvement of the calcium-calmodulin pathway, although sufficient, is not an obligatory prerequisite for the activation of eNOS, and argue in favor of protein tyrosine kinase activation as an intermediate mechanism coupling the ET_B receptor and constitutive NOS. Using two distinct model systems and physiological concentrations of ET-1 (10^{-10}-10^{-8} M), the above data indicate that the receptor/effector coupling is not governed solely by the elevation of cytosolic Ca^{2+}. The most plausible explanation for this apparent failure of calcium/calmodulin inhibitors to affect ET-1-induced NOS activation is that the partial occupancy of the ET_B receptors by the endogenously synthesized and released ET-1 in HUVEC produces a tonic stimulation of the calcium/calmodulin pathway. This would result in the permanent activation of calmodulin and explain the observed failure of calcium chelators and calmodulin inhibitors to affect the ET-1-induced NO production. Hence, we hypothesize that the constitutive endogenous production of ET-1 by HUVEC may represent a cardinal difference in the receptor-enzyme coupling in HUVEC, as compared to CHO cells.

It should be mentioned that recent data on the increased production of NO by the endothelial cells stimulated with insulin-like growth factor-I, a known ligand for the receptor tyrosine kinase, also support the possibility that an alternative, tyrosine kinase-dependent pathway is involved in activation of constitutive NOS.[8] A panel of tyrphostin derivatives has recently been shown to attenuate NO release from activated murine macrophages, human monocytes, and endotoxin-stimulated rat aortic smooth muscle cells.[7,8] Collectively, these data implicate a tyrosine kinase-dependent pathway in the activation of both the constitutive and the inducible forms of NOS.

The above observations provide strong evidence of direct coupling between the ET_B receptor and eNOS in the genetically-engineered CHO cells and outline the existing similarities and differences in HUVEC signaling between the receptor and the enzyme, thus supporting the hypothesis that the ET_B receptor serves as a storage pool for ET-1 and is coupled to NO synthase. Such a structure of the regulatory network and intercommunication between the two potent endothelium-derived vasoactive moieties is physiologically meaningful in providing a cooperative and coordinated system for the local control of vascular tone. The next question to be addressed is concerned with the possibility of NO modulation of ET-1 signaling in vascular smooth muscle cells, as discussed below.

Role of Nitric Oxide in Termination of ET-1 Signal

Effects of nitric oxide on ET-1-induced $[Ca^{2+}]_i$ transients. To examine any possible effects of NO on ET-1 signaling via ET_A receptor, again a simplified genetically engineered system, CHO cells stably transfected with ET_A receptor, was utilized. The reason behind this selection of a model was that ET-1-induced $[Ca^{2+}]_i$ transients can be obscured by the activation of nonselective cationic stretch-activated channels in vascular smooth muscle cells contracting in response to ET-1.

ET-1 caused a robust $[Ca^{2+}]_i$ response in CHO-ET_A cells (but not in wild-type CHO cells), as previously described.[9] Effects of ET-1 on CHO-ET_A cell $[Ca^{2+}]_i$ were characterized by (1) a profound homologous desensitization persisting for at least 30 min after removal of the agonist and (2) a slow rate of recovery of elevated $[Ca^{2+}]_i$ (Fig. 5.3). Application of NO-donor, 3- morpholino-sydnonimine HCl, SIN-1, shortly before (25 sec or less) or after (50-100 sec) stimulation with ET-1 resulted in accelerated recovery of $[Ca^{2+}]_i$. Furthermore, exposure to SIN-1 (or sodium nitroprusside [SNP]) prior to or immediately after stimulation with ET-1 resulted in abrogation of homologous desensitization. However, prolonged exposure to SIN-1 before application of ET-1 (100 sec) caused a decline in the amplitude of $[Ca^{2+}]_i$ responses to the agonist, not affecting the amplitude of subsequent responses. Hence, the results demonstrate that NO affects (1) the amplitude of $[Ca^{2+}]_i$ responses to ET-1, (2) the rate of recovery of $[Ca^{2+}]_i$, and (3) the development of homologous desensitization.

When the sequential pulses of ET-1 stimulation were separated by a 1 min wash-out period, the second response did not occur. Forty percent recovery of the second response was detected after a 3 min wash-out period, and full-scale responses were registered after a 10 min wash-out period.

The possible role of cyclic GMP in modulation of ET-1-induced $[Ca^{2+}]_i$ transients was examined by substituting SIN-1 with a membrane-permeant 8-bromo-cyclic GMP.[9] Neither the development of homologous desensitization nor the amplitude of $[Ca^{2+}]_i$ responses to ET-1 were affected by cyclic GMP in CHO-ET_A cells. These data suggest that the observed effects of NO on parameters of $[Ca^{2+}]_i$ response to ET-1 in CHO-ET_A cells were not mediated by the soluble guanylate cyclase-cyclic GMP system, which is the only known intracellular signaling system for SIN-1 or SNP.

Receptor-mediated effects of nitric oxide: To overcome existing obstacles for intravital studies of ET-1/ET_A interaction, we have synthesized a biotinylated analog of endothelin which provided the benefit of vital fluorescence "mapping" of ET_A receptors and their dynamics.[9] Fluorescence microscopy and image analysis demonstrated minimal decay of fluorescence intensity over 1 hr of observation. Distribution of fluorescence was unchanged during this period of time. These observations suggested a unique stability of ligand-receptor complexes, consistent with the previous finding of poor dissociation of ET-1 from its receptor.

Exposure of CHO-ET cells stained with ET-1[BtK9] to SIN-1 or SNP, donors of NO, resulted in a rapid decay of fluorescence (Fig. 5.4). When SIN-1 or SNP were removed, the cells regained the ability to bind ET-1[BtK9], demonstrating the

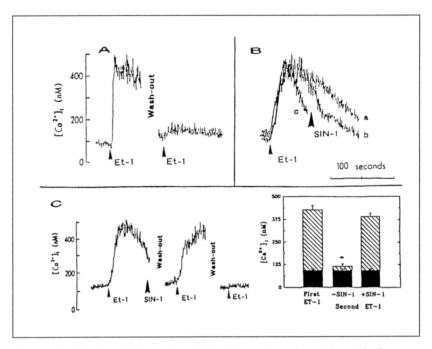

Fig. 5.3. Modulation of ET-1-induced [Ca^{2+}]$_i$ transients by nitric oxide donor 3-morpholino- sydnonimine HCl (SIN-1). (A-C) application of 1 nM ET-1 (small arrowheads) to CHO-ET$_A$ receptor cells results in development of homologous desensitization which is alleviated by a brief treatment with 10 uM SIN-1 (large arrowhead). This concentration of SIN-1 also results in the accelerated recovery of [Ca^{2+}]$_i$ transients, as depicted in B (a-control tracing, b-SIN-1 was added at the arrowhead, c-SIN-1 was added 25 sec prior to ET-1). Reprinted with permission of John Wiley & Sons, Inc. from reference 9.

reversibility of NO action. Endocytosis of ET-1[BtK9] contributed less than 15% to the total cell-associated fluorescence intensity, as judged from the residual fluorescence insensitive to trypsin treatment. This endocytosed fraction of the probe did not exhibit changes in fluorescence intensity upon exposure to SIN-1.

The EC$_{50}$ of the SNP-induced decrease in fluorescence of biotinylated ET-1 averaged 75 µM. SIN-1 modified the affinity of the receptor to the ligand with the EC$_{50}$ approximating 6x10^{-6} M.[9] Cyclic GMP failed to decrease the fluorescence of CHO-ET cells labeled with biotinylated ET-1. This finding does not support the possibility that NO dissociates the ET-1-receptor complex in CHO-ET cells via generation of cyclic GMP. Collectively, these data indicate that NO modifies the affinity of the ET$_A$ receptor to ET-1.

Postreceptor effects of nitric oxide. The effects of mastoparan, a wasp venom compound directly activating pertussis toxin-sensitive GTP-binding proteins, thus bypassing the receptor-mediated activation, on [Ca^{2+}]$_i$ were examined in CHO cells expressing ET$_A$ receptors.[9] Application of mastoparan evoked an elevation of [Ca^{2+}]$_i$. Pretreatment of cells with SIN-1 for 100 sec resulted in a dose-dependent attenuation of mastoparan-induced [Ca^{2+}]$_i$ transients. These data suggest that NO has an

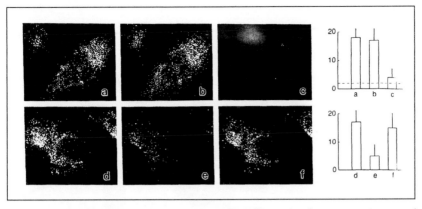

Fig. 5.4. Mapping of ET_A receptors in CHO-ET_A cells. (a-c)—fluorescence images of cells stained with 1 nM ET-1[BtK9] under control conditions, again after 60 min incubation, and 5 min following the application of 10 uM SIN-1, respectively. (d-f)—reversibility of SIN-1 effect: initial image of 1 nM ET-1[BtK9] labeled cells, 5 min after exposure to SIN-1 (as in c), and following repeated labeling with 1 nM ET-1[BtK9], respectively. Bars on the right summarize the relative fluorescence intensity. Reproduced with permission of John Wiley & Sons, Inc. from reference 9.

additional target(s) along the pathway between GTP-binding proteins and calcium mobilization. To examine the possibility that activation of GTP-binding proteins *per se* leads to the dissociation of the ligand-receptor complexes, cells labeled with ET-1[BtK9] were exposed to mastoparan. Monitoring of fluorescence intensity over the labeled cells, however, did not show any effect of mastoparan on the binding of ET-1[BtK9]. These observations suggest that, although NO interferes with postreceptor events leading to the mobilization of $[Ca^{2+}]_i$, its accelerated dissociation of ET-1-ET_A receptor complexes is distinct from the effect on intracellular messengers.

A working hypothesis on the cellular targets of NO is presented schematically in Figure 5.5. NO acts on a target cell by a) modifying the affinity of receptors to their respective ligands and b) interfering with the intracellular signaling cascade leading to mobilization of intracellular calcium. The latter effect may be related to the activation of guanylyl cyclase.

"Ping-pong" Hypothesis of NO/ET-1 Interactions

This hypothesis is based on the above demonstration of (1) the expression of mRNA for the ET_B receptor by endothelial cells, (2) the expression of the ET_B receptor on endothelial cells, (3) the coupling of the receptor to the NO synthase and (4) NO-induced dissociation of ET-1 from the ET_A receptor. Using several genetically-engineered CHO cells, we have demonstrated a direct coupling, not requiring any paracrine or autocrine intermediates, between the ET_B receptor and eNOS; dissociation of ET-1 from the ET_A receptor and postreceptor actions of NO. Extrapolating these findings to endothelial cell-smooth muscle cell interaction, one can envisage the following scenario (Fig. 5.6). ET-1, in addition to acting in a paracrine fashion and contracting smooth muscle cells, acts as an autocrine mediator of NO release from endothelial cells. The latter acts on smooth muscle cells

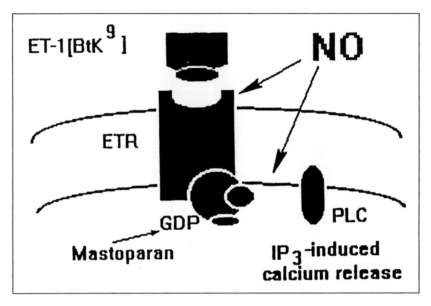

Fig. 5.5. The possible targets for nitric oxide: accelerated dissociation of ET-1 from the ET$_A$ receptor and inhibition of Ca^{2+} mobilization from the intracellular stores.

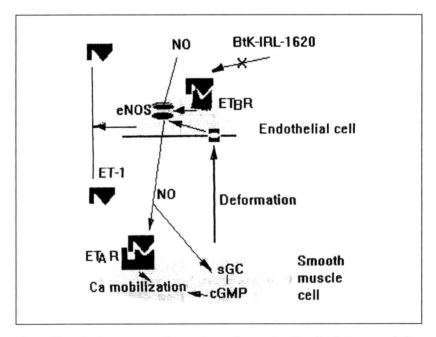

Fig. 5.6. Hypothetic summary of autocrine and paracrine signaling between endothelial and smooth muscle cells through coordinated regulation of ET-1 and NO release and action. See text for the details.

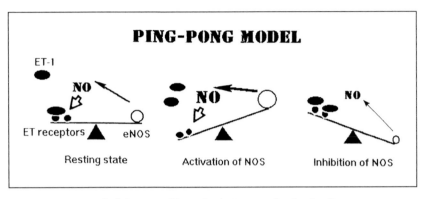

Fi.e 5.7. Summary of a "ping-pong" hypothesis. See text for the details.

to dissociate ET-1 from ET_A receptors and by interfering with ET-induced mobilization of $[Ca^{2+}]_i$; it relaxes the cell. During this process, on the one hand, previously occupied ET_A receptors become again available for ET-1, and on the other hand, endothelial cells undergo deformational activation, turning on the release of ET-1. It is obvious that the above closed-loop regulatory cycle contains in itself the capability for auto-oscillations, an intrinsic property of vascular wall. The "ping-pong" hypothesis, schematically depicted in Figure 5.7, allows one to make a prediction that inhibition of NO production should not only eliminate its tonic vasorelaxing action, but should also potentiate the effect of ET-1 (due to the defect in termination of ET-1 signaling) and other Ca-mobilizing vasoconstrictors (due to the unopposed Ca-mobilization).

ET-1 and NO in Hemodynamic Adaptations

Regulation of Renal Medullary Circulation

The mammalian renal medulla is a major source of production both of ET-1 and of NO.[10,11] Both agents have been shown to exert important paracrine/autocrine actions in this region of the kidney, related to salt and water transport in the collecting duct and PGE_2 production.[12,13] Recently, we provided evidence suggesting an important interaction of both agents in the modulation and maintenance of renal medullary blood flow.[14]

Under normal physiological conditions only 10-15% of the renal blood flow circulates through the medulla. This limited blood supply together with the unique hairpin like structure of the medullary blood vessels, the vasa recta, are essential for preservation of a normal concentrating ability by the kidney, but at the same time markedly decreasing oxygen supply to the medulla.[15] The finding of high immunoreactive levels of ET in the renal inner medulla could be of crucial importance during pathophysiological activation of the renal ET system, since such a potent vasoconstrictor could further diminish the limited medullary blood supply and accentuate the hypoxia present in this region even under physiologic conditions.[15] We therefore speculated that adaptive mechanisms must exist to protect medullary structures from the potent vasoconstrictor effect of ET. Indeed, systemic infusion of ET-1 at a dose of 1.0 nmol/kg which decreased cortical blood flow

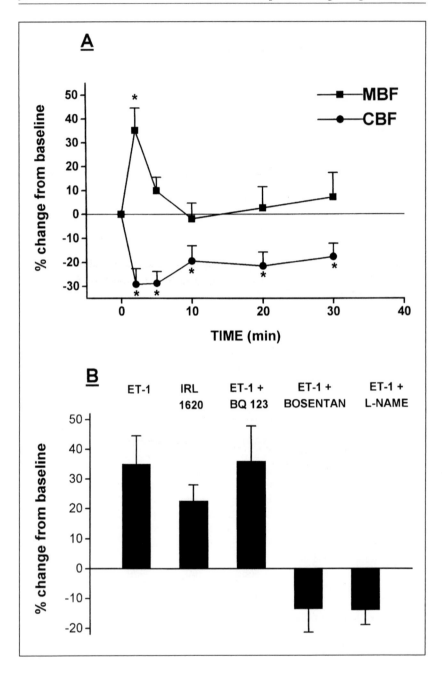

Fig. 5.8. A. Effects of bolus injection of ET-1 (1.0 nmol/kg) on renal cortical blood flow (CBF) and medullary blood flow in normal rats. Data are expressed as percent change from baseline. B. Effects of pretreatment with various agonists and antagonists of ET receptors and of NO synthase blockade, on ET-1 induced medullary vasodilatation (peak response). See Text for further explanation (Modified from ref. 14).

by more than 30% was associated with a transient increase in medullary blood flow by 35% (Fig. 5-8A). The medullary vasodilatation induced by ET-1 was dose-dependent and was observed with doses lower than that required to produce cortical vasoconstriction, suggesting that the medullary circulation is highly sensitive to vasodilatory action of ET-1. As demonstrated in Figure 5.8 B, the ET-1 related medullary vasodilatation was not affected by pretreatment with the ET_A receptor antagonist BQ-123, but was completely blocked by the mixed ET_A/ET_B receptor antagonist bosentan and was mimicked by IRL 1620, a potent and selective ET_B agonist. Moreover, in rats chronically pretreated with the NO synthase blocker L-NAME, ET-1 administration failed to increase medullary blood flow and, in fact, produced a slight vasoconstrictor response.[14] These data clearly suggest that the vasodilatory effect of ET-1 in the renal medulla is mediated by activation of the ET_B receptor and is apparently linked to increased production of NO. Thus, it might be speculated that the renal medulla, by adopting primarily the vasodilator pathway of ET action, maintains adequate perfusion and may protect itself from the potent vasoconstrictor effect of the high concentrations of ET present in this region of the kidney. The finding of a higher preponderance of the ET_B receptor in the renal medulla as compared to the renal cortex is in concert with the latter proposal.[16]

Vascular Remodeling and Angiogenesis: Permissive Role of NO in ET-induced Endothelial Cell Migration

The recent finding that ischemia results in increased production of ET-1 by several tissues raises an important question on the biological significance of such a response. If vasoconstriction is the only function of ET, overexpression of ET-1 would necessarily aggravate local hemodynamics and oxygen supply. It seems, however, that under these pathophysiological conditions other functions of ET-1 predominate. One of such functions may be an adaptive stimulation of vascular remodeling.

Formation of new blood vessels, driven by the morphogenetic program and/or by the functional demand for increased blood supply, is initiated by the budding of endothelial cells off the microcirculatory bed. Tissue hypoxia represents a physiologic stimulus for angiogenesis, and several autocrine angiogenic mediators produced by endothelial cells have been recognised.[17] Endothelin production is also enhanced by hypoxia.[18,19] Two investigative teams have recently demonstrated that ET-1 and ET-3, acting via the ET_B receptor, stimulate endothelial cell migration and proliferation.[20,21] As discussed above, the occupancy of ET_B receptors in endothelial cells is accompanied by the activation of constitutive eNOS,[7] raising the question whether the observed motogenic effects of members of ET family may in fact be mediated by the release of NO. Indeed, there is emerging evidence that NO release serves as a prerequisite for epithelial and endothelial cell motility. Leibovich et al have shown that production of angiogenic activity by activated monocytes (assayed by chemotaxis of endothelial cells and corneal angiogenesis) is absolutely dependent on L-arginine and NO synthase.[22] These observations are in concert with findings reported by Ziche et al who detected the potentiation by sodium nitroprusside of the angiogenic effect of substance P in the rabbit cornea.[23] Our own observations expand this function to the classical angiogenic signal, vascular endothelial growth factor. We demonstrated that endogenous NO production by the endothelial cells is a prerequisite for the motogenic and angiogenic effects of this factor.[24]

To address this question, three different approaches were used (migration and wound healing by endothelial cells, wound healing by Chinese hamster ovary cells expressing ET$_B$ receptor with or without eNOS, and application of antisense oligodeoxynucleotides targeting eNOS) to provide evidence that the effect of ET-1 on cell migration is mediated via ET$_B$ receptor and requires functional enzymatic machinery for NO generation.

Motogenic effect of endothelin is mediated via ET$_B$ receptor and is NO-dependent. Renal microvascular EC used in this study displayed sensitivity to ET-1 within the physiologic concentration range. ET-1 stimulated NO production by rat microvascular endothelial cells (RMVEC), and L-NAME blunted NO release in a dose-dependent manner with 2 mM L-NAME resulting in complete inhibition of NO release. Trans-well migration of RMVEC, studied in a modified Boyden chamber, showed the similar exquisite sensitivity to ET-1. Statistically significant almost 50% increase in the number of migrated cells was detected with ET-1 concentrations of 10-100 pM, and it almost doubled with further elevation of ET-1 concentration to 1 nM. Inhibition of NO synthase with 2 mM L-NAME completely abrogated the ET-1-induced cell migration.

To elucidate the type of ET receptor(s) involved in the observed stimulation of RMVEC migration, trans-well migration experiments were performed with different endothelins and in the presence of selective inhibitors of their receptors.[25] ET-1 and ET-3 were equipotent in stimulating endothelial cell migration. A selective ET$_B$ agonist, IRL 1620 exhibited similar stimulatory effect, suggesting that the observed phenomena were mediated via ET$_B$ receptors. This conclusion is further supported by the observations that a selective ET$_A$ antagonist BQ-123 did not affect ET-1-induced endothelial cell transmigration, whereas a selective ET$_B$ antagonist BQ-788 completely abrogated the stimulatory action of ET-1. It is therefore concluded that ET-1 induces endothelial cell migration via the ET$_B$ receptor, and that the previously demonstrated coupling of this receptor to eNOS is critical for NO production and cell migration.[7] A cell-permeant analog of cyclic GMP, 8-bromo-cyclic GMP, restored ET-1-induced locomotion of endothelial cells pretreated with L-NAME, suggesting that the effect of NO on cell migration is mediated via cyclic GMP signaling pathway.

Effects of endothelin on the rate of endothelial wound healing. The rate of endothelial wound healing was examined by electroporation-induced RMVEC denudation from the surface of a flat microelectrode, as previously detailed.[25] ET-1 accelerated wound healing at 10 h by 66% compared to control medium 40%. The rate of restitution of endothelial integrity was equally enhanced by IRL 1620 and by ET-3, and this effect was abolished by coapplication of ET$_B$ receptor antagonist BQ-788, further implicating ET$_B$ receptor in the observed phenomena. The addition of L-NAME to ET-1- or ET-3-stimulated endothelial cells virtually abrogated this response.

Effects of endothelin on wound healing in CHO cells. To further test the above premise that NO production is required for ET-1-induced RMVEC migration, experiments were performed in a model system of CHO cells expressing ET$_B$ receptors and eNOS, as verified and detailed previously.[7] ET-1 was ineffective in stimulating wound healing in CHO cells transfected with ET$_B$ receptors alone. However, coexpression of both ET$_B$ receptors and eNOS in CHO cells imparted on them the migratory response to ET-1 and resulted in acceleration of wound healing. The data further confirm the permissive role of NO production in ET-1-induced migration in a simplified model of CHO cells which are devoid of complex autocrine signaling characteristic for endothelial cells.

Effects of endothelin in the presence of antisense oligodeoxynucleotides targeting eNOS. A selective downregulation of eNOS was performed using antisense S-ODNs.[25] To avoid any possible species-specific variations in the sequence, experiments were performed in HUVEC using an antisense construct directed against the initiation codon of human eNOS cDNA. While intact HUVEC, as well as HUVEC pretreated with control sense or scrambled constructs showed immunodetectable eNOS, cells pretreated with the antisense S-ODN revealed only a weak staining with monoclonal antibodies to eNOS. Both sense and scrambled S-ODN-treated cells responded to ET-1 with increased NO release, similar to that observed in intact control cells. In contrast, antisense S-ODN-treated HUVEC failed to produce NO in response to ET-1 (Fig. 5.9). These data confirm the validity of the utilized S-ODN antisense construct. To study the effects of ET-1 on motility of S-ODN-treated HUVEC, experiments were performed in a modified Boyden apparatus. Pretreatment of HUVEC with antisense S-ODN resulted in a dramatic deceleration of transmigration, as compared to intact control cells and cells exposed to sense or scrambled constructs.

Hence, two different endothelial cell preparations, derived from rats and humans, showed a consistent acceleration of motility in response to members of endothelin family. Through application of specific agonists and antagonists of known endothelin receptors, ET_A and ET_B, it was possible to implicate ET_B receptor in the observed responses. However, these effects of ET were mediated by endogenous NO production, as asserted to by the results of three independent experimental approaches. In the first series of experiments, L-NAME prevented ET-1-induced endothelial cell trans-well migration. In the second, CHO-ET_B/NOS cell, a model cell system devoid of complexity inherent to endothelial cells, showed NO-dependency of migration induced by ET-1. Finally, application of eNOS isoform-selective antisense S-ODNs, but not sense or scrambled constructs, to endothelial cells suppressed their migratory responsiveness to ET-1. Therefore, stimulated NO production serves a permissive role in ET-induced acceleration of endothelial cell motility and wound healing.

The exact mode of NO action on endothelial cell migration is presently unknown. We have recently demonstrated that locomoting epithelial cells, also dependent on NO production, is accompanied by polarized distribution of NOS, thus suggesting a gradient in NO release during cell migration.[26] Furthermore, endothelial cell locomotion triggered by the vascular endothelium growth factor showed the similar requirement for the functional NOS.[24] In these cells NO stimulated micromotions, which we termed podokinesis, indicative of accelerated turnover of focal adhesions. Based on these observations and the results presented above, we propose that endothelial cell migration requires simultaneous action of two signals: (1) NO to agitate podokinesis and (2) guidance cues, e.g., ET-1 or vascular endothelial growth factor (VEGF), to secure the necessary direction of movement.

The described phenomenon of NO production, serving as a prerequisite for endothelial cell locomotion in response to activation of ET_B receptor, may explain a host of pathophysiologic observations on inadequate angiogenesis despite enhanced generation of ET-1. In hypercholesterolemic pigs and in humans with atherosclerotic lesions, ET-1 generation is augmented.[27-29] The reason for the

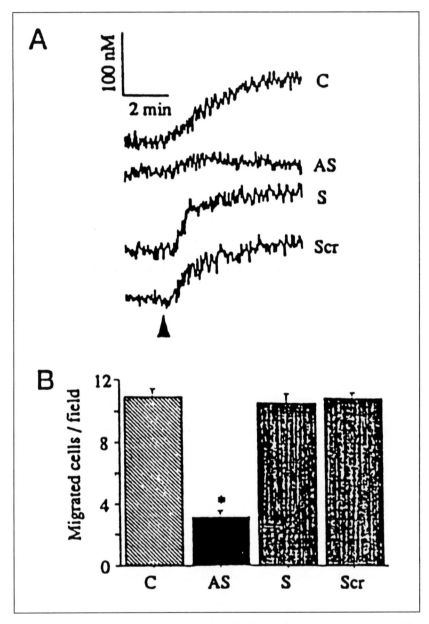

Fig. 5.9. Effects of antisense phosphorothioated oligonucleotides targeting the initiation codon of human eNOS cDNA on the release of nitric oxide (A) and transwell migration (B) of human umbilical vein endothelial cells. Abbreviations—AS, S, Scr indicate antisense, sense and scramble constructs, respectively. * denotes $p < 0.05$ versus control. An arrowhead in A shows the time of ET-1 application. Reprinted with permission of the American Society for Biochemistry and Molecular Biology from reference 25.

insufficient angiogenesis towards ischemic sites could be conceptually explained by the fact that NO production by the endothelium is suppressed under these conditions.[30-32] Similarly, endothelial wound healing after balloon angioplasty is retarded, neointimal formation by proliferating smooth muscle cells is enhanced, and endothelin receptor antagonist SB 209670 protects angioplastic vessels against neointimal formation.[33] Interpretation of these observations again rests on the above presented findings indicating the permissive role of NO in endothelial cell migration and wound healing. Based on the contribution of endothelial cell migration to angiogenesis, these data may implicate insufficient NO production in pathologic states, e.g., atherosclerosis, heart failure and hypertension, in the inappropriate response to angiogenic stimuli.

Pathophysiology of ET and NO Coordination

Congestive Heart Failure

Congestive heart failure (CHF) is a clinical syndrome characterized, among its many facets, by activation of several neurohumoral systems with potent vasoactive properties.[34] These include vasoconstrictor systems such as the sympathetic nervous system, renin angiotensin and antidiuretic hormone as well increased production of vasodilator agents such prostaglandin E_2 and the natriuretic peptide family. Evidence for participation of the paracrine ET and NO systems in the pathophysiology of CHF is steadily accumulating. However, their exact role in the pathogenesis of CHF, as well as their relative contribution to the vascular alterations in this syndrome is far from being established.

Kaiser et al originally documented that acetylcholine infusion directly into the femoral artery of dogs with CHF, resulted in a blunted vasodilatory response.[35] Similar observations were later reported in other vascular beds such as the thoracic and abdominal aorta[36,37] and peripheral arteries,[38-40] both by in vivo infusions of acetylcholine and NO synthase blockers, as well as in isolated vessels studied in vitro. The mechanisms responsible for the impaired vasodilatory response to acetylcholine in CHF have not been elucidated at present. A blunted response to endothelial-dependent vasodilators has been generally equated with a decrease in NOS activity and NO generation. Thus, it has been speculated that such a decrease in NO production may be secondary to a reduced shear stress associated with the decline in cardiac output in CHF.[41] Another hypothesis suggests that increased activity of angiotensin converting enzyme in blood vessels in CHF is associated with accelerated degradation and lower local levels of bradykinin.[42] The latter is thought to be a potent stimulus of NO release and therefore chronically reduced bradykinin levels may result in downregulation of the NO synthase.[42] The notion of impaired NO generation in CHF is supported by the finding of low cGMP content in vascular segments of rats with CHF.[43] However, additional experimental evidence suggest that basal levels of NO release may be preserved or even increased in CHF. This is based on the finding of an exaggerated vasoconstriction in response to NO synthase blockade[44,45] and high circulating levels of nitrate, the stable end product of NO, in patients with CHF.[46] Taken together these data suggest a dissociation between a preserved or enhanced basal NO release and an impaired stimulated NO production. Thus, it is the vasodilatory activity of NO, rather than its production, which is impaired in CHF. The issue is further complicated by the demonstration that endothelial dysfunction in heart failure is a progressive time-dependent process which may develop in an advanced stage of cardiac failure.[37] Recent data from our laboratory suggest that increased activity of the renin-angiotensin system may also contribute to the impaired vasodilatory action of NO in rats with experimental CHF induced by aorto-caval fistula,[47] supporting the findings of Mulder et al in rats with coronary ligation model.[48] In summary, the mechanism of endothelial dysfunction in CHF appears to be complicated and may differ at stages of the development of cardiac dysfunction. Evidently, many questions regarding the role of NO in CHF remain unanswered at present. It is clear, however, that such an impaired activity of the NO system could contribute to the development of the systemic vasoconstriction in CHF.

Several reports have raised the notion that CHF is associated with increased activity of the ET system, and that ET could contribute to the increased vascular tone in CHF. Thus, plasma levels of ET-1 and its precursor, Big ET-1, have been shown to be elevated in patients as well as in experimental animals with CHF.[49-51] The high circulating levels of the peptide in CHF apparently reflect its increased production and spill-over from vascular wall to plasma. Moreover, Wei et al were able to demonstrate a good correlation between circulating ET levels and various parameters reflecting the severity of heart failure, such as, cardiac index, left ventricular end diastolic volume index and NYHA classification.[51] Other studies have documented that CHF is associated with several other alterations in the ET system. These include changes in the myocardial ir-ET levels and ET receptor binding,[52] increased expression of the mRNAs of ET-1 and the ET_A receptor subtype in the myocardium,[53] and a blunted vasoconstrictor response to exogenous administration of ET-1.[49]

Because of its mitogenic properties, cardiac ET-1, acting in autocrine/paracrine fashion, has been implicated in the pathogenesis of cardiac hypertrophy in CHF.[54,55] Indeed, it has been demonstrated that chronic administration of BQ-123, a specific ET_A receptor antagonist was able to prevent cardiac hypertrophy induced in rats by aortic banding, a model of pressure overload CHF.[56] Finally, activation of the ET system may also participate in the induction of the structural changes in blood vessels in CHF. Thus, ET-1 exerts a potent mitogenic activity on cultured vascular smooth muscle cells and has been shown to play an important role in the vascular remodeling and neointimal formation following carotid artery balloon angioplasty.[57] Perhaps the most compelling evidence for the involvement of ET-1 in the pathogenesis of CHF has emerged from the development of potent, orally active ET receptor antagonists. The acute administration of bosentan (Ro 47-0203), a nonselective, nonpeptide $ET_{A/B}$ antagonist to rats with experimental CHF and to patients suffering from severe CHF resulted in a significant decrease in mean arterial pressure.[58,59] Moreover, in patients with CHF, bosentan caused a considerable improvement in cardiac function apparently by unloading the failing myocardium.[59] These data clearly suggest that the ET system plays an important adaptive role in the maintenance of blood pressure in CHF, similar to other neurohumoral vasoconstrictor systems activated to support blood pressure in the face of deteriorating cardiac output. In summary, current evidence supports the notion that CHF is associated with both activation of the ET system and impaired activity of the vasodilatory NO system. This imbalance could contribute, among other factors, to the increase in vascular resistance in heart failure.

Liver Cirrhosis

Liver cirrhosis, especially in advanced stages, is characterized by a hyperdynamic circulation, the manifestations of which include peripheral arterial vasodilatation, hypotension and increased cardiac output as well as hyporeactivity of blood vessels from cirrhotic animals to vasoconstrictors.[60,61] The peripheral vasodilatation, which occurs initially in the splanchnic circulation, may lead to compensatory activation of vasoconstrictor systems such as the sympathetic nervous system, renin-angiotensin-aldosterone axis, and ADH release, which may be involved in the renal complications of cirrhosis and eventually result in salt and water retention and ascites formation.[61] In the most advanced stages of liver cirrhosis with ascites, the development of hepato-renal syndrome with the characteristic severe renal vasoconstriction may further complicate the clinical presentation of cirrhosis.[62] Thus, paradoxically, while the systemic circulation exhibits profound vasodilatation, intense vasoconstriction prevails in the renal vasculature. It is not surprising, therefore, that recent studies suggest the involvement of both NO and endothelin in the pathophysiology of cirrhosis and its complications.

In 1991, Vallance and Moncada suggested that overproduction of NO in vascular wall, induced by endotoxins, may account for the arterial vasodilatation in cirrhosis.[63] Since then several studies have confirmed that vascular production of NO or cGMP is increased both in human and in experimental cirrhosis.[64-67] Moreover, inhibition of NO formation in cirrhotic animals is associated with normalization of peripheral vascular resistance and cardiac output as well as the vascular hyporeactivity to vasoconstrictors.[66,68,69] Although these data strongly support the concept that increased production of NO is involved in mediating the hyperdynamic circulation in cirrhosis the mechanisms responsible for the activation of the NO system have not been fully elucidated.[70] The Vallance-Moncada hypothesis implicated the endotoxemia, present in cirrhosis, as a mechanism causing the increased NO formation by the inducible NO synthase (iNOS) isoform, which is known to be regulated by cytokines and endotoxins. The finding that aminoguanidine, a preferential blocker of the iNOS, was able to reverse the hyperdynamic circulation in rats with portal hypertension tended to support this assumption.[71] However, direct evaluation of the NOS isoform in cirrhotic rats by Western blotting analysis, using specific monoclonal antibodies against iNOS and the constitutive endothelial NOS (eNOS), indicated that upregulation of the later isoform rather than iNOS was responsible for the increased NO production.[72] Although one should be cautious in extrapolating from data obtained in CCl_4 cirrhotic rat model to human cirrhosis, these findings are interesting and raise once again the question regarding the cause of the activation of this isoform in cirrhosis. Thus, although the notion of increased NO production in cirrhosis is fairly established at present, the source of NO and the mechanisms mediating this phenomenon remain to be elucidated.

In addition to activation of the NO system several reports also raised the possibility of increased activity of the ET system in cirrhosis. Plasma ET-1 levels were reported to be elevated in cirrhotic patients, with higher circulating levels in patients with cirrhosis complicated by ascites or renal failure.[73-77] Whether these increased levels have any pathophysiological significance remains unresolved at present. Likewise, the source of the increased plasma level of ET in cirrhosis is currently unknown. Because of the potent vasoconstrictor effect of the peptide it is tempting to speculate that the increased production of ET in cirrhosis may be responsible for the sustained and intense renal vasoconstriction of the hepato-renal syndrome. However, the evidence supporting the latter postulate is far from convincing. Thus, the increments in plasma ET levels in patients with hepato-renal syndrome, although statistically significant, are modest in their magnitude. Moreover, an elevation in plasma ET-1 levels may be related to a variety of complicating events in this syndrome such as renal failure, endotoxic shock and disseminated intravascular coagulation, casting further doubt on a cause and effect relationship. The mechanism responsible for the increased plasma levels of ET-1 in liver dysfunction is another debated issue. Recent data by Bernardi et al suggest that this elevation is not mediated by hemodynamic factors since in contrast to plasma renin activity and aldosterone, ET-1 levels may not be suppressed by hemodynamic manipulations.[77] Theoretically, endotoxemia may serve as a common stimulus both for the induction of NO synthesis as well as of ET-1 in cirrhosis, since endotoxin has been reported to stimulate ET synthesis.[78] However, no correlation between plasma endotoxin and ET-1 levels was found in a recent study by Salo et al.[79] It is also possible that the endothelial cells respond by an augmented release of ET-1 to the high circulating concentrations of angiotensin II, norepinephrine and vasopressin found in advanced stages of liver dysfunction. It should be noted that the kidney itself may be an important source of ET in patients with advanced cirrhosis with and without hepatorenal syndrome.[74] Recently, Ackerman et al provided evidence for increased immunoreactive ET-1 levels in the renal medullary and papillary tissues of bile duct ligated and portal vein ligated rats.[80] In view of the paracrine nature of ET-1 action and the high sensitivity of renal vasculature to its vasoconstrictor effect, increased renal production of the peptide in liver cirrhosis may be of critical importance in mediating the renal vascular alterations in hepatorenal syndrome. This assumption, however, requires further validation.

Finally, increased NO production may be responsible, in part, for the vascular hyporesponsiveness to ET-1 reported by Hartleb et al in bile duct ligated rats.[81] This group demonstrated a blunted vasopressor response to ET-1 administration in this experimental model of cirrhosis, which could be partly reversed by NO synthesis blockade. The finding that NO blockade restored the responsiveness only to high doses but not to low dose of ET-1 suggests that other factors may be involved in this phenomenon. Nevertheless, this interesting observation depicts an additional mode of interaction between NO and ET-1 in modulating vascular reactivity in hepatic cirrhosis.

Cyclosporine-induced Hypertension and Nephrotoxicity

Cyclosporine A (CsA) is a potent, widely used, immunosuppressive agent that has revolutionized the field of organ transplantation.[82] However, its long-term clinical use has been hampered by adverse side effects such as acute and chronic nephrotoxicity and hypertension.[83,84] The precise mechanisms by which CsA exerts these side effects are not entirely clear. Yet, the systemic and renal vasoconstriction induced by the drug appear to be of primary importance in mediating these phenomena. CsA-induced vasoconstriction has been attributed to several factors including activation of the renin-angiotensin system, activation of the sympathetic nervous system and increased calcium uptake by vascular smooth muscle cells.[85-77] However, the possibility that the endothelial factors ET and NO, in particular ET-1, may be involved in the pathogenesis of the nephrotoxicity and hypertension is gaining increasing support in recent years, based on the following observations:

First, CsA stimulates ET production by several cell types, including endothelial, epithelial, mesangial and smooth muscle cells.[88-90] Second, CsA induced nephrotoxicity is associated with increased urinary excretion of ET-1 and its precursor Big-ET-1.[91] Third, CsA related renal hypoperfusion and cellular proliferation may be ameliorated by antiiET antibodies and by selective ET antagonists.[92-94] Finally, Abassi et al have recently demonstrated an increased expression of ET-1 mRNA in the renal medulla of rats chronically treated with CsA.[95] Using the technique of reverse transcription followed by polymerase chain reaction, these authors demonstrated a selective increase in ET-1 mRNA level but not in that of ET-3 or in endothelin converting enzyme, following CsA treatment. This suggests that CsA increases renal synthesis of ET-1 by regulating its production at the transcriptional level.[95] Taken together these data clearly support the notion that ET-1 plays an important role in mediating CsA induced nephrotoxicity and vasoconstriction.

Less obvious is the role of NO in the pathogenesis of CsA-induced nephrotoxicity and hypertension. Rego et al demonstrated an impaired vasodilatory response to acetylcholine and sodium nitroprusside as well as low cGMP content in preconstricted aortic rings taken from CsA treated rats.[96] Other studies have confirmed the presence of an impaired endothelial dependent vasorelaxation in CsA treated rats, whereas the endothelial independent vasorelaxation, induced by nitroprusside, was preserved.[97,98] De-Nicola et al further demonstrated that the preglomerular vasoconstriction in CsA treated rats could be partially restored by L-arginine feeding, suggesting that the renal hemodynamic alterations in these animals were mediated in part by diminished NO production in the kidney.[99] In contrast, Bobadilla et al failed to demonstrate a defect in the NO system in experimentally induced CsA nephrotoxicity.[100] In their study, despite of the characteristic increase in preglomerular resistance that resulted in a marked fall in glomerular filtration in CsA treated rats, the ability of the renal endothelium to produce NO was maintained. Moreover, in a recent in vitro study utilizing cultured bovine aortic cells that were exposed to CsA, Lopez-Ongil et al reported an increased production of NO, measured by nitrite concentrations.[101] This augmentation in NO production was associated with a threefold increase in the constitutive eNOS mRNA, as evaluated by Northern blotting. This finding suggests that CsA, in the setting of latter experiment, has the capacity to induce NO production by upregulating eNOS gene expression. This interesting phenomenon may represent a counter-regulatory homeostatic response to the generation of other potent vasoconstrictor agents, such as ET-1, by CsA.

Conclusions

In conclusion, in this last section evidence has been provided that an imbalance in the physiological activities and regulation of NO and ET systems may play an important role in the pathogenesis of diverse clinical situations of which we described three, namely, CHF, liver cirrhosis and the complications of CsA therapy. Two of these situations, CHF and CsA nephrotoxicity are characterized by activation of the ET system and impaired activity of NO, whereas in liver cirrhosis activation of the NO system seems to predominate. As pointed out, many gaps still exist in our understanding of the exact interactions between these two potent vasoregulatory systems. Similarly, the relative importance of this imbalance in the overall context of activation of other local and circulating neurohumoral systems which may occur in these clinical conditions is far from elucidated. Nevertheless, we believe that the concept of a balance between the two systems playing an important role in the regulation of vascular tone, as supported by diverse experimental and clinical findings, should prove to be fertile ground for understanding the physiology and pathophysiology of circulation.

Acknowledgments

The studies from authors' laboratories were supported in part by NIH grants DK-41573, DK-45695 and DK-45462 (MSG) and grant 3256 from the Israeli Ministry of Health. (JMW).

References

1. Furchgott R, Zawadzki J. The obligatory role of endothelial cells in the relaxation of arterial smooth muscle by acetylcholine. Nature 1984; 288:373-376.
2. Palmer RM, Ferridge AG, Moncada S. Nitric oxide release accounts for the biological activity of endothelium-derived relaxing factor. Nature 1987; 327:524-526.
3. Ignarro LJ, Buga GM, Woods KS et al. Endothelium-derived relaxing factor produced and released from artery and vein is nitric oxide. Proc Natl Acad Sci USA 1987; 84:9265-9269.
4. Papapetropoulos A and Sessa WC. Regulation of the nitric oxide synthase gene family. In: Goligorsky MS, Gross SS ed. Nitric Oxide and the Kidney: Physiology and Pathophysiology. Chapman & Hall, 1997.
5. Huang P, Huang Z, Mashimo H et al. Hypertension in mice lacking the gene for endothelial nitric oxide synthase. Nature 1995; 377:239-242.
6. Fleming I and Busse R. Vascular effects of NO. In: Nitric Oxide and the Kidney: Physiology and Pathophysiology. Edited by Goligorsky MS and Gross SS, Chapman & Hall, 1997.
7. Tsukahara H, Ende H, Magazine HI et al. Molecular and functional characterization of the non- isopeptide-selective ET receptor in endothelial cells. Receptor coupling to nitric oxide synthase. J Biol Chem 1994; 269, 21778-21785.
8. Tsukahara H, Gordienko DV, Tonschoff B et al. Direct demonstration of insulin-like growth factor- I-induced nitric oxide production by endothelial cells. Kidney Int 1994; 45:598-604.
9. Goligorsky MS, Tsukahara H, Magazine HI et al. Termination of endothelin signaling: role of nitric oxide. J Cell Physiol 1994; 158:485-494.
10. Kitamura K, Tanaka T, Kato J et al. Immunoreactive endothelin in rat kidney inner medulla: marked decrease in spontaneously hypertensive rats. Biochem Biophys Res Comm 1989; 162:38-44.
11. Mattson DL, Roman RJ, and Cowley Jr. AC. Role of Nitric Oxide in renal papillary blood flow and sodium excretion. Hypertension 1992; 19:766-769.
12. Kohan D.E. Endothelins in the kidney: physiology and pathophysiology. Am J Kid Dis 1993; 22:493-510.
13. Bachmann S, Mundel P. Nitric oxide in the kidney: synthesis, localization, and function. Am J Kid Dis 1994; 24:112-129.
14. Gurbanov K, Rubinstein I, Hoffman A et al. Differential regulation of renal regional blood flow by endothelin-1. Am J Physiol 1996; 271:F1166-F1172.
15. Brezis M, Rosen S. Hypoxia of the renal medulla: its implications for disease. N Engl J Med 1995; 269:115-121.
16. Gellai,M, DeWolf R, Pullen M et al. Distribution and functional role of renal ET receptor subtypes in normotensive and hypertensive rats. Kidney Int 1994; 46:1287-1294.
17. Ziche M, Mirabidelli L, Donnini S. Angiogenesis. Exp Nephrol 1996; 4:1-14.

18. Shweiki D, Itin A, Soffer D et al. Vascular endothelial growth factor induced by hypoxia may mediate hypoxia-initiated angiogenesis. Nature 1992; 359:843-845.
19. Levy A, Levy N, Wegner S et al. Transcriptional regulation of the rat vascular endothelial growth factor gene by hypoxia. J Biol Chem 1995; 270:13333-13400.
20. Wren A, Hiley C, Fan T-P et al. Endothelin-3 mediated proliferation in wounded human umbilical vein endothelial cells. Biochem Biophys Res Comm 1993; 196:369-375.
21. Morbidelli L, Orlando C, Maggi C et al. Proliferation and migration of endothelial cells is promoted by endothelins via activation of ET_B receptors. Am J Physiol 1995; 269:H686-H695.
22. Leibovich SJ, Polverini PJ, Fong TW et al. Production of angiogenic activity by human monocytes requires an L-arginine/nitric oxide-synthase-dependent effector mechanism.Proc Natl Acad Sci USA 1994; 91, 4190-4194.
23. Ziche M, Morbidelli L, Masini E et al. Nitric oxide mediates angiogenesis in vivo and endothelial cell growth and migration in vitro promoted by substance P. J Clin Invest 1994; 94:2036-2044.
24. Noiri E, Lee E, Testa J et al. Podokinesis in endothelial cell migration: Role of nitric oxide. Am J Physiol 198; 274:C236-C244.
25. Noiri E, Hu Y, Bahou WF et al. Permissive role of nitric oxide in endothelin-induced migration of endothelial cells. J Biol Chem 1997; 272:1747-1752.
26. Noiri E, Peresleni T, Srivastava N et al. Nitric oxide is necessary for a switch from stationary to locomoting phenotype in epithelial cells. Am J Physiol 1996; 270, C794-C802.
27. Lerman A, Webster M, Chesebro J et al. Circulating and tissue endothelin immunoreactivity in hypercholesterolemic pigs. Circulation 1993; 88:2923-2928.
28. Lerman A, Edwards B, Hallet J et al. Circulating and tissue endothelin immunoreactivity in advanced atherosclerosis. N Engl J Med 1991; 325:997-1001.
29. Winkles J, Alberts G, Brogi E et al. Endothelin-1 and endothelin receptor m-RNA expression in normal and atherosclerotic human arteries. Biochem Biophys Res Comm 1993; 191:1081-1088.
30. Chester A, O'Neal G, Moncada S et al. Low basal and stimulated release of nitric oxide in atherosclerotic epicardial coronary arteries. Lancet 1990; 336:897-900.
31. Bossaler C, Habib G, Yamamoto H et al. Impaired muscarinic endothelium-dependent relaxation and cyclic guanosine 5'-monophosphate formation in atherosclerotic human coronary artery and rabbit aorta. J Clin Invest 1987; 79:170-174.
32. Shimokawa H and Vanhoutte PM. Impaired endothelium dependent relaxation to aggregating platelets and related vasoactive substances in porcine coronary arteries in hypercholesterolemia and atherosclerosis. Circ Res 1989; 64:900-914.
33. Douglas S, Louden C, Vickery-Clark L et al. A role for endogenous endothelin-1 in neointimal formation after rat carotid artery ballon angioplasty. Protective effects of the novel nonpeptide endothelin receptor antagonist SB 209670. Circ Res 1994; 75:190-197.
34. Dzau VJ. Renal and circulatory mechanisms in congestive heart failure. Kidney Int 1987; 31:1402-1415.
35. Kaiser L, Spickard RC, Olivier NB. Heart failure depresses endothelium-dependent responses in canine femoral artery. Am J Physiol 1989; 256:H962-H967.
36. Ontkean M, Gay R, Greenberg B. Diminished endothelium-derived relaxing factor activity in an experimental model of chronic heart failure. Circ Res 1991; 69:1088-1096.
37. Teerlink JR, Clozel M, Fischli W et al. Temporal evolution of endothelial dysfunction in a rat model of chronic heart failure. J Am Coll Cardiol 1993; 22:615-620.

38. Kubo SH, Rector TS, Bank AJ et al. Endothelium-dependent vasodilation is attenuated in patients with heart failure. Circulation 1991; 84:1589-1596, 1991.

39. Drexler H, Lu W. Endothelial dysfunction of hindquarter resistance vessels in experimental heart failure. Am J Physiol 1992; 262:H1640-H1645.

40. Katz SD, Schwartz M, Yuen J et al. Impaired acethylcholine-mediated vasodilation in patients with congestive heart failure: Role of endothelium-derived vasodilating and vasoconstricting factors. Circulation 1993; 88:55-61.

41. Vanhoutte PM. Endothelium-dependent responses in congestive heart failure. J Mol Cell Cardiol 1996; 28:2233-2240.

42. Drexler H. Holtz J. Endothelium dependent relaxation in chronic heart failure. Cardiovasc Res 1994; 28:720-721.

43. Teerlink JR, Gray GA, Clozel M et al. Increased vascular responsiveness to norepinephrine in rats with heart failure is endothelium dependent: Dissociation of basal and stimulated nitric oxide release. Circulation 1994; 89:393-401.

44. Drexler H, Hayoz D, Munzel T et al. Endothelial function in chronic congestive heart failure. Am J Cardiol 1992; 69:1596-1601.

45. Habib F, Dutka D, Crossman D et al. Enhanced basal nitric oxide production in heart failure: another failed counter-regulatory vasodilator mechanism? Lancet 1994; 344:371-373.

46. Winlaw DS, Smythe GA, Keogh AM et al. Increased nitric oxide production in heart failure. Lancet 1994; 344:373-374.

47. Potlog K, Hoffman A, Gurbanov K et al. Angiotensin II (AII) receptor blockade restores endothelial dependent renal vasodilation in rats with experimental congestive heart failure (CHF). J Am Soc Nephrol 1994; 5:610.

48. Mulder P, Elfertak L, Richard V et al. Peripheral artery structure and endothelial function in heart failure: effect of ACE inhibition. Am J Physiol 1996; 271:H469-H477.

49. Cavero PG, Miller WL, Heublein DM et al. Endothelin in experimental congestive heart failure in the anesthetized dog. Am J Physiol 1990; 259:F312-F317.

50. McMurray JJ, Ray SG, Abdullah I et al. Plasma endothelin in chronic heart failure. Circulation 1992; 85:1374-1379.

51. Wei CM, Lerman A, Rodeheffer RJ et al. Endothelin in human congestive heart failure. Circulation 1994; 89:1580-1586.

52. Loeffler BM, Roux S, Kalina B et al. Influence of congestive heart failure on endothelin levels and receptors in rabbits. J Mol Cell Cardiol 1993; 25:407-416.

53. Brown LA, Nunez DJ, Brookes CIO et al. Selective increase in endothelin-1 and endothelin A receptor subtype in the hypertrophied myocardium of the aortovenacaval fistula rat. Cardiovasc Res 1995; 29:768-774.

54. Fareh J, Touyz RM, Schiffrin EL et al. Endothelin-1 and Angiotensin II receptors in cells from rat hypertrophied heart. Receptor regulation and intracellular Ca^{2+} modulation. Circ Res 1996; 78:302-311.

55. Ito H, Hirata Y, Adachi S et al. Endothelin. An autocrine/paracrine factor in the mechanism of angiotensin II-induced hypertrophy in cultured rat cardiomyocytes. J Clin Invest. 1993; 92:398-403.

56. Ito H, Hiroe M, Hirata Y et al. Endothelin ET_A receptor antagonist blocks cardiac hypertrophy provoked by hemodynamic overload. Circulation 1994; 89:2198-2203.

57. Wang X, Douglas SA, Louden C et al. Expression of Endothelin-1, Endothelin-3, Endothelin- converting enzyme-1, and Endothelin-A and Endothelin-B receptor mRNA after angioplasty-induced neointimal formation in the rat. Circ Res 1996; 78:322-328.

58. Teerlink JR, Loeffler BM, Hess P et al. Role of Endothelin in the maintenance of blood pressure in conscious rats with chronic heart failure. Circulation 1994; 90:2510-2518.

59. Kiowski W, Sutsch G, Hunziker P et al. Evidence for endothelin-1-mediated vaso-constriction in severe chronic heart failure. Lancet 1995; 346:732-736.
60. Bomzon A. Vascular reactivity in liver disease. In: Bomzon A, Blendis LM, eds. Cardiovascular complications of liver disease. Boca Raton, FL: CRC Press, 1990:207-224.
61. Schrier RW, Arroyo V, Bernardi M et al. Peripheral arterial vasodilation hypothesis: a proposal for the initiation of renal sodium and water retention in cirrhosis. Hepatology 1988; 8:1151-1157 .
62. Epstein M. Hepatorenal Syndrome In: Epstein M, ed. The Kidney in Liver Disease. 4th edition Philadelphia: Hanley & Belfus Inc. 1996; 75-108.
63. Vallance P, Moncada S. Hyperdynamic circulation in cirrhosis: a role for nitric oxide? Lancet 1991; 337:776-778.
64. Claria J, Jimenez W, Ros J et al. Pathogenesis of arterial hypotension in cirrhotic rats with ascites: role of endogenous nitric oxide. Hepatology 1994; 20:1615-1621.
65. Guarner C, Soriano G, Tomas A et al. Increased serum nitrite and nitrate levels in patients with cirrhosis: relationship to endotoxemia. Hepatology 1993; 18:1139-1143.
66. Sieber CC, Lopez-Talavera JC, Groszmann RJ. Role of nitric oxide in the in vitro splanchnic vascular hyporeactivity in ascitic cirrhotic rats. Gastroenterology 1993; 104:1750-1754.
67. Niederberger M, Gines P, Tsai P et al. Increased aortic cyclic guanosine monophosphate concentration in experimental cirrhosis in rats: evidence for a role of nitric oxide in the pathogenesis of arterial vasodilation in cirrhosis. Hepatology 1995; 21:1625-1631.
68. Niederberger M, Martin PY, Gines P et al. Normalization of NO production corrects arterial vasodilation and hyperdinamic circulation in cirrhotic rats. Gastroenterology 1995; 109:1624- 1630.
69. Pizcueta P, Pique JM, Fernandez M et al. Modulation of the hyperdynamic circulation of cirrhotic rats by nitric oxide inhibition. Gastroenterology 1992; 103:1909-1915.
70. Bomzon A, Blendis LM. The nitric oxide hypothesis and the hyperdynamic circulation in cirrhosis. Hepatology 1994; 20:1343-1350.
71. Lopez-Talavera JC., Groszmann RJ. Treatment with aminoguanidine, a specific inhibitor of the inducible nitric oxide synthase, ameliorates the hyperdynamic syndrome in portal hypertensive rats (Abstract). Hepatology 1994; 20:143.
72. Martin P-Y, Xu DL, Niederberger M et al. Upregulation of endothelial constitutive NOS: a major role in the increased NO production in cirrhotic rats. Am J Physiol 1996; 270:F494-F499.
73. Uchihara M, Izumi N, Sato C et al. Clinical significance of elevated plasma endothelin concentration in patients with cirrhosis. Hepatology 1992; 16:95-99.
74. Moore K, Wendon J, Frazer M et al. Plasma endothelin immunoreactivity in liver disease and the hepatorenal syndrome. N Eng J Med 1992; 327:1774-1778.
75. Asbert M, Gines A, Gines P et al. Circulating levels of endothelin in cirrhosis. Gastroenterology 1993; 104:1485-1491.
76. Moller S, Emmeluth C, Henriksen JH. Elevated circulating plasma endothelin-1 concentrations in cirrhosis. J Hepatol 1995; 22:389-398.
77. Bernardi M, Gulberg V, Colantoni A et al. Plasma endothelin-1 and -3 in cirrhosis: relationship with systemic hemodynamics, renal function and neurohumoral systems. J Hepatol 1996; 24:161-168.
78. Sugiura M, Inagami T, Kon V. Endotoxin stimulates endothelin release in vivo and in vitro as determined by radioimmunoassay. Biochem Biophys Res Comm 1989; 161:1220-1227.

79. Salo J, Francitorra A, Follo A et al. Increased endothelin in cirrhosis. Relationship with systemic endotoxemia and response to changes in effective blood volume. J Hepatol 1995; 22:389-398.

80. Ackerman Z, Karmeli F, Amir G et al. Renal vasoactive mediator generation in portal hypertensive and bile duct ligated rats. J Hepatol 1996; 24:478-486.

81. Hartleb M, Moreau R, Cailmail S et al. Vascular hyporesponsiveness to endothelin-1 in rats with cirrhosis. Gastroenterology 1994; 107:1085-1093.

82. Kahan BD. Medical intelligence: cyclosporine. N Engl J Med 1989; 321:1725-1738.

83. Myers BD. Cyclosporine nephrotoxicity. Kidney Int 1986; 30:964-974.

84. Mimran A, Mourad G, Ribstein J et al. Cyclosporine-associated hypertension. In: Laragh JH, Brenner BM. Hypertension: Pathophysiology, Diagnosis, and Management. 2nd edition, New-York: Raven Press Ltd 1995; 2459-2469.

85. Barros EJG, Boim MA, Ajzen H et al. Glomerular hemodynamics and hormonal participation in cyclosporine nephrotoxicity. Kidney Int 1987; 32:19-25.

86. Murray, BM, Paller MS, Ferris T. Effect of cyclosporine administration on renal hemodynamics in conscious rats. Kidney Int 1985; 28:767-774.

87. Pfeilschifter J, Ruegg UT. Cyclosporine A augments angiotensin II-stimulated rise in intracellular free calcium in vascular smooth muscle cells. Biochem J 1987; 248:883-887.

88. Bunchman TE, Brookshire CA. Cyclosporine-induced synthesis of endothelin by cultured human endothelial cells. J Clin Invest 1991; 88:310-314.

89. Ong ACM, Jowett TP, Scoble JE et al. Effect of cyclosporine A on endothelin synthesis by cultured human renal cortical epithelial cells. Nephrol Dial Transplant 1993; 8:748-753.

90. Takeda Y, Itoh Y, Yoneda T et al. Cyclosporine A induces endothelin-1 release from cultured rat vascular smooth muscle cells. Eur J Parmacol 1993; 233:299-301.

91. Benigni A, Perico N, Ladny JR et al. Increased urinary excretion of endothelin-1 and its precursor, Big-endothelin-1, in rats chronically treated with cyclosporine. Transplantation. 1991; 52:175- 177.

92. Kon V, Sugiura M, Inagami T et al. Role of endothelin in cyclosporine-induced glomerular dysfanction. Kidney Int 1990; 37:1487-1491.

93. Fogo A, Hellings SE, Inagami T et al. Endothelin receptor antagonism is protective in in vivo acute cyclosporine toxicity. Kidney Int 1992; 42:770-774.

94. Kivlighn SD, Gabel RA, Siegl PKS. Effect of BQ-123 on renal function and acute cyclosporine-induced renal dysfunction. Kidney Int 1994; 45:131-136.

95. Abassi ZA, Pieruzzi F, Nakhoul F et al. Effects of cyclosporine A on the synthesis, excretion, and metabolism of endothelin in the rat. Hypertension 1996; 27:1140-1148.

96. Rego A, Vargas R, Wroblewska B et al. Atenuation of vascular relaxation on cyclic GMP responses by cyclosporine A. J Pharmacol Exp Ther. 1989; 25:165-170.

97. Diederich D, Yang Z, Luscher TF. Chronic cyclosporine therapy impairs endothelium-dependent relaxation in the renal artery of the rat. J Am Soc Nephrol 1992; 2:1291-1297.

98. Gallego MJ, Lopez Farre A, Riesco A et al. Blockade of endothelial dependent responses in conscious rats by cyclosporine A: effect of L-arginine. Am J Physiol 1993; 264:H708-H714.

99. De Nicola L, Thomson SC, Wead LM et al. Arginine feeding modifies cyclosporine nephrotoxicity in rats. J Clin Inves 1993; 92:1859-1865.

100. Bobadilla NA, Tapia E, Franco M et al. Role of nitric oxide in th renal hemodynamics abnormalities of cyclosporine nephrotoxicity. Kidney Int 1994; 46:773-779.

101. Lopez-Ongil S, Saura M, Rodriguez-Puyol D et al. Regulation of endothelial NO synthase expression by cyclosporine A in bovine aortic endothelial cells. Am J Physiol 1996; 271:H1072-H1078.

CHAPTER 6

Molecular and Structural Biology of Endothelin Receptors

Maria L. Webb and Stanley R. Krystek Jr.

Introduction

The discovery of the vasodilatory factors prostacyclin and nitric oxide that are secreted by, and act on, the endothelium and underlying vascular smooth muscle, led to the search for counter-regulatory vasoconstrictory substances.[1-4] In 1988, Yanagisawa and colleagues reported the discovery of an endothelium-derived, 21 amino acid peptide with potent and long-lasting vasoconstrictive and pressor actions.[5] This peptide was termed endothelin-1 (ET-1) and is one of a family of three related mammalian peptides. The three peptides are closely related to the sarafotoxin peptides found in the venom of the Israeli burrowing asp, *Actractaspic engaddensis* and it has become part of the lore surrounding ET-1 that the extremely potent vasoconstrictory effects of the sarafotoxins led to the demise of Cleopatra.

The actions of the endothelin peptides are many and include potent and prolonged vasoconstriction of smooth muscle, vasodilation, pressor and depressor effects, positive myocardial inotropy and chronotropy, aldosterone and prostaglandin release, mitogenicity prolactin secretion and neuromodulation.[5-17] The multiple actions of endothelins are widely attributed to the existence of at least two endothelin receptor subtypes with discrete and regulated cellular distributions and functions. This review will focus on the molecular biology of endothelin receptors and will utilize current data to speculate upon the molecular structure of these membrane proteins.

Molecular Biology of ET_A and ET_B Receptors

Evidence for Multiple Receptor Subtypes

Intravenous administration of ET-1 to animals causes an acute biphasic pressor response: an initial and transient depressor effect is followed by a sustained pressor effect.[18-20] In humans, infusion of small doses of ET-1 through the brachial artery causes significant local vasoconstrictory responses.[21] As with animal studies, the vasoconstrictor response is slow to develop, but is sustained for 2 hours after cessation of infusion. Similarly, systemic administration of ET-1 in man causes

Endothelin Receptors and Signaling Mechanisms, edited by David M. Pollock and Robert F. Highsmith. © 1998 Springer-Verlag and R.G. Landes Company.

elevated blood pressure that is maintained after clearance of endothelin from the circulation.[22] Haynes et al observed similar effects of ET-1 in man and extended the earlier observations to include venoconstriction as well as vasoconstriction.[23]

The kinetics of the slow-to-develop but sustained contractions, can be explained in part by endothelin receptor trafficking. Intravenously administered ET-1 is rapidly cleared from the circulation, due largely to a first-pass effect in the pulmonary and renal beds, with a half-life of 7 minutes.[24] Despite this rapid clearance, exogenous endothelin-induced pressor responses are maintained for hours. Specific [125I]ET-1 binding was slowly and partially reversible, even in the presence of GTP analogs or high ET-1 concentration.[25-28] Initially, the slow dissociation of agonist-receptor complexes was given as the explanation for the long-lasting actions of ET-1.[29] However, endothelin receptors are rapidly internalized in the presence of agonist and Marsault et al suggested that sustained ET-1-induced contractions are more likely attributed to a late intracellular signaling event rather than to slow off-rate of ET-1 from receptor.[30-32] Chun et al has since demonstrated that endothelin receptors are associated with caveolae at or near the plasma membrane, that ET-1 induced rapid endothelin receptor endocytosis, and that ET-1 remains intact and bound to receptors for up to 2 hours after endocytosis.[33,34] Thus, it appears that enduring ET-1 induced contractions are likely due to a sustained activation of a signal-transducing G-protein.

The vasodilatory actions of ET-1 were dependent upon the presence of an intact endothelium.[8,35] Moreover, cyclooxygenase and NO inhibitors reduced the endothelin-induced vasodilatory response, consistent with the view that the vasodilatory response was mediated by endothelium-derived prostanoids and NO.

An important advance was made when several groups reported differential structure-activity relationships among the endothelin peptides for the vasoconstrictory and vasodilatory responses. The rank order of potency for the vasoconstrictory response was ET-1 ≥ ET-2 >> ET-3 while for the vasodilatory response it was ET-1 = ET-2 = ET-3.[20] Warner et al noted that ET-3 was equipotent to ET-1 with regard to vasodilatory responses.[8] Similarly, Inoue et al reported that although ET-3 has only 25-50% of the in vivo pressor activity of ET-1, it produces a more potent vasodepressor response in rats.[18] These data indicated that the vasoconstrictory and vasodilatory action of ETs were mediated through different receptor systems.

The biphasic nature of ET-1 actions in the vasculature taken with the different rank order of potency among the endothelin peptides led to the hypothesis that multiple receptors for ETs existed. This was supported by radioligand binding studies that showed that the binding sites for [125I]ET-1 and [125I]ET-3 were distinct. The contribution of the earliest radioligand binding studies was to demonstrate the presence of high affinity (100-200 pM) [125I]ET-1 sites on cultured rat and human vascular smooth muscle cells that were consistent with the in vivo and in vitro potency of ET-1.[25,26] Many studies verified the existence of high affinity ET-1 binding sites in tissues and cells from other species including porcine atria, arteries and veins, Swiss 3T3 fibroblasts, chick cardiac membranes, rat lung, and human placenta.[27,36-39] Following these studies however, came a closer examination of the binding of ET-3. [125I]ET-3 bound with picomolar affinity to sites in brain microvessel endothelial cells, bovine adrenomedullary chromaffin cells, and porcine artery.[40-42] However, the receptor density for the [125I]ET-3 population was significantly lower

Table 6.1 Binding affinities of the endothelin isopeptides at recombinant human endothelin-A (ET_A) and endothelin-B (ET_B) receptors

Peptide Ligand	ET_A Receptor (Ki nM)	ET_B Receptor (Ki nM)
ET-1	0.4±0.2	0.2±0.0
ET-2	0.4±0.04	0.2±0.0
ET-3	820±260	0.4±0.2
SFX-6c	29,000±4,000	0.3±0.1

Data (means ± S.E.M., n≥3) from Rose et al using recombinant human ET_A and ET_B cDNAs cloned from a human placental cDNA library and expressed i n CHO cells.[44] Abbreviations: SFX, sarafotoxin.

than that for the [125I]ET-1 population of sites in porcine coronary artery (80 vs. 280 fmol/mg) and in some smooth muscle preparations, [125I]ET-3, but not [125I]ET-1, binding was undetectable.[41,42] In such preparations, i.e., those with low or undetectable [125I]ET-3 binding, [125I]ET-1 binding was inhibited with a rank order of potency of ET-1 >> ET-3 undetectable.[36,41,42] In contrast, in preparations where [125I]ET-3 binding was detected, the rank order of potency was ET-1 = ET-3. These data indicated that [125I]ET-1 was labeling two populations of binding sites, each with high affinity, but that [125I]ET-3 was labeling a subset of the overall ET-1 binding sites. While these studies showed that differential affinity for ET-3 could distinguish these binding sites, Williams et al reported that sarafotoxin S6c could do the same in rat and human uterine smooth muscle but not in rat or human hippocampal tissue.[43] In sum, these studies clearly showed that two binding sites exist, one with high affinity for ET-1 and the second with high affinity for both ET-1 and ET-3.

Thus, three lines of data supported the view that at least two endothelin receptors mediated the effects of the endothelin peptides: (1) biphasic vascular response to ET-1, (2) ET-1-selectivity for the vasoconstrictory effect and ET-1/ET-3 equipotency for the vasodilatory effects, and (3) distinct ET-1 and ET-3 binding sites with ET-1 preferring sites on VSMC and equipotent ET-1/ET-3 sites on endothelium. These data led to the notion that vasoconstrictory, ET-1 selective receptors were found on vascular smooth muscle while vasodilatory, nonselective receptors were found on the endothelium. These receptors were termed ET_A and ET_B, respectively. The pharmacological differentiation of the ET_A and ET_B receptors by the endothelin isopeptides and sarafotoxin 6c are summarized in Table 6.1.

Another breakthrough in the field occurred when it was recognized that activation of nonselective endothelin receptors also caused contraction of vascular smooth muscle in select circulatory beds. Williams et al showed that sarafotoxin S6c (SFX-S6c) increased blood pressure in pithed rats and Harrison and colleagues found that ET-3 and sarafotoxin SFX-6c stimulated force development in endothelium-denuded pig coronary arteries.[45,46] The efficacy of these ETB agonist peptides in endothelium-denuded preparations was consistent with the presence of these endothelin receptors on smooth muscle.

Additional evidence that ET_B receptors could mediate vasoconstriction came from work with receptor antagonists. The vasoconstrictory responses to sarafotoxin SFX-6c in rabbit saphenous vein and endothelium-denuded porcine coronary artery were not totally reversed by high concentrations of the ET_A receptor antagonist, BQ-123 (D-Trp-D-Asp-Pro-D-Val-Leu), consistent with the suggestion that contraction was mediated by non-ET_A receptors.[47,48] Warner and colleagues studied the effects of PD 142893 (Ac-D-diphenylalanine -Leu-Asp-Ile-Ile-Trp), an antagonist of both ET_A and ET_B receptors.[8] Contractions mediated by non-ET_A receptors were not blocked by PD 142893 in the rabbit pulmonary artery but were blocked by the same compound in the rat mesentery. These data contributed to the hypothesis that a third, perhaps molecularly distinct but ET_B-like, receptor subtype mediates some contractions of isolated smooth muscle.

Cloning of the Endothelin Receptor cDNAs

The isolation of cDNA clones that encoded proteins with selective (ET_A) and nonselective (ET_B) peptide responses was reported in 1990.[49,50] As anticipated, the cloning of the endothelin receptors revealed significant homology ($\approx$60%) between the ET_A and ET_B receptors. The ET_A cDNA identified by Arai and colleagues was isolated from a bovine lung cDNA library by expression cloning. The deduced amino acid sequence encoded a protein of 427 amino acids with a molecular mass of 48.5 kilodaltons and seven hydrophobic regions of 22 to 27 residues each. The ET_B cDNA initially identified by Sakurai et al was isolated from a rat lung expression library.[50] The encoded protein was predicted to be 441 amino acids, a molecular mass of 47 kilodaltons and seven hydrophobic regions.

The ET_A and ET_B receptor cDNAs encode proteins that have clear sequence (Fig. 6.1) homology. Both cDNAs encode for proteins with 7 hydrophobic regions of 20-25 amino acids that are thought to be transmembrane spanning domains. Each receptor has an amino-terminal signal sequence. In addition, when compared to other GPC receptors, the amino terminus is relatively long (70-100 residues); that may reflect a binding site for the 21-amino acid peptide as is the case with GPC receptors for large protein hormones. Both receptors have extracellular sites for N-linked glycosylation (Asn-X-Ser/Thr) and an intracellular site just distal to the seventh transmembrane domain for palmitoylation (Cys-Leu-Cys-Cys-X-Cys). The third intracellular loop and cytoplasmic tail have numerous sites that are substrates for kinase phosphorylation. When compared to other members of the GPC receptor superfamily, the sequence homology is generally on the order of 20-25% until one compares the transmembrane regions where the homology is higher.[51]

Following the initial coexpression cloning of the bovine ET_A and rat ET_B receptor cDNAs, many subsequent cDNAs encoding ET_A and ET_B receptors from different species were been identified. The conservation across species is high, ranging from 91-94 % for ET_A and 83-88% for ET_B (Fig. 6.2). The ET_A receptor has been cloned from bovine lung, rat vascular smooth muscle cells, human placenta, human lung, and human heart.[49,52-58] The ET_B receptor has been cloned from rat lung, human jejunum, human liver, and human placenta, porcine cerebellum, human lung, and human prostate.[50,57-63] These mammalian ET_A and ET_B cDNAs are all nearly identical by nucleic acid and amino acid sequences and when expressed in mammalian cells displayed an agonist rank order of potency of ET-1 = ET-2 >> ET-3 at ET_A receptors and ET-1 = ET-2 = ET-3 for ET_B receptors.

Nonmammalian Endothelin Receptor cDNAs

Endothelin receptors have been identified in *Xenopus laevis* that are distinct from the mammalian subtypes.[66,67] Karne et al cloned a novel endothelin receptor from *Xenopus laevis* dermal melanophores that has 47% and 52% identity with the ET_A and ET_B receptors.[66] The pharmacology of this receptor suggests that ET-3 and ET-1 have K_i values of 45 and 114 nM, respectively. The EC_{50}, measured by pigment dispersion of granules in melanophores, was ET-3 24 nM but 10 μM for ET-1, an order of magnitude different from the binding affinity of ET-1 for this site. These data suggest that this receptor, termed ET_C, is selective for ET-3.

Kumar et al also cloned a novel endothelin receptor from *Xenopus laevis* heart that is 74%, 60%, and 51% identical to the human ET_A, ET_B, *and Xenopus laevis* ET_C receptors.[67] When expressed in COS-7 cells, this receptor displays Ki values for ET-1 and ET-3 of 0.15 and 100 nM, respectively, but sarafotoxin S6c and BQ-123 were inactive in competing [^{125}I]ET-1. These data are not consistent with typical ET_A receptor pharmacology and suggest that Xenopus contains at least two novel forms of endothelin receptors.

Alternate Transcripts

A novel ET_B receptor cDNA was cloned from rat brain that was found to represent an alternative transcript of the ET_B gene.[68] Compared to the rat lung cDNA, the rat brain transcript has divergent 5' and 3' untranslated sequences and three extra nucleotides in the coding sequence.[50] Consequently, the encoded protein is predicted to have four amino acid substitutions and to be one amino acid longer than the rat lung ET_B receptor. The four substitutions were Ser-Ser-Ala-Pro instead of Phe-Ala-Thr localized at position 60 in the amino terminus. Expression of this cDNA in COS-1 cells resulted in a binding profile consistent with that observed in rat lung. Since Southern blots of the rat genome indicated that the ET_B cDNA was present as a single copy gene, these results indicate that the two cDNAs are derived from the same gene.[68]

Receptor Genes

Southern blots of human genomic DNA using the ET_A and ET_B receptor cDNAs as probes are consistent with the presence of two endothelin receptor genes. Restriction digests of genomic DNA were probed with either the ET_A or ET_B cDNA and in each case a single product hybridized with the probe.[69,70] These results indicate that the ET_A and ET_B loci exist as a single copy in the human genome. Similar results have been reported in other species for the ET_B gene. Cheng et al detected a single band in restriction digests of rat genomic DNA and Mizuno et al identify a single hybridization product in bovine genomic DNA.[68,71,72] These results consistently indicate that the human, bovine and rat and endothelin receptors are present as single copy genes.

The genomic organization of the ET_A and ET_B genes has been established.[69-72] The human ET_A gene spans 40 kb and contains 8 exons and 7 introns.[69] The human ET_B spans 24 kb and contains 7 exons and 6 introns.[70] Thus, with the exception that intron 1 of ET_A, which is in the 5' noncoding region, is lacking in ET_B gene, the gene structures are similar with introns 1-6 of ET_B corresponding to introns 2-7 of ET_A. In addition, the intron-exon borders correspond to the borders of the seven putative transmembrane domains. These findings are consistent with the ET_A and ET_B genes originating from the same ancestral gene.[70]

```
                                                              TM-1
hETA:  MET---LCLRASFWLALVGCVISDNP-E-R-YSTNLSNHVDDFTFRGTELSFLVTTHQPTNLV--LPSNGSMHNYCPQQTKITSA-----------
bETA:  MET---FWLRLSFWVALVGGVISDNP-E-S-YSTNLSIHVDSVATFHGTELSFVVTTHQPTNLA--LPSNGSMHNYCPQQTKITSA-----------
rETA:  MGV---LCFLASFWLALVGGAIADNA-E-R-YSANLSSHVEDFTPFPGTEFDFLGTTLRPPNLA--LPSNGSMHGYCPQQTKITTA-----------
hETB:  MQPPSLCGRALVALVLA-CGLSRIWGEERGFPPDRATP-LLQTAEIMTPPTKTLWPKGSNASLARSLAPAEVPKGDRTAGSPPRTISPPCQGPIEIKE
bETB:  MQPLPSLCGRALVALILA-CGVAGIQAEEREFPPAGATQPLPGTGEVMETPETSWPGRSNASDPRSAATPQIPRGGRVAGIPPRT--PPCDGPIEIKE
rETB:  MQSSASRCGRALVALILLA-CGLLGVWGEKRGFPPAQATPSLLGTKEVMTPPTKTSWTRGSNSSLMRFRTAEVT-KGGRVAGVPPRSF-PPCQRKIEINK
pETB:  MQPLRSLCGRALVALIFA-CGVAGVQSEERGFPPAGATPPALRTGEIVAPPTKTFWPRGSNASLPRSSSPPQMPKGGRMAGPPARTLTPPPCEGPIEIKD
       ==========                                                                          ==========

                                                              TM-2                                      TM-3
hETA:  -FKYINTVISCTFIVGMVGNATLLRIIYQNKYMRNGPNALIASLALGDLIYVVIDLPINVFKLLAGRWPFDHNDFGVFLCKLFPFLQKSSVGITVLNLC
bETA:  -FKYINTVISCTFIVGMVGNATLLRIIYQNKCMRNGPNALIASLALGDLIYVVIDLPINVFKLLAGRWPFEQNDFGVFLCKLFPFLQKSSVGITVLNLC
rETA:  -FKYINTVISCTIFIVGMVGNATLLRIIYQNKCMRNGPNALIASLALGDLIYVVIDLPINVFKLLAGRWPFDHNDFGVFLCKLFPFLQKSSVGITVLNLC
hETB:  TFKYINTVVSCLVFVLGIIGNSTLLRIIYKNKCMRNGPNILIASLALGDLLHIIIDIPINTYKLLAKDWPFGVE----MCKLVPFIQKASVGITVLSLC
bETB:  TFKYINTVVSCLVFVLGIIGNSTLLRIIYKNKCMRNGPNILIASLALGDLLHIVIDIPINVYKLLAEDWPFGAE----MCKLVPFIQKASVGITVLSLC
rETB:  TFKYINTVVSCLVFVLGIIGNSTLLRIIYKNKCMRNGPNILIASLALGDLLHIIIDIPINAYKLLAGDWPFGAE----MCKLVPFIQKASVGITVLSLC
pETB:  TFKYINTVVSCLVFVLGIIGNSTLLRIIYKNKCMRNGPNILIASLALGDLLHIIIDIPINVYKLLAEDWPFGVE----MCKLVPFIQKASVGITVLSLC
       ==============================================                           =======================

                                                              TM-4                                      TM-5
hETA:  ALSVDRYRAVASWSRVQGIGIPLVTAIEIVSIWILSFILAIPEAIGFVMVPFEYRGEQHKTCMLNATSK--FMEFYQDVKDWWLFGFYFCMPLVCTAIFY
bETA:  ALSVDRYRAVASWSRVQGIGIPLVTAIEIVSIWILSFILAIPEAIGFVMVPFEYKGAQHRTCMLNATSK--FMEFYQDVKDWWLFGFYFCMPLVCTAIFY
rETA:  ALSVDRYRAVASWSRVQGIGIPLITAIEIVSIWILSFILAIPEAIGFVMVPFEYKGEQHRTCMINATTK--FMEFYQDVKDWWLFGFYFCMPLVCTAIFY
hETB:  ALSIDRYRAVASWSRIKGIGVPKWTAVEIVLIWVVSVVLAVPEAIGFDIITMDYKGSYLRICLLHPVQKTAFMQFYKTAKDWWLFSFYFCLPLAITAFFY
bETB:  ALSIDRYRAVASWSRIKGIGVPKWTAVEIVLIWVVSVVLAVPEAVGFDIITSDHIGNKLRICLLHPTQKTAFMQFYKTAKDWWLFSFYFCLPLAITALFY
rETB:  ALSIDRYRAVASWSRIKGIGVPKWTAVEIVLIWVVSVVLAAPEAIGFDVTTSDYKGKPLRVCMLNPFQKTAFMQFYKTAKDWWLFSFYFCLPLAITAIFY
pETB:  ALSIDRYRAVASWSRIKGIGVPKWTAVEIVLIWVVSVVLAVPEALGFDMITTDYKGNRLRICLLHPTQKTAFMQFYKTAKDWWLFSFYFCLPLAITAFFY
       ==========================================================               =======================
```

Fig. 6.1. See full caption on following page.

```
                       TM-6                                                           TM-7
        ==  ===============================                     ===================================
hETA:   TLMTCEMLNRRNGSLRIALSEHLKQRREVAKTVFCLVVIFALCWFPLHLSRILKKTVYNEMDKNRCELLSFLLLMDYIGINLATMNSCINPIALYFVSKK
bETA:   TLMTCEMLNRRNGSLRIALSEHLKQRREVAKTVFCLVVIFALCWFPLHLSRILKKTVYDEMDTNRCELLSFLLLMDYIGINLATMNSCINPIALYFVSKK
rETA:   TLMTCEMLNRRNGSLRIALSEHLKQRREVAKTVFCLVVIFALCWFPLHLSRILKKTVYDEMDKNRCELLSFLLLMDYIGINLATMNSCINPIALYFVSKK
hETB:   TLMTCEMLRKKSGMQ-IALNDHLKQRREVAKTVFCLVLVFALCWLPLHLSRILKLTLYNQNDPNRCELLSFLLVLDYIGINMASLNSCINPIALYLVSKR
bETB:   TLMTCEMLRKKSGMQ-IALNDHLKQRREVAKTVFCLVLVFALCWLPLHLSRILKLTLYDQHDPRRCEFLSFLLVLDYIGINMASLNSCINPIALYLVSKR
rETB:   TLMTCEMLRKKSGMQ-IALNDHLKQRREVAKTVFCLVLVFALCWLPLHLSRILKLTLYDQSNPQRCELLSFLLVLDYIGINMASLNSCINPIALYLVSKR
pETB:   TLMTCEMLRKKSGMQ-IALNDHLKQRREVAKTVFCLVLVFALCWLPLHLSRILKLTLYDQNDSNRCELLSFLLVLDYIGINMASLNSCINPIALYLVSKR

hETA:   FKNCFQSCLCC-CC-YQSKSLMTSVPMNGTSIQWKNHDQNNHNTDRSSHKDMN--
bETA:   FKNCFQSCLCC-CC-YQSKSLMTSVPMNGTSIQWKNHEQNNHNTERSSHKDIN--
rETA:   FKNCFQSCLCC-CC-HQSKSLMTSVPMNGTSIQWKNQEQN-HNTERSSHKDMN--
hETB:   FKNCFKSCLCCWCQSFEEKQSLEEKQSC---LKFKANDHG-YDNFRSSNKYSSSS
bETB:   FKNCFKSCLCCWCQSFEEKQSLEEKQSC---LKFKANDHGYDN-FRSSNKYSSSS
rETB:   FKNCFKSCLCCWCQTFEEKQSLEEKQSC---LKFKANDHGYDN-FRSSNKYSSSS
pETB:   FKNCFKSCLCCWCQSFEEKQSLEEKQSC---LKFKANDHGYDN-FRSSNKYSSSS
```

Fig. 6.1. Sequence alignment of the human, bovine, rat and porcine endothelin receptor cDNA sequences. The transmembrane regions (TM) are identified by the double bar (===) above the sequence.

	hETA	bETA	rETA	hETB	bETB	rETB	pETB
hETA	--	--	--	--	--	--	--
bETA	94	--	--	--	--	--	--
rETA	92	91	--	--	--	--	--
hETA	55	55	55	--	--	--	--
bETB	55	55	56	85	--	--	--
rETB	55	56	56	83	84	--	--
pETB	56	57	57	84	88	83	--

Fig. 6.2. Homology of the endothelin receptors across species. (A). Percent homology of human (h), bovine (b), rat (r), and porcine (p) ET_A and ET_B receptors. (B). Serpentine schematic of the ET_A receptor. Each circle represents an amino acid in he sequence of ET_A as it is compared to the other endothelin receptors, including ET_B receptors. The coloring is as follows: white ; identity, green = conserved across species and subtype, red = not conserved across receptor subtype or species, blue = insertion of gap into sequence alignment. The extracellular and intracellular loops are labeled EL-# or IL-#, respectively. The transmembrane regions are also labeled 1 through 7.

The 5'-flanking regions of the human ET_A and ET_B genes are also similar. Both receptor genes lack conventional TATA and CCAAT boxes but contain SP1 and GATA elements. As with some other genes that lack TATA and CCAAT boxes, the ET_B gene has two major initiation sites found at positions -258 and -229, with a third potential but minor transcription initiation site at -283. The transcription start site of the ET_A gene is at -502 bp upstream of the methionine initiation codon. The genes map to different chromosomal locations with the ET_A on chromosome 4 and the ET_B on chromosome 13. The identification of the ET_B and ET-3 genes as recessive susceptibility loci for Hirschsprung's disease gives impetus to the search for genetic disorders associated with the endothelin receptor system.[73-75]

Tissue Distribution of Endothelin Receptors

The presence of receptors in human tissues is critical to our understanding of the role of the endothelin receptor system in disease. ET_A and ET_B receptors or their mRNA have been identified in numerous tissues.[76] Molenaar and colleagues showed that in human myocardium, ET_A and ET_B receptors were localized to atrial and ventricular myocardium, the atrioventricular conducting system, and endocardial cells.[77] These investigators found that the ET_A receptor density exceeded that for the ET_B receptor in human left ventricle (60% ET_A to 40% ET_B) as well as in right atrial myoctyes (90% ET_A to 10% ET_B). In rat and pig heart, ET_A is the predominant endothelin receptor subtype expressed.[78,79] These studies demonstrate that the ET_A receptor is the primary endothelin receptor subtype expressed in human ventricular myocardium.

Studies examining the endothelin receptor distribution in human vasculature also indicate that the ET_A receptor is the predominant subtype expressed.[80-82] Moreover, the relative contribution of the ET_A receptor to vasoconstriction of human vessels supersedes that of the ET_B subtype by approximately two orders of magnitude.[81,83-85] These data are consistent with the suggestion that, in human vasculature the ET_A receptor mediates vasoconstriction.

Molecular Structure

Agonist Structure

The endothelin peptides, ET-1, ET-2, and ET-3, are highly homologous (Fig. 6.3). The three mature endothelin polypeptides, as well as a related family of highly similar peptides found in cobra venom known as the sarafotoxins consist of 21 amino acids and contain fully conserved motifs of two Cys-Cys disulfide bridges (^{1}Cys-^{15}Cys and ^{3}Cys-^{11}Cys), an internal highly-charged tripeptide ^{8}Asp-Lys-^{10}Glu, and a carboxyl-terminal pentapeptide ^{16}His-Leu-Asp-Ile-Ile-^{21}Trp. The establishment of the two disulfide bonds are essential for tight binding of the ligands to the ET_A receptor, but less so for the ET_B receptor.[86,87] Alanines can be substituted for ^{3}Cys and ^{11}Cys to replace one of the disulfide loops, which results in a slight loss in agonist activity, while alanine substitution for the other disulfide loop (^{1}Cys and ^{15}Cys) nearly eliminates activity.[88] The free carboxylic form of the C-terminal Trp residue in the endothelins is required for biological activity. Systemic replacement of peptide residues within endothelin-1 has revealed that ^{8}Asp, ^{10}Glu, ^{13}Tyr, ^{14}Phe, ^{20}Ile and ^{21}Trp are all essential for binding of the ligand to its receptor.[88-90]

The three-dimensional structure of human endothelin-1 has been determined by both X-ray crystallography and by solution nuclear magnetic resonance (NMR) methods.[91-100] The NMR structures of endothelin analogs and endothelin-3 have also been determined.[101-104] The NMR studies described a well-defined core region (residues 1-15) while the C-terminal region (residues 16-21) was shown to have no defined conformation due to a lack of specific interactions between the core region the C-terminal tail region. Fig. 6.4A shows the results from "a typical" NMR study where the models have similar core regions and the C-terminus exhibits conformational isomerism.[94] Within the core region two consensus structural features were found: (1) a turn from ^{5}Serto ^{8}Asp and (2) an α-helical region from ^{9}Lys to ^{15}Cys. Similar features have been noted for the related snake toxin sarafotoxin-6b.[105,106] Figure 6.4B shows the X-ray structure determined in the laboratory of Wallace et al with side chains colored by amino acid type.[107] The X-ray structure

ET-1	Cys-Ser-Cys-Ser-Ser-Leu-Met-Asp-Lys-Glu-Cys-Val-Tyr-Phe-Cys-His-Leu-Asp-Ile-Ile-Trp
ET-2	Cys-Ser-Cys-Ser-Ser-Trp-Leu-Asp-Lys-Glu-Cys-Val-Tyr-Phe-Cys-His-Leu-Asp-Ile-Ile-Trp
ET-3	Cys-Thr-Cys-Phe-Thr-Tyr-Lys-Asp-Lys-Glu-Cys-Val-Tyr-Tyr-Cys-His-Leu-Asp-Ile-Ile-Trp
SFX-6a	Cys-Ser-Cys-Lys-Asp-Met-Thr-Asp-Lys-Glu-Cys-Leu-Asn-Phe-Cys-His-Gln-Asp-Val-Ile-Trp
SFX-6b	Cys-Ser-Cys-Lys-Asp-Met-Thr-Asp-Lys-Glu-Cys-Leu-Tyr-Phe-Cys-His-Gln-Asp-Val-Ile-Trp
SFX-6c	Cys-Thr-Cys-Asn-Asp-Met-Thr-Asp-Glu-Glu-Cys-Leu-Asn-Phe-Cys-His-Gln-Asp-Val-Ile-Trp
SFX-6d	Cys-Thr-Cys-Lys-Asp-Met-Thr-Asp-Lys-Glu-Cys-Leu-Tyr-Phe-Cys-His-Gln-Asp-Ile-Ile-Trp

Fig. 6.3. Primary amino acid sequence of the endothelin (ET) peptides and the related sarafotoxin (SFX) peptides from snake venom.

has a core somewhat similar to the NMR models that contain an extended β-strand at the N-terminus followed by a bulge between residues 5 and 7, which is followed by a hydrogen-bonded loop between residues 7 and 11. An irregular helix extends from residues 11 until the end of the molecule. A comparison of the X-ray and NMR structures describes the differences seen in side chain conformers, conformation of the C-terminal region as well as the central core region.[107] The overall folds of X-ray and NMR structures are similar; however, the two regions of greatest difference (the loop between residues 6 and 10) and the C-terminus are most critical for binding specificity and biological activity (as noted below). Structure activity relationships (SARS) of endothelin peptides have demonstrated the importance of a free amino terminus and a carboxylic acid at the C-terminus.[108] Other important SARs are: a negative charged side chain at positions 8, 10 and 18 as well as aromatic side chains at positions 13, 14 and 21.[108-111] The importance of residues within the core region and the C-terminal tail region which are greater than 17 angstroms apart (CA of [10]Glu to the CA of [21]Trp) suggests that there are multiple discrete points of contact which are distal to one another in the membrane bound receptor. The interaction sites on the receptor may be both in the transmembrane regions and the extracellular loops which extend into the solvent.

Characterization of Agonist Binding Site

Since endothelin receptors are members of the superfamily of G protein-coupled receptors for which no experimentally determined three dimensional structures are available to date, 3-D models of the endothelin receptors have been used to generate hypotheses concerning ligand-receptor interactions. These models are based upon orientation of the transmembrane helices of either bacteriorhodopsin or rhodopsin.[112-115] In turn, model-building is based on the ability to define the location of the seven transmembrane helices and develop a 3-D model of the heptahelical domains.[116] There are no structural templates from which to model the extracellular or intracellular loops connecting the transmembrane helices. There are several databases which have been published and contain sequence and structural information as well as experimental data and a variety of molecular models.[117-120] These databases in their most useful form exist on the world wide web (WWW).

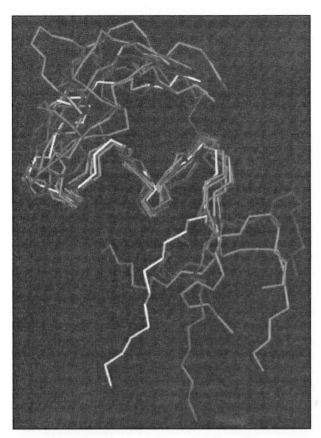

A

Fig. 6.4. NMR and X-ray crystal structure of ET-1. (A) NMR of the peptide backbone for eight low energy conformers of endothelin-1 superimposed for best fit of the helix (residues 9-15). Each structure is displayed in a different color with the disulfides shown in one structure as a thin red line. (B) X-ray structure of endothelin. The backbone trace is in magenta and the side chains are colored by functional group as follows: aromatic = orange, basic = blue, acidic = red, disulfides = yellow, aliphatic and hydrophobic = green, neutral and hydrophilic = light blue. See Color Figures on page 217.

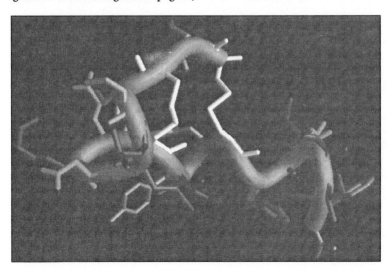

B

To predict the transmembrane domains, the sequence alignments and species comparisons of the ET_A and ET_B receptors were utilized. As previously described the endothelin receptors consist of two subtypes. Figure 6.1 shows the sequences of the endothelin receptors with the predicted transmembrane regions indicated. From the sequence alignment (Fig. 6.1) and species comparison matrix (Fig. 6.2), it can be seen that the sequences show high identity within subtypes (about 90%) and less sequence similarity between subtypes (about 60%). Based upon these similarities and differences the transmembrane regions were identified and the sequence predictions were translated to three-dimensional models as described.[121]

In addition, the computer-generated receptor models were used to predict the amino acid residues likely to be near the extracellular surface of the putative binding cavity. Figure 6.5 shows a model of the ET_A receptor with the C-terminus of ET-1 inserting into the transmembrane region while the bicyclic core would be involved with several interactions with the N-terminal extracellular portion of the receptor and the extracellular loops. Based upon molecular modeling, single amino acid mutations were targeted for alanine replacement by site-directed mutagenesis in order to explore ligand binding and selectivity. The first and most obvious task was to predict specific amino acids involved in subtype selective binding. In two such studies, mutation of [129]Tyr in transmembrane region 2 of ET_A was found to alter the binding profile of the native endothelin peptides.[121,122] Indeed, the affinity of subtype-selective agonists (and antagonists as discussed below) were altered to that of the opposing subtype. These results indicated that [129]Tyr is critically involved in the subtype-selective binding observed for the endothelin isopeptide agonists.[121-123] In addition, Lee and coworkers showed that [182]Lys of ET_B was also critical for agonist binding.[124]

Several groups have used chimeric receptors to delineate functional domains of ET_A and ET_B receptors that are involved in agonist binding. These studies suggested that two distinct regions (transmembrane regions and connecting loops) of the receptor, transmembrane regions 1, 2, 3, and 7 and transmembrane regions 4, 5, and 6 are distinct subdomains involved in ligand binding and isopeptide subtype selectivity.[125-128] These results are especially interesting as they are consistent with the potential binding cavities that are seen in the molecular models based upon rhodopsin.[115,116]

Characterization of Antagonist Binding Site

Recently, the mechanism underlying antagonist-ET_A receptor complex formation was studied. Four chemically distinct antagonists, BQ-123, BMS-182874, Ro, and SB 209670, were studied. The data supported a role of aromatic interactions in the binding of four chemically distinct antagonists to ET_A receptors. This hypothesis was supported by structure-activity data with analogs of BMS-182874 that varied the C-5 dimethylamino substituent on the naphthalene ring. On the basis of these data, a model of the docked conformation of BMS-182874 in the ET_A receptor (Fig. 6.6) was proposed in which a putative hydrophobic binding pocket is formed by residues in transmembrane domains 2, 3, and 7. A direct contact between the C-5 dimethylamino group and naphththalene ring of BMS-182874 and [129]Tyr in transmembrane domain 2 was supported by structure-activity data.[123]

An alternative hypothesis for the interaction of antagonists and agonists with endothelin receptors suggests that binding into a cleft formed between transmembrane domains 2, 3, and 7.[129] In this model, the orientation of ET-1 would be similar

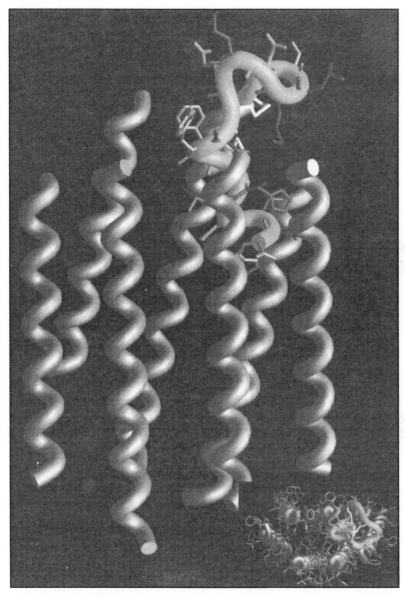

Fig. 6.5. Cross-sectional view of the X-ray structure of ET-1 in the ET$_A$ receptor. The backbone trace of the receptor is colored blue and the sidechains of the receptor were removed for clarity. The side chains and backbone trace of ET-1 are colored as in Figure 6.4. The inset is a view of the receptor from the extracellular surface. The C-terminal tail of endothelin inserts into the cavity formed by transmembrane domain 1, 2, 3, and 7. See Color Figures on page 218.

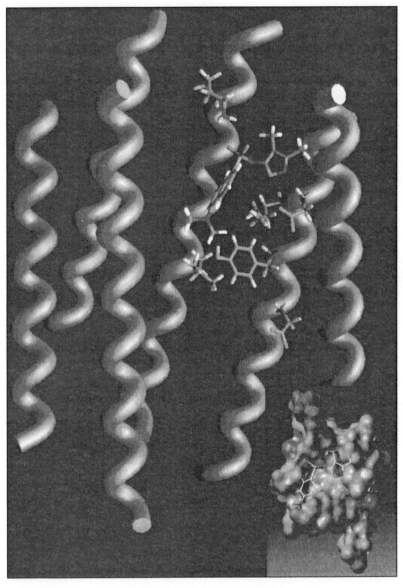

Fig. 6.6. Molecular model of the ET_A receptor with BMS-182874 docked into the putative binding cavity. The backbone trace of the transmembrane regions are colored blue with transmembrane region 7 removed for clarity. Selected side chains of the receptor that interact with BMS-182874 are displayed. BMS-182874 is colored as follows: C = green, N = blue, S = orange, O = red, and H = yellow. The inset shows the molecular surface of the binding pocket in the ET_A receptor colored by lipophilic potential (green & orange = lipophilic, red & blue = polar). The cavity has a complementary shape to the ligand and contains both hydrophilic and hydrophobic interactions. See Color Figures on page 219.

to Figure 6.5, however the C-terminal region would be in an extended conformation, rather than in a helical conformation as in the X-ray structure. Based upon this scheme, [18]Asp of ET-1 forms a salt bridge with [166]Lys on transmembrane domain 3. [165]Gln has several contacts to the main chain of ET-1. The amino acids in the C-terminus of ET-1 are involved in several hydrophobic interactions while the C-terminal carboxylate was proposed to contact [126]Asp on transmembrane domain 2. Mutagenesis of [126]Asp was shown to abolish transmembrane signaling and since [126]Asp corresponds to the highly conserved aspartate present in many GPC receptors and shown to be critical for agonist efficacy, this model attempts to rationalize binding and receptor activation.[130] In contrast to ET-1, when the pseudopeptide FR-139317 was placed in the extracellular binding pocket, it could not contact [126]Asp and was predicted correctly to be an antagonist.

Conclusion

In the relatively short time since the discovery of endothelin peptides (1988) and the cloning of their receptors (1990), much has been learned of the biology of the endothelin-receptor system. The primary underlying biology, that of vascular pressor regulation, has broad implications for numerous diseases and has captured the interest of the pharmaceutical industry. Numerous selective and nonselective small molecules, mostly antagonists, have been generated and will undoubtedly shed light on the pathophysiological roles of endothelin-receptor interactions. Much is yet to be learned of the molecules that interact with the endothelin receptors. This will start to close the circle so that as we learn about the molecules that block endothelin function, will we learn more of endothelin biology.

References

1. Moncada S, Gryglewski R, Bunting S et al. An enzyme isolated from arteries transforms prostaglandin endoperoxides to an unstable substance that inhibits platelet aggression. Nature 1976; 263:663-665.
2. Furchgott RF, Zawadzki JV. The obligatory role of endothelial cells in the relaxation of arterial smooth muscle by acetylcholine. Nature 1980; 288:373-376.
3. Palmer RM, Ferrige AG, Moncada S. Nitric oxide release accounts for the biological activity of endothelium-derived relaxing factor. Nature 1987; 327:524-526.
4. Hickey KA, Rubanyi GM, Paul RJ et al. Characterization of a coronary vasoconstrictor produced by cultured endothelial cells. Am J Physiol 1985; 248:C550-C556.
5. Yanagisawa M, Kurihara H, Kimura S et al. Primary structure, synthesis, and biological activity of rat endothelin, an endothelium derived vasoconstrictor peptide. Nature 1988; 332:411-415.
6. Simonson M, Dunn M. Cellular signaling by peptides of the endothelin gene family. J Clin Invest 1990; 85:790-797.
7. Secrest RJ, Cohen ML. Endothelin: differential effects in vascular and nonvascular smooth muscle. Life Sci 1989; 45:1365-1372.
8. Warner TD, de Nucci G, Vane JR. Rat endothelin is a vasodilator in the isolated perfused mesentary of the rat. Eur J Pharm 1989; 159:325.
9. Cristol JP, Warner TD, Thiemermann C et al. Mediation via different receptors of the vasoconstrictor effects of endothelins and sarafotoxins in the systemic circulation and renal vasculature of the anesthetized rat. Br J Pharmacol 1993; 108:776-779.
10. Ishikawa T, Yanagisawa M, Kimura S et al. Positive chronotropic effects of endothelin, a novel endothelium derived vasoconstrictor peptide. Am J Physiol 1988; 255:H970-973.

11. Stier Jr CT, Quilley CP, McGriff JC. Endothelin 3 effects on renal function and prostanoid release in the rat isolated kidney. J Pharm Exp Ther 1992; 262:252-262.

12. Morishita R, Higaki J, Ogihara T. Endothelin stimulates aldosterone biosynthesis by dispersed rabbit adrenocapsular cells. Biochem Biophys Res Comm 1989; 160:628-632.

13. Komuro I, Kurihara H, Sugiyama T et al. Endothelin stimulated c-fos and c-myc expression and proliferation of vascular smooth muscle cells. FEBS Lett 1988; 238:249-252.

14. Takuwa N, Takuwa Y, Yanagisawa M et al. A novel vasoactive peptide endothelin stimulates mitogenesis through inositol lipid turnover in swiss 3T3 fibroblasts. J Biol Chem 1989; 264:7856-7861.

15. Weber H, Webb ML, Serafino R et al. Endothelin-1 and angiotensin II induce additive but delayed mitogenic stimulation in cultured rat aortic vascular smooth muscle cells: Evidence for common signaling pathways. Mol Endocrinol 1994; 8:148-158.

16. Samson WK, Skala KD, Alexander BD et al. Hypothalamic endothelin: presence and effects related to fluid and electrolyte homeostasis. Biochem Biophys Res Comm 1990; 169:737-743.

17. Matsumura K, Abe I, Tsuchihashi T et al. Central effect of endothelin on neuro-hormonal responses in conscious rabbits. Hypertension 1991; 17:1192-1196.

18. Inoue A, Yanagisawa M, Kimura S et al. Endothelin augments unitary calcium channel currents on the smooth muscle cell membrane of guinea pig portal vein. Proc Natl Acad Sci. USA 1989; 86:2863-2867.

19. Yanagisawa M, Inoue A, Ishikawa T et al. Primary structure, synthesis, and bio-logical activity of rat endothelin, an endothelium derived vasoconstrictor pep-tide. Proc Natl Acad Sci USA 1988; 85:6964-6967.

20. Spokes RA, Ghatei MA, Bloom SR. Studies with endothelin 3 and endothelin 1 on rat blood pressure and isolated tissues: Evidence for multiple endothelin receptor subtypes. J Cardiovasc Pharmacol 1989; 17:S427-S429.

21. Clarke JG, Benjamin N, Larkin SW et al. Endothelin is a potent long-lasting vaso constrictor in men. Am J Physiol 1989; 257:H2033-H2035.

22. Vierhapper H, Wagner O, Nowotny P et al. Effect of endothelin-1 in man. Circu-lation 1990; 81, 1415-1418.

23. Haynes WG, Clarke JG, Cockcroft JR et al. Pharmacology of endothelin 1 in vivo in humans. J Cardiovasc Pharmacol 1991; 17:S284-S286.

24. Shiba R, Yanagisawa M, Miyauchi T et al. Elimination of intravenously injected endothelin-1 from the circulation of the rat. J Cardiovasc Pharmacol 1989; 13:S98-S101.

25. Clozel M, Fishli W, Guilly C. Specific binding of endothelin on human vascular smooth muscle cells in culture. J Clin Invest 1989; 83:1758-1761.

26. Hirata Y, Yoshimi H, Takaichi S et al. Binding and receptor down regulation of a novel vasoconstrictor endothelin in cultured rat vascular smooth muscle cells. FEBS Lett 1988; 239:13-17.

27. Kanse PA, Ghatei MA, Bloom SR. Endothelin binding sites in porcine aortic and rat lung membranes. Eur J Biochem 1989; 182:175-179.

28. Waggoner WG, Genova SL, Rash VA. Kinetic analyses demonstrate that the equi-librium assumption does not apply to [125I]endothelin-1 binding data. Life Sci 1992; 51:1869-1876.

29. Bolger GT, Liard F, Krogsrud R et al. Tissue specificity of endothelin binding sites. J Cardiovascular Pharmacol 1990; 16:367-375.

30. Resink TJ, Scott-Burden T, Boulanger C et al. Internalization of endothelin by cultured human vascular smooth muscle cells: Characterization and physiological significance. Mol Pharmacol 1990; 38:244-252.

31. Marsault R, Vigne P, Frelin C. The irreversibility of endothelin action is a property of a late intracellular signalling event. Biochem Biophys Res Comm 1991; 179:1408-1413.

32. Marsault R, Vigne P, Breittmayer JP et al. Astrocytes are target cells for endothelins and sarafotoxin. Am J Physiol 1991; 261:C986-C993.

33. Chun M, Liyanage UK, Lisanti MP et al. Signal transduction of a G protein-coupled receptor in caveolae: Colocalization of endothelin and its receptor with caveolin. Proc Natl Acad Sci USA 1994; 91:11728-11732.

34. Chun M, Lin HY, Henis YI et al. Endothelin-induced endocytosis of cell surface ET_A receptors: Endothelin remains intact and bound to the ET_A receptor. J Biol Chem 1995; 270:10855-10860

35. De Nucci G, Thomas R, D'Orleans-Juste PJ et al. Pressor effects of circulating endothelin are limited by its removal in the pulmonary circulation and by release of prostacyclin and endothelium-derived relaxing factor. Proc Natl Acad Sci USA 1988; 85:9797-9800.

36. Takayanagi R, Hashiguchi T, Ohashi M et al. Presence of nonselective type of endothelin receptor on vascular endothelium and its linkage to vasodilation. Reg Peptides 1990; 27:247-255.

37. Devesly P, Phillips PE, Johns A et al. Receptor kinetics differ for endothelin 1 and endothelin 2 binding to Swiss 3T3 fibroblasts. Biochem Biophys Res Comm 1990; 172:126-134.

38. Watanabe H, Miyazaki H, Kondoh M et al. Two distinct types of endothelin receptors are present on chick cardiac membranes. Biochem Biophys Res Comm 1989; 161:1252-1259.

39. Fischli W, Clozel M, Guilly C. Specific receptors for endothelin on membranes from human placenta. Characterization and use in a binding assay. Life Sci 1989; 44:1429-1436.

40. Vigne P, Ladoux A, Frelin C. Endothelins activate $Na^+/^+$ exchange in brain capillary endothelial cells via a high affinity endothelin 3 receptor that is not coupled to phospholipase. J Biol Chem 1991; 266:5925-5928.

41. Wilkes LC, Boarder MR. Characterization of the endothelin binding site on bovine adrenomedullary chromaffin cells: Comparison with vascular smooth muscle cells. Evidence for receptor heterogeneity. J Pharmacol Exp Ther 1991; 256:628-633.

42. Ihara M, Saeki T, Funabashi K et al. Two endothelin receptor subtypes in porcine arteries. J Cardiovasc Pharmacol 1991; 17:S119-S121.

43. Williams DL Jr, Jones KL, Colton CD et al. Identification of high affinity endothelin-1 receptor subtypes in human tissues. Biochem Biophys Res Comm 1991; 180:475-480.

44. Rose PM, Krystek SR Jr, Patel PS et al. Aspartate mutation distinguishes ET_A but not ET_B receptor subtype-selective ligand binding while abolishing phospholipase C activation in both receptors. FEBS Letters 1995; 361:243-249.

45. Williams Jr DL, Jones KL, Pettibone DJ et al. Sarafotoxin S6c: an agonist which distinguishes between endothelin receptor subtypes. Biochem Biophys Res Comm 1991; 175:556-561.

46. Harrison VJ, Randriantsoa A, Schoeffter P. Heterogeneity of endothelin sarafotoxin receptors mediating contraction of pig coronary artery. Br J Pharmacol 1992; 105:511-513.

47. Moreland M, McMullen DM, Delaney CL et al. Venous smooth muscle contains vasoconstrictor ET_B-like receptors. Biochem Biophys Res Comm 1992; 184:100-106.
48. Ihara M, Noguchi M., Saeki T et al. Biological profiles of highly potent novel endothelin antagonists selective for the ET_A receptor. Life Sci 1992; 50:247-250.
49. Arai H, Hori S, Aramori I et al. Cloning and expression of a cDNA encoding an endothelin receptor. Nature 1990; 348:730-732.
50. Sakurai T, Yanagisawa M, Takuwa Y et al. Cloning of a cDNA encoding a non isopeptide selective subtype of the endothelin receptor. Nature 1990; 348:732-735.
51. Probst WC, Snyder LA, Schuster DJ et al. Sequence alignment of the G-protein coupled receptor superfamily. DNA Cell Biol 1992; 11:1-20.
52. Lin HY, Kaji EH, Winkel GK et al. Cloning and functional expression of a vascular smooth muscle endothelin-1 receptor. Proc Natl Acad Sci USA 1991; 88:3185-3189.
53. Adachi M, Yan Y-Y, Furuichi Y et al. Cloning and characterization of cDNA encoding human A-type endothelin receptor. Biochem Biophys Res Comm 1991; 180:1265-1272.
54. Cyr C, Huebner K, Druck T et al. Cloning and chromosomal localization of a human endothelin ETA receptor. Biochem Biophys Res Comm 1991; 181:184-190.
55. Hosoda K, Nakao K, Arai H et al. Cloning and expression of human endothelin-1 receptor cDNA. FEBS Lett 1991; 287:23-36.
56. Hayzer D, Rose P, Lynch J et al. Cloning and expression of a human endothelin receptor: Subtype A. J Am Med Sci 1992; 304:231-238.
57. Williams DL Jr, Jones KL, Alves K et al. Characterization of cloned human endothelin receptors. Life Sci 1993; 53:407-414.
58. Elshourbagy NA, Korman DR, Wu H-L et al. Molecular cloning and regulation of the human endothelin receptors. J Biol Chem1993; 268:3873-3879.
59. Sakamoto A, Yanagisawa M, Sakurai T et al. Cloning and functional expression of human cDNA for the ET_B endothelin receptor. Biochem Biophys Res Comm 1991; 178:656-663.
60. Nakamuta M, Takayanagi R, Sakai Y et al. Cloning and sequence analysis of a cDNA encoding human nonselective type of endothelin receptor. Biochem Biophys Res Comm 1991; 177:34-39.
61. Ogawa Y, Nakao K, Arai H et al. Molecular cloning of a nonisopeptide-selective human endothelin receptor. Biochem Biophys Res Comm 1991; 178:248-255.
62. Elshourbagy NA, Lee JA, Korman DR et al. Molecular cloning and characterization of the major endothelin receptor subtype in porcine cerebellum. Mol Pharmacol 1992; 41:465-473.
63. Webb ML, Chao CC, Rizzo M et al. Cloning and expression of an endothelin receptor subtype B (ET_B) from human prostate that mediates contraction. Mol Pharmacol 1995; 47:730-737.
64. Fukuroda T, Fujikawa T, Ozaki S et al. Clearance of circulating endothelin-1 by ET receptors in rats. Biochem Biophys Res Comm 1994; 199:1461-1465.
65. Iwasaki S, Homma T, Matsuda Y et al. Endothelin receptor subtype B mediates autoinduction of endothelin-1 in rat mesangial cells. J Biol Chem 1995; 270:6997-7003.
66. Karne S, Jayawickreme CK, Lerner MR. Cloning and characterization of an endothelin-3 specific receptor (ETC receptor) from Xenopus laevis dermal melanophores. J Biol Chem 1993; 268:19126-19133.
67. Kumar C, Mwangi V, Nuthalaganti P et al. Cloning and characterization of a novel endothelin receptor from Xenopus heart. J Biol Chem 1994; 269:13414-13420.
68. Cheng H-F, Su Y-M, Yeh J-R et al. Alternative transcript of the nonselective-type endothelin receptor from rat brain. Mol Pharmacol 1993; 44:533-538.

69. Hosoda K, Nakao K, Tamura N et al. Organization, structure, chromosomal assignment, and expression of the gene encoding the human endothelin A receptor. J Biol Chem 1992; 267:18797-18804.

70. Arai H, Nakao K, Takaya K et al. The human endothelin B receptor gene. Structural organization and chromosomal assignment. J Biol Chem 1993; 268:3463-3470.

71. Mizuno T, Saito Y, Itakura M et al, Structure of the bovine ET$_B$ endothelin receptor gene. Biochem J 1992; 287:305-309.

72. Mizuno T, Imai T, Itakura M et al Structure of the bovine endothelin-B receptor gene and its tissue-specific expression revealed by Northern analysis. J Cardiovasc Pharmacol 1992; 20:S8-S10.

73. Puffenberger EG, Hosoda K, Washington SS et al. A missense mutation of the endothelin-B receptor gene in multigenic Hirschsprung's disease. Cell 1994; 79:1257-1266.

74. Hosoda K, Hammer RE, Richardson JA et al. Targeted and natural (Piebald-lethal) mutations of endothelin-B receptor gene produce megacolon associated with spotted coat color in mice. Cell 1994; 79:1267-1276.

75. Baynash AG, Hosoda K, Giaid A et al. Interaction of endothelin-3 with endothelin-B receptor is essential for development of epidermal melanocytes and enteric neurons. Cell 1994; 79:1277-1285.

76. Hori S, Komatsu Y, Shigemoto R et al. Distinct tissue distribution and cellular localization of two messenger ribonucleic acids encoding different subtypes of rat endothelin receptors. Endocrinology 1992; 130:1885-1895.

77. Molenaar P, O'Reilly G, Sharkey A et al. Characterization and localization of endothelin receptor subtypes in the human atrioventricular conducting system and myocardium. Circ Res 1993; 72:526-538.

78. Sargent C, Liu ECK, Chao CC et al. Role of endothelin receptor subtypes in myocardial ischemia. Life Sci 1994; 55:1833-1844.

79. Thomas PB, Liu ECK, Webb ML et al. Evidence of an endothelin-1 autocrine loop in cardiac myocytes: Relation to contractile function with congestive heart failure. Am J Physiol 1996; 271:H2629-H2637.

80. Davenport AP, O'Reilly G Kuc RE. Endothelin ET$_A$ and ET$_B$ mRNA and receptors expressed by smooth muscle in the human vasculature: Majority of the ET$_A$ subtype. Br J Pharmacol 1995; 114:1110-1116.

81. Maguire JJ, Davenport AP. Endothelin-induced vasoconstriction in human isolated vasculature is mediated predominantly via activation of ET$_A$ receptors. Br J Pharmacol 1993; 110:47P.

82. Fukuroda T, Kobayashi M, Ozaki S et al. Endothelin receptor subtypes in human versus rabbit pulmonary arteries. J Appl Physiol 1994; 76:1976-1982.

83. Buchan KW, Magnusson H, Rabe KF et al. Characterisation of the endothelin receptor mediating contraction of human pulmonary artery using BQ-123 and Ro 46-2005. Eur J Pharmacol 1994; 260:221-226.

84. Davenport AP, Maguire JJ. Is endothelin-induced vasoconstriction mediated only by ET$_A$ receptors in humans? Trends Pharmacol 1994; 15:9-11.

85. Deng L-Y, Li J-S, Schriffrin EL. Endothelin receptor subtypes in resistance arteries from humans and rats. Cardiovascular Res 1995; 29:532-535.

86. Heyl DL, Cody WL, He JX et al. Truncated analogues of endothelin and sarafotoxin are selective for the ET$_B$ receptor subtype. Peptide Res 1993; 6:238-41.

87. Sakurai T, Yanagisawa M, Masaki T. Molecular characterization of endothelin receptors. Trends Pharmacol 1992; 13:103-108.

88. Nakajima K, Kubo S, Kumagaye S-I et al. Structure-activity relationship of endothelin: importance of charged groups. Biochem Biophys Res Comm 1989; 163:424-429.

89. Hunt JT, Lee VG, Stein PD et al. Structure-activity studies of endothelin leading to novel peptide ET_A antagonists. Bioorg Med Chem 1993; 1:59-65.

90. Hunt JT. Drug News Perspect 1992; 5:78-82

91. Janes RW, Peapus DH, Wallace BA. The crystal structure of human endothelin. Nature Struct Biol 1994; 1:311-319.

92. Saudek V, Hoflack J, Pelton JT. 1H-NMR study of endothelin, sequence-specific assignment of the spectrum and a solution structure. FEBS Letters 1989; 257:145-148.

93. Endo S, Inooka H, Ishibashi Y et al. Solution conformation of endothelin determined by nuclear magnetic resonance and distance geometry. Solution conformation of endothelin determined by nuclear magnetic resonance and distance geometry. FEBS Letters 1989; 257:149-154.

94. Krystek SR Jr, Bassolino DA, Novotny J. Conformation of endothelin in aqueous ethylene glycol determined by 1H-NMR and molecular dynamics simulations. Confomration of endothelin in aqueous ethylene glycol determined by 1H NMR and molecular dynamics simulations. FEBS Letters 1991; 281:212-218.

95. Munro S, Craik D, McConville C et al. Solution conformation of endothelin, a potent vaso-constricting bicylic peptide. Solution conformation of endothelin, a potent vasoconstricting bicyclic peptide. A combined use of 1H NMR spectroscopy and distance geometry calculations. FEBS Letters 1991; 278:9-13.

96. Reily MD, Dunbar Jr JB, The conformation of endothelin-1 in aqueous solution: NMR-derived constraints combined with distance geometry and molecular dynamics calculations. Biochem Biophys Res Comm 1991; 178:570-577.

97. Aumelas A, Chiche L, Mahe E et al. Determination of the structure of [Nle]-endothelin by 1H-NMR. Int J Pept Protein Res 1991; 37:315-324.

98. Tamaoki H, Kobayashi Y, Nishimura S et al. Solution conformation of endothelin determined by means of 1H-NMR spectroscopy and distance geometry calculations. Protein Engineering 1991; 4, 509-518.

99. Dalgarno DC, Slater L, Chackalamannil S et al. Solution conformation of endothelin and point mutants by nuclear magnetic resonance spectroscopy. Int J Pept Protein Res 1992; 40:515-523.

100. Andersen NH, Chen C, Marschner TM. Conformational isomerism of endothelin in acidic aqueous media: a quantitative NOESY analysis. Biochemistry 1992; 31:1280-1295.

101. Saudek V, Hoflack J, Pelton JT. Solution conformation of endothelin-1 by 1H-NMR, CD, and molecular modeling. Int J Pept Protein Res 1991; 37:174-179.

102. Coles M, Munro SLA, Craik DJ. The solution structure of monocyclic analogue of endothelin [1,15 Aba]-ET-1, determined by 1H NMR spectroscopy. J Med Chem 1994; 37:656-664.

103. Mills RG, O'Donoghue SI, Smith R et al. Solution structure of endothelin-3 determined using NMR spectroscopy. Biochemistry 1992; 31:5640-5645.

104. Bortmann P, Hoflack J, Pelton JT et al. Solution conformation of endothelin-3 by 1H NMR and distance geometry calculations. Neurochem Int 1991; 18:491-496.

105. Mills RG, Atkins AR, Harvey T et al. Conformation of sarafotoxin-6b in aqueous solution determined by NMR spectroscopy and distance geometry. FEBS Letters 1991; 282, 247-252.

106. Atkins AR, Martin RC, Smith R. 1H NMR studies of sarafotoxin SRTb, a nonselective endothelin receptor agonist, and IRL 1620, an ET_B receptor-specific agonist. Biochemistry 1995; 34:2026-2033.

107. Wallace BA, Janes RW, Bassolino DA et al. A comparison of X-ray and NMR structures for endothelin-1. Protein Sci 1995; 4:75-83.

108. Kimura S, Kasuya Y, et al. Structure-activity relationships of endothelin: Importance of the C-terminal moiety. Biochem Biophys Res Comm 1988; 156:1182-1186.

109. Nakajima K, Kubo S, Kumagaye S, Nishio H, Tsunemi M, Inui T, Kuroda H, Chino N, Watanabe TX, Kimura T et al. Structure-activity relationship of endothelin: importance of charged groups. Biochem Biophys Res Comm 1989; 163:424-429.

110. Hunt JT, Lee VG, McMullen D et al. Structure-activity of endothelin leading to novel peptide ET_A antagonists. Bioorganic Med Chem 1993; 1:59-65.

111. Tam JP, Liu W, Zhang JW et al. Alanine scan of endothelin: importance of aromatic residues. Peptides 1994; 4:703-708

112. Henderson R, Baldwin JM, Ceska TA et al. Model for the structure of bacteriorhodopsin based upon high resolution electron cryomicroscopy. J Mol Biol 1990; 213:899-929.

113. Schertler GF, Villa C, Henderson R. Projection structure of frog rhodopsin in two crystal forms. Proc Natl Acad Sci USA 1995; 92:11578-11582.

114. Unger VM, Schertler GF. Low resolution structure of bovine rhodopsin determined by electron cryo-microscopy. Biophys J 1995; 68:1776-1786.

115. Schertler GF, Hargrave PF. Projection structure of frog rhodopsin in two crystral forms. Proc Natl Acad Sci USA 1995; 92:11578-11582.

116. Baldwin JM. The probable arrangement of the helices in G protein-coupled receptors. EMBO J 1993; 12:16931703.

117. Kristiansen K, Dahl SG, Edvardsen O. A database of mutants and effects of site-directed mutagenesis experiments on G protein-coupled receptors. Proteins 1996; 26:81-94. (http://wwwgrap.fagmed.uit.no/GRAP/homepage.html)

118. van Rhee AM, Jacobson KA. Molecular architecture of G protein-coupled receptors. Drug Dev Res 1996; 37:1. (http://mgddk1.niddk.nih.gov:8000/GPCR.html)

119. GCRDb-WWW, The G protiein-coupled receptor database world-wide-web site. (http://receptor.mgh.harvard.edu/GCRDBHOME.html)

120. GPCRDB: Information system for the G protein-coupled receptors. (http://swift.embl-heidelberg.de/7tm)

121. Krystek S, Patel PS, Rose PM et al. Mutation of a peptide binding site in a transmembrane region of a G protein-coupled receptor accounts for endothelin receptor subtype selectivity. J Biol Chem 1994; 269:12383-12386.

122. Lee JA, Elliott JD, Sutiphong JA et al. Tyr-129 is important to the peptide ligand affinity and selectivity of human endothelin type A receptor. Proc Natl Acad Sci USA 1994; 91:7164-7168.

123. Webb ML, Patel PS, Rose PM et al. Mutational analysis of the endothelin type A receptor (ET_A): Interactions and model of the selective ET_A antagonist BMS-182874 with the putative ET_A receptor binding cavity. Biochemistry 1996; 35:2549-2556.

124. Lee JA, Brinkmann JA, Longton ED et al. Lysine 182 of endothelin B receptor modulates agonist selectivity and antagoniust affinity: Evidence for the overlap of peptide and nonpeptide ligand binding sites. Biochemistry 1994; 33:14543-14549.

125. Adachi M, YangY-Y, Trzeciak A et al. Identification of a domain of ET_A receptor required for ligand binding. FEBS Letters 1992; 311:179-183.

126. Sakamoto A, Yanagisawa M, Sawamura T et al. Distinct subdomains of human endothelin receptors determine their selectivity to endothelin A-selective antagonist and endothelin B-selective agonists. J Biol Chem 1993; 268:8547-8553.

127. Sakamoto A, Yanagisawa M, Sakurai T et al. The ligand receptor interactions of the endothelin systems are mediated by distinct "message" and "address" domains. J Cardio Pharmacol 1993; 22:S113-S116.

128. Adachi M, Hashido K, Trzeciak A et al. Functional domains of human endothelin receptor. J Cardio Pharmacol 1993; 22:S121-S124.

129. von Geldern TW, Hutchins C, Kester JA et al. Azole endothelin antagonists. 1. A receptor model explains an unusual structure-activity profile. J Med Chem 1996; 39:957-967.
130. Rose PM, Krystek SR Jr, Patel PS et al. Aspartate mutation distinguishes ET_A but not ET_B receptor subtype-selective ligand binding while abolishing phospholipase C activation in both receptors. FEBS Letters 1995; 361:243-249.

CHAPTER 7

Cytokine Regulation of Endothelin Action

Timothy D. Warner

Introduction

In healthy humans or animals endothelin-1 (ET-1) is generally reported to circulate at particularly low concentrations (e.g., ref 1). This suggests that ET-1 may only have minor roles within the healthy circulation, even bearing in mind its particularly short circulating half-life.[2] This proposal is supported by the observations that inhibitors of the production or actions of ET-1 rarely produce effects in healthy, mature animals. (Although it is important to mention that in the human there is evidence that ET-1 may provide some contribution to the regulation of basal blood pressure[3]). In marked comparison to the healthy animal, substantial elevations in the circulating and/or tissue levels of ET-1 have been noted in pathological states. These states include coronary vasospasm and cerebral vasospasm following subarachnoid hemorrhage, Raynaud's disease, certain forms of systemic and pulmonary hypertension, ischemia reperfusion injury and organ infarctions, congestive heart failure, various forms of shock, and hypercholesterolaemia and atherosclerosis (see refs. 4, 5 for review). Clearly such increases can be secondary rather than causative to the pathological processes. However, this does not appear to be the case for ET-1. Experiments in animal models have demonstrated ET-1 to be an active participant in many of the above disease states, for antibodies directed against ET-1 are protective, as are the more recently developed endothelin receptor antagonists (see refs. 4-7). Taken together these observations suggest that the endothelin system is largely dormant in the healthy adult body, and particularly within the healthy adult circulation, but that it becomes activated in a range of pathological conditions. The marked damping down of the production of ET-1 under normal circumstances may well be to the body's benefit for, with its very long duration of action, its particular potency as a vasoconstrictor and its promitogenic activities, ET-1 is an autocoid that should not be produced in an ungoverned fashion. What, however, is the stimulus, or stimuli, that turns this inactive system into an active participant in so many disease states?

Endothelin Receptors and Signaling Mechanisms, edited by David M. Pollock and Robert F. Highsmith. © 1998 Springer-Verlag and R.G. Landes Company.

Kinetics of Endothelin-1 Release

In reviewing the stimuli that may regulate the synthesis and release of ET-1 it is useful first of all to consider what is known about the kinetics of these processes. It is often thought that the production of ET-1 is exclusively regulated at the level of gene transcription. In agreement with this view the production of ET-1 by cultured endothelial cells is increased by stimulants such as growth factors and vasoactive substances over the course of hours rather than minutes. ET-1 has, therefore, not been seen as a mediator of rapid responses within the vasculature. There are, however, certain stimuli that will quickly increase the circulating levels of ET-1.

Stimuli of Rapid Increases in Endothelin Release

One of the first reports noting that ET-1 levels in the circulation could increase rapidly was from studies in the human in which it was shown that such changes followed rapid postural changes.[8] Surgery also appears to cause fairly rapid increases in the circulating levels of ET-1 both in humans[9] and in animals.[10] Interestingly, in these latter experiments the increases in ET-1 were not affected by treatment with phosphoramidon, an endothelin converting enzyme inhibitor. From these experiments it is not possible to determine the sites from which ET-1 is released but clearly it is tempting to suggest that it is from the endothelium. Possibly, there could also be release of endothelin into the circulation from the hypothalamus and related brain regions, where ET-1 and ET-3 are present in large amounts.[11-13]

It appears clear, therefore, that in vivo the circulating levels of ET-1 can be regulated over short periods, i.e., minutes rather than hours. Although it is generally much more difficult to stimulate the release of ET-1 from endothelial cells in culture over such short periods there are a few stimuli that can be employed to produce such an effect. Physical stretching of endothelial cells, for example, can stimulate a quick release of ET-1, such that after 20 min the medium bathing stretched endothelial cells contains more than twice as much ET-1 as that bathing identical but stationary cells.[14] Clearly as these experiments employ mono-cultures of cells the results indicate that endothelial cells contain intracellular stores of ET-1 which can be released rapidly in response to certain stimuli. This conclusion is supported by studies which show within endothelial cells staining for ET-1[15] the presence of ET-1-rich structures following subcellular fractionation[16] and, of course, strong staining for ET-1 within the endothelium of tissue sections.[13]

From the above it appears safe to conclude that ET-1 can be rapidly released into the circulation, possibly from stores held within the endothelium. However, to establish whether such a release is of any significance it is also necessary to determine any functional effects that are dependent upon this rapidly produced ET-1. Such effects can be found. For instance, rapid intravenous injection of a high dose of endotoxin (LPS) to anesthetized rats causes within 5-10 min both a dramatic fall in blood pressure and a marked increase in hematocrit. Treatment with endothelin receptor antagonists active at the ET_A receptor limit both the fall in blood pressure and the increase in hematocrit.[17] Clearly the speed of these responses does not allow for the expression and synthesis of ET-1 and we must conclude that it is dependent upon the activity of performed ET-1.

The above experiments start to lead us in the direction of agents more selective than LPS that may regulate the production of ET-1. Obviously, it is difficult to use LPS as a fine tool to understand the regulation of ET-1 production for it is a widely active agent that increases the production and/or expression of a large number of autacoids, and particularly cytokines. To elucidate the roles of this latter class of agents in regulating the production of ET-1 we have been examining the effects of cytokines on the production of ET-1.

Intravenous administration of tumor necrosis factor-α (TNF-α) to rats causes quite rapid increases in the circulating levels of ET-1, that are unaffected by pretreatment of the animals with phosphoramidon,[18] as has been noted previously for surgery (see above), suggestive of release from preformed stores. As before, however, we need to ask whether this increase in circulating ET-1 is linked to any functional changes. In particular, bearing in mind ET-1's very marked effects on vascular smooth muscle tone, we might expect to find marked vasoconstrictions. Interestingly, this does indeed appear to be the case, an effect that we have demonstrated following ex vivo measurements of blood vessel reactivity. In detail, we have found that when hearts are removed from rats treated 15 min previously with TNF-α and perfused at constant flow ex vivo with Krebs' buffer a profound coronary vasoconstriction is revealed.[19] This vasoconstriction is seen functionally as an increase in coronary perfusion pressure to more than double that seen in hearts removed from sham-treated animals. The coronary vasoconstriction can be positively correlated to the accompanying increase in the circulating level of ET-1 for it is greatly decreased when rats are pretreated with an endothelin ET$_A$ receptor antagonist or with antibodies directed against ET-1. From these experiments, therefore, we may conclude that exogenous cytokines can rapidly increase the production of ET-1 and so promote coronary vasoconstriction. This suggests that cytokines, and in particular TNF-α, may precipitate the release of ET-1 from the in vivo storage sites indicated by the experiments discussed previously.

Stimuli of Medium Term and Prolonged Increases in Endothelin Synthesis and/or Release

Extending the above argument further we might also suggest that cytokines, and in particular TNF-α or interleukin-1 (which causes similar changes to TNF-α) are endogenous 'turn-on' switches for ET-1 production. To support such a suggestion we need, however, to know the effects that we see following injection of exogenous TNF-α can be precipitated by TNF-α derived from endogenous sources. Our studies indicate that they can. Treatment of rats with IL-2 causes in the next few hours a time-dependent increase in the concentration of TNF-α within the circulation which is shadowed shortly afterwards by an increase in the coronary perfusion pressure.[19] More importantly, the coronary vasoconstriction seen in these IL-2-treated rats is reduced by antibodies directed against TNF-α or by endothelin ET$_A$ receptor antagonists, underscoring the link between TNF-α and ET-1.

Is there other evidence supporting the idea that cytokines may increase the production of ET-1 over the short to medium term (minutes to hours)? The answer is that such evidence does exist. In the rat, for example, sepsis resulting from cæcal perforation is associated with an increase in circulating ET-1 that is reduced by treatment with a monoclonal anti-endotoxin antibody.[20] Similarly, intravenous infusion of live bacteria into young swine causes after 4 h increases in circulating levels of ET-1 which are markedly reduced if the animals are treated with anti-TNF-α antibody.[21] In an animal more closely related to the human, the baboon, the increase in the circulating concentration of big ET-1 caused by endotoxemia is also reduced by treatment with an anti-TNF-α antibody.[22] Furthermore, in humans suffering with sepsis there is a clear positive correlation between the circulating levels of TNF-α and ET-1.[23] Thus, it would appear that cytokines, and in particular TNF-α can promote both an increase in the immediate release of ET-1, and also an up-regulation in the expression of ET-1 over periods of hours.

What about vascular cells in culture? Here too it has been widely reported that cytokines increase the production of endothelin-1, usually over a period of time consistent with an up-regulation of the expression of ET-1. For example, treatment of bovine aortic endothelial cells with either TNF-α or IL1-β causes an up to 2-fold increase in the release of ET-1 after 6 h.[24] However, it is not only the endothelium within the vascular wall that may respond to cytokines by up-regulating their production of ET-1. The vascular smooth muscle, which is not considered to be a producer of ET-1 under normal conditions, can also be induced to synthesis and release ET-1. Treatment of human vascular smooth muscle cells in culture, for instance, with the vasoconstrictor hormones angiotensin II or arginine-vasopressin, or the growth factors transforming growth factor β, platelet derived growth factor or epidermal growth factor stimulates the expression of ET-1.[25] These results suggest that ET-1 may be formed within the vascular smooth muscle and act in an autocrine manner to influence both vascular tone and function. Indeed, in cultured vascular smooth muscle cells derived from atherosclerotic coronary arteries there is a marked increase in the basal production of ET-1 compared to that from vascular smooth muscle cells taken from macroscopically normal vessels.[26] Interestingly, atherosclerosis is a condition associated with cytokine up-regulation, particularly within the blood vessel wall (see below). Outside of the blood vessel wall, other studies have indicated that cytokines, particularly TNF-α and IL1-β possess the ability to stimulate the synthesis of ET-1 by a range of endothelial, epithelial and other cell types (Table 7.1).

Table 7.1 Cytokines that stimulate endothelin synthesis in various cells in culture

Cell	Species	Cytokine	Ref
endothelial	porcine	IL-1, TGF-β, TNF-α	27,28
	bovine	TNF-α, INF-γ, IL-1β,	24,29-32
	HUVEC	IL-1β	33
	rat	IL-1a, IL-1, TNF-α	34
amnion	human	IL-1β, EGF, TNF-α,	35,36
	IL-6		
LLCPK1 (renal epithelium)	porcine	TGF-β, TNF-α, IL-1β	37
airway epithelium	human	IL-1, TNF-α, IL-2	38,39
	guinea-pig	IL-8, TNF-α, TGF-β	40
mesenteric artery	rat	IL-2	41
vascular smooth muscle	human	TGF-β	42
masangial	human	TGF-β	43
	rat	TNF-α	44
breast cancer	human	IL-6	45

Many of the above studies report predominantly pharmacological investigations, albeit interesting ones, that rely heavily upon the application of exogenous cytokines to provoke the changes seen. A relevant cytokine-ET-1 axis may be better demonstrated by following changes in a pathological model where cytokine production is entirely driven by endogenous changes within the animal. Adjuvant polyarthritis in the rat, which precipitates marked joint swelling and has been reported to involve elevated circulating levels of cytokines,[46,47] provides such a model. Using our model of coronary vasoconstriction described above we have found that there are dramatic increases in the ex vivo coronary perfusion pressures of hearts removed from adjuvant polyarthritic rats that follow closely the development of the joint swelling. For instance, at day 12 when the joint swelling is most marked the ex vivo coronary perfusion pressure is increased by almost three-fold.[48] Most interestingly, this coronary vasoconstriction is absent when rats are treated for the preceding 48 h with an endothelin receptor antagonist.

What can be inferred from the above observations? Clearly, that cytokine production is associated with a marked up-regulation of the production and/or effects of ET-1, and in the rat with an increased drive towards coronary vasoconstriction. However, we should not assume that this implies that inflammation must be associated with coronary vasoconstriction and myocardial dysfunction. It is rather that in the rat coronary perfusion pressure serves as a very good bioassay for ET-1 activity, as ET-1 is a strong and almost irreversible constrictor of rat coronary vessels in vitro.[49] It is still noteworthy, however, that IL-2 therapy in humans is accompanied by myocardial dysfunction (see ref. 19).

Evidence for a Link Between Cytokines and ET-1 Production

In Human Disease States

In looking for a link between cytokines and ET-1 in a wider variety of disease states we need to look for more indirect evidence, as there have been few trials that have combined measurements of cytokines, ET-1 and the effectiveness of endothelin receptor antagonists in such conditions; although there is evidence that ET-1 is often increased in the circulation in conditions in which cytokines, and in particular TNF-α are elevated (Table 7.2). However, we can ask whether or not our experience with disease states in which ET-1 appears involved also accord with the suggestion that cytokines are a turn-on signal for ET-1? The answer appears to be yes in many instances (Table 7.3). Much experimental evidence from models in animals including rats, rabbits and dogs has strongly implicated endothelin as a causative player in the injuries that follow ischemia/reperfusion (see ref. 76), and here there is good evidence that in all species there are increases in cytokine production.[77] Following an ischemic period, for instance, the hindquarters of rats release an early, transient burst of TNF-α on reperfusion, and treatment with antibodies against TNF-α maintains postischemic flow.[78] Similarly, focal cerebral ischemia in rats increases TNF-α production by ischemic neurones[79] supporting a role for ET-1 in stroke. Interestingly, an ET_A receptor antagonist is found to be protective in a gerbil model of transient global brain ischemia, which acts as a model for stroke.[80] As might be expected there is also a strong cytokine activation in human vasculitis,[81,82] and interestingly in congestive heart failure[83] (in which human condition ET-1 receptor antagonists are beneficial[7]), and possibly hypercholesterolemia.[84] In association with this latter condition, it is also interesting to note that endothelin

Table 7.2 Human pathologies in which elevated levels of both ET-1 and cytokines have been noted

Pathology	Site of Endothelin	Associated Cytokine	Ref
Kawasaki disease	plasm	IL-6	50
systemic sclerosis	fibroblasts	IL-1IL-β	51
HIV + retinal microangiopathic syndrome	plasma		
sepsis	plasma	TNF-α	52
disseminated intravascular coagulation & sepsis	plasma	TNF-α	53
>20% burn	plasma		54,55
asthma	bronchial epithelial cells	IL-1	56
malaria	plasma	TNF-α	57
organ rejection	plasma	TNF-α, IL-1	58
burn wound infection	plasma	TNF-α	59

Table 7.3 Various human pathologies associated with cytokine activation of the endothelium

Pathology	Ref
ulcerative colitis	60
cutaneous vasculitis	61
ischemic heart disease	62
dilated cardiomyopathy	63
preeclampsia	64
lung fibrosis	65
necrotizing vasculitis	66
dilated cardiomyopathy	67
aortic aneurysm	68
Kawasaki disease	69
atherosclerosis; occlusive vascular disease	70,71
asthma	72
psoriasis	73
Crohn's disease and ulcerative colitis	74
Raynaud's phenomenon	75

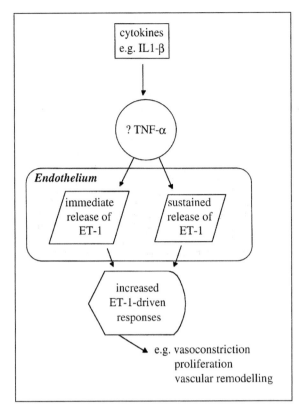

Fig. 7.1. Cytokines such as IL-1β, acting possibly via TNF-α as a final mediator, promote both a rapid release of ET-1 and a more prolonged upregulation of ET-1 production. This may well underpin the vaso-constriction, proliferation and vascular remodelling seen in a number of cyto-kine and endothelin related disease states (Tables 7.2 and 7.3).

receptor antagonists are found to reduce neointimal proliferation in models of restenosis following percutaneous transluminal angioplasty.[85] In accordance with this anti-proliferative effect at the level of the vascular smooth muscle, endothelin receptor antagonists have also been demonstrated to decrease the thickening of pulmonary arteries seen in animal models of pulmonary hypertension.[86] Clearly there are substantial increases in the production and circulating level of TNF-α in systemic inflammatory response syndrome (septic shock) and related pathologi-cal states,[87] and here animal studies indicate that endothelin production is increased for endothelin receptor antagonists potentiate endotoxemia-associated hypoten-sion[88] and increase mortality, at least acutely.[89] Aside from these conditions there are also a number of other disease states that have been reported to be associated with cytokine activation of the endothelium in which we might predict that ET-1 could have an involvement (Table 7.3).

Conclusion

In conclusion, the ET-1 system should be viewed as being in the larger part latent, but in numerous disease states rapidly activatable. This rapid activation may well depend upon activation of the endothelium by cytokines, and in particular by TNF-α (Fig. 7.1). The current state-of-the-art in the area of endothelin research gives much support to this idea. In particular, endothelin receptor antagonists are found to be beneficial in pathological disease states such as ischemia reperfusion injury, congestive heart disease, (some forms of) hypertension, restenosis/athero-sclerosis and septic shock, all of which are associated with cytokine activation of the endothelium. Numerous other disease states in which there is also known to be cytokine activation of the endothelium may also offer additional targets at which endothelin antagonists could be aimed. The presence of this cytokine-endothelin axis, and other axes in disease states such as hypertension where ET-1 has been implicated but elevated cytokine production does not appear to be involved, would then explain the relative lack of effects of endothelin receptor antagonists under physiological conditions but their beneficial actions in pathological states. We might therefore regard ET-1 as not being purely a vasoconstrictor peptide but in fact an integral part of the cytokine network orchestrated by the endothelium.

References

1. Cernacek P, Stewart DJ. Immunoreactive endothelin in human plasma: marked elevations in patients in cardiogenic shock. Biochem Biophys Res Commun 1989; 161:562-567.
2. Änggård EE, Galton S, Rae G et al. The fate of radioiodinated endothelin-1 and endothelin-3 in the rat. J Cardiovasc Pharmacol 1990; 13(Suppl 5):S46-S49.
3. Haynes WG, Webb DJ. Contribution of endogenous generation of endothelin-1 to basal vascular tone. Lancet 1994; 344:852-854.
4. Rubanyi GM, Polokoff MA. Endothelins: molecular biology, biochemistry, pharmacology, physiology and pathophysiology. Pharmacol Rev 1994; 46:325-415.
5. Warner TD, Battistini B, Doherty AM et al. Endothelin receptor antagonists: actions and rationale for their development. Biochem Pharmacol 1994; 48:625-635.
6. Warner TD. Endothelin receptor antagonists. Pharmacol Drug Rev 1994; 12:105-122.
7. Kiowski W, Sütsch G, Hunziker P et al. Evidence for endothelin-1-mediated vaso-constriction in severe chronic heart failure. Lancet 1995; 346:732-736.
8. Kaufmann H, Oribe E, Oliver JA. Plasma endothelin during upright tilt: relevance for orthostatic hypotension. Lancet 1991; 338:1542-1545.
9. Haak T, Matheis G, Kohleisen M et al. Endothelin during cardiovascular surgery: the effect of diltiazem and nitroglycerin. J Cardiovasc Pharmacol 1995; 26(Suppl 3):S494-S496.
10. Pollock DM, Divish BJ, Opgenorth TJ. Stimulation of endogenous endothelin release in the anesthetized rat. J Cardiovasc Pharmacol 1993; 22(Suppl. 8), S295-S298.
11. Matsumoto H, Suzuki N, Onda H et al. Abdunance of endothelin-3 in rat intestine, pituitary gland and brain. Biochem Biophys Res Commun 1989; 164:74-80.
12. Yoshizawa Y, Shinmi O, Giaid A et al. Endothelin: a novel peptide in the posterior pituitary system. Science 1990; 247:462-464.
13. Lee M-E, de la Monte SM, Ng S-C et al. Expression of the potent vasoconstrictor endothelin in the human central nervous system. J Clin Invest 1990; 86:141-147.
14. Macarthur H, Warner TD, Wood EG et al. Endothelin-1 release from endothelial cells in culture is elevated both acutely and chronically by short periods of mechanical stretch. Biochem Biophys Res Commun 1994; 200:395-400.
15. Loesch A, Bodin P, Burnstock G. Colocalization of endothelin, vasopressin and serotonin in cultured endothelial cells of rabbit aorta. Peptides 1991; 12:1095-1103.
16. Harrison VJ, Corder R, Änggård EE et al. Evidence for vesicles that transport endothelin-1 in bovine aortic endothelial cells. J Cardiovasc Pharmacol 1993; 22(Suppl 8):S57-S60.
17. Allcock GH, Warner TD. Activation of ET_A receptors is partially responsible for the rapid increase in haematocrit induced by bacterial lipopolysaccharide in the rat. Life Sci; in press.

18. Vemulapalli S, Chiu PJS, Griscti K et al. Phosphoramidon does not inhibit endogenous endothelin-1 release stimulated by hemorrhage, cytokines and hypoxia in rats. Eur J Pharmacol 1994; 257:95-102.

19. Klemm P, Warner TD, Hohlfeld T et al. Endothelin 1 mediates *ex vivo* coronary vasoconstriction caused by exogenous and endogenous cytokines. Proc Natl Acad Sci USA 1995; 92:2691-2695.

20. Lundblad R, Giercksky FE. Effect of volume support, antibiotic therapy, and monoclonal antiendotoxin antibodies on mortality rate and blood concentrations of endothelin and other mediators in fulminant intraabdominal sepsis in rats. Crit Care Med 1995; 23:1382-1390.

21. Han JJ, Windsor A, Drenning DH, Leeper-Woodford S et al. Release of endothelin in relation to tumor necrosis factor-a in porcine Pseudomonas aeruginosa-induced septic shock. Shock 1994; 1:343-346.

22. Redl H, Schlag G, Bahrami S et al. Big-endothelin release in baboon bacteremia is partially TNF dependent. J Lab Clin Med 1994; 124:796-801.

23. Takakuwa T, Endo S, Nakae H et al. Plasma levels of TNF-α, endothelin-1 and thrombomodulin in patients with sepsis. Res Commun Chem Pathol Pharmacol 1994; 84:261-269.

24. Corder R, Carrier M, Khan N et al. Cytokine regulation of endothelin-1 release from bovine aortic endothelial cells. J Cardiovasc Pharmacol 1995; 26(Suppl 3):S56-S58.

25. Resink TJ, Hahn AWA, Scott-Burden T, Powell J, Weber E, Bühler F. Inducible endothelin mRNA expression and peptide secretion in cultured human vascular smooth muscle cells. Biochem Biophys Res Commun 1990; 168:1303-1310.

26. Haug C, Voisard R, Lenich A et al. Increased endothelin release by cultured human smooth muscle cells from atherosclerotic coronary arteries. Cardiovasc Res 1996; 31:807-813.

27. Yoshizumi M, Kurihara H, Morita T et al. Interleukin 1increases the production of endothelin-1 by cultured endothelial cells. Biochem Biophys Res Commun 1990; 166:324-329.

28. Maemura K, Kurihara H, Morita T. et al. Production of endothelin-1 in vascular endothelial cells is regulated by factors associated with vascular injury. Gerontology 1992; 38(Suppl 1):29-35.

29. Kanse SM, Takahashi K, Lam HC et al. Cytokine stimulated endothelin release from endothelial cells. Life Sci 1991; 48:1379-1384.

30. Marsden PA, Brenner BM. Transcriptional regulation of the endothelin-1 gene by TNF-α. Am J Physiol 1992; 262:C854-C861.

31. Lamas S, Michel T, Collins T et al. Effects of interferon-γ on nitric oxide synthase activity and endothelin-1 production by vascular endothelial cells. J Clin Invest 1992; 90:879-887.

32. Estrada C, Gomez C, Martin C. Effects of TNF-α on the production of vasoactive substances by cerebral endothelial and smooth muscle cells in culture. J Cereb Blood Flow Metab 1995; 15:920-928.

33. Katabami T, Shimizu M, Okano K et al. Intracellular signal transduction for interleukin-1β-induced endothelin production in human umbilical vein endothelial cells. Biochem Biophys Res Commun 1992; 188:565-570.

34. Golden CL, Kohler JP, Nick HS et al. Effects of vasoactive and inflammatory mediators on endothelin-1 gene expression in pulmonary endothelial cells. Am J Respir Cell Mol Biol 1995; 12:503-512.

35. Sunnergren KP, Word RA, Sambrook JF et al. Expression and regulation of endothelin precursor mRNA in avascular human amnion. Mol Cell Endocrinol 1990; 68:R7-R14.

36. Mitchell MD, Lundin-Schiller S, Edwin SS. Endothelin production by amnion and its regulation by cytokines. Am J Obstet Gynecol 1991; 165:120-124.

37. Ohta K, Hirata Y, Imai T et al. Cytokine-induced release of endothelin-1 from porcine renal epithelial cell line. Biochem Biophys Res Commun 1990; 169:578-584.

38. Nakano J, Takizawa H, Ohtoshi T et al. Endotoxin and pro-inflammatory cytokines stimulate endothelin-1 expression and release by airway epithelial cells. Clin Exp Allergy 1994; 24:330-336.

39. Calderon E, Gomez-Sanchez CE, Cozza EN et al. Modulation of endothelin-1 production by a pulmonary epithelial cell line. I. Regulation by glucocorticoids. Biochem Pharmacol 1994; 48:2065-2071.

40. Endo T, Uchida Y, Matsumoto H et al. Regulation of endothelin-1 synthesis in cultured guinea-pig airway epithelial cells by various cytokines. Biochem Biophys Res Commun 1992; 186:1594-1599.

41. Miyamori I, Takeda Y, Yoneda T et al. Interleukin-2 enhances the release of endothelin-1 from the rat mesenteric artery. Life Sci 1991; 49:1295-1300.

42. Resink TJ, Hahn AWA, Scott-Burden T et al. Inducible endothelin mRNA expression and peptide secretion in cultured human vascular smooth muscle cells. Biochem Biophys Res Commun 1990; 168:1303-1310.

43. Zoja C, Orisio C, Perico N et al. Constitutive expression of endothelin gene in cultured human mesangial cells and its modulation by transforming growth factor-β, thrombin and a thromboxane A_2 analogue. Lab Invest 1991; 64:16-20.

44. Kohan DE. Production of endothelin-1 by rat mesangial cells: regulation by tumor necrosis factor. J Lab Clin Med 1992; 119:477-484.

45. Yamahita J, Ogawa M, Nomura K et al. Interleukin 6 stimulates the production of immunoreactive endothelin 1 in human breast cancer cells. Cancer Res 1993; 53:464-467.

46. Arai K, Lee F, Miyajima A, Miyatake S et al. Cytokines: coordinators of immune and inflammatory responses. Ann Rev Biochem 1990; 59:783-794.

47. Arend WP, Dayer J-M. Cytokines and cytokine inhibitors or antagonists in rheumatoid arthritis. Arthritis Rheum 1990; 33:305-315.

48. Klemm P, Warner TD, Willis D et al. Coronary vasoconstriction in vitro in the hearts of polyarthritic rats: effectiveness of *in vivo* treatment with the endothelin receptor antagonist SB 209670. Br J Pharmacol 1995; 114:1327-1328.

49. Baydoun AR, Peers SH, Cirino G et al. Effects of endothelin-1 on the rat isolated heart. J Cardiovasc Pharmacol 1989; 13(Suppl. 5):S193-S196.

50. Morise T, Takeuchi Y, Takeda R et al. Increased plasma endothelin levels in Kawasaki disease: a possible marker for Kawasaki disease. Angiology 1993; 44:719-723.

51. Kawaguchi Y, Suzuki K, Hara M et al. Increased endothelin-1 production in fibroblasts derived from patients with systemic sclerosis. Ann Rheum Dis 1994; 53:506-510.

52. Takakuwa T, Endo S, Nakae E et al. Plasma levels of TNF-α, endothelin-1 and thrombomodulin in patients with sepsis. Res Commun Chem Pathol Pharmacol 1994; 84:261-269.

53. Endo S, Inada K, Nakae H et al. Blood levels of endothelin-1 and thrombomodulin in patients with disseminated intravascular coagulation and sepsis. Res Commun Mol Pathol Pharmacol 1995; 90:277-288.

54. Huribal M, Cunningham ME, D'Aiuto ML et al. Endothelin levels in patients with burns covering more than 20% body surface area. J Burn Care Rehabil 1995; 16:23-26.

55. Huribal M, Cunningham ME, D'Aiuto ML et al. Endothelin-1 and prostaglandin E2 levels increase in patients with burns. J Am Coll Surg 1995; 180:318-322.

56. Ackerman V, Carpi S, Bellini A et al. Constitutive expression of endothelin in bronchial epithelial cells of patients with symptomatic and asymptomatic asthma and modulation by histamine and interleukin-1. J Allergy Clin Immunol 1995; 96:618-627.

57. Wenisch C, Wenisch H, Wilairatana P et al. Big endothelin in patients with complicated Plasmodium falciparum malaria. J Infect Dis 1996; 173:1281-1284.

58. Watschinger B, Sayegh MH. Endothelin in organ transplantation. Am J Kidney Dis 1996; 27:151-161.

59. Takakuwa T, Endo S, Nakae H et al. Relationship between plasma levels of type II phospholipase A_2, PAF acetylhydrolase, endothelin-1, and thrombomodulin in patients with infected burns. Res Commun Mol Pathol Pharmacol 1994; 86:335-340.

60. Nielsenn OH, Brynsov J, Vainer B. Increased mucosal concentrations of soluble intercellular adhesion molecule-1 (sICAM-1), sE-selectin, and interleukin-8 in active ulcerative colitis. Dig Dis Sci 1996; 41:1780-1785.

61. Jurd KM, Stephens CJ, Black MM et al. Endothelial activation in cutaneous vasculitis. Clin Exp Dermatol 1996; 21:28-32.

62. Galea J, Armstrong J, Gadsdon P et al. Interleukin-1β in coronary arteries of patients with ischemic heart disease. Arterioscler Thromb Vasc Biol 1996; 16:1000-1006.

63. Habib FM, Springall DR, Davies GJ et al. Tumour necrosis factor and inducible nitric oxide synthase in dilated cardiomyopathy. Lancet 1996; 347:1151-1155.

64. Chen G, Wilson R, Wang SH et al. Tumour necrosis factor-α (TNF-α) gene polymorphism and expression in preeclampsia. Clin Exp Immunol 1996; 104:154-59.

65. Vaillant P, Menard O, Vignaud JM et al. The role of cytokines in human lung fibrosis. Monaldi Arch Chest Dis 1996; 51:145-152.

66. Teofoli P, Lotti T. Cytokines, fibrinolysis and vasculitis. Int Angiol 1995; 14:125-129.

67. Kuhl U, Noutsias M, Schultheiss HP. Immunohistochemistry in dilated cardiomyopathy. Eur Heart J 1995; 16 Suppl O:100-106.

68. Koch AE, Kunkel SL, Pearce WH et al. Enhanced production of the chemotactic cytokines interleukinnn-8 and monocyte chemoattractant protein-1 in human abdominal aortic aneurysms. Am J Pathol 1993; 142:1423-1431.

69. Leung DY. The immunologic effects of IVIG in Kawasaki disease. Int Rev Immunol 1939; 5:197-202.

70. Libby P, Sukhova G, Lee RT et al. Cytokines regulate vascular functions related to stability of the atherosclerotic plaque. J Cardiovasc Pharmacol 1995; 25 Suppl 2:S9-S12.

71. Brody JI, Pickering NJ, Capuzzi DM et al. Interleukin-1α as a factor in occlusive vascular disease. Am J Clin Pathol 1992; 97:8-13.

72. Howarth PH. The airway inflammatory response in allergic asthma and its relationship to clinical disease. Allergy 1995; 50(Suppl 22):13-21.

73. Das PK, de Boer OJ, Visser A et al. Differential expression of ICAM-1, E-selectin and VCAM-1 by endothelial cells in psoriasis and contact dermatitis. Acta Derm Venereol Suppl 1994; 186:21-22.

74. Murch SH, Braegger CP, Walker-Smith JA et al. Location of tumour necrosis factor alpha by immunohistochemistry in chronic inflammatory bowel disease. Gut 1993; 34:1705-1709.

75. Blann AD, Illingworth K, Jayson MI. Mechanisms of endothelial cell damage in systemic sclerosis and Raynaud's phenomenon. J Rheumatol 1993; 20:1325-1330.

76. Warner TD. Endothelin receptor antagonists. Cardiovasc Drug Rev 1994; 12:105-122.

77. Nose PS. Cytokines and reperfusion injury. J Cardiac Surg 1993; 8(Suppl. 2):305-308.

78. Sternbergh WC, Tuttle TM, Makhoul RG et al. Postischemic extremities exhibit immediate release of tumor necrosis factor. J Vasc Surg 1994; 20:474-481.

79. Liu T, Clark RK, McDonnell PC et al. Tumor necrosis factor-alpha expression in ischemic neurons. Stroke 1994; 25:1481-1488.

80. Feuerstein G, Gu J-L, Ohlstein EH et al. Peptidic endothelin-1 receptor antagonist, BQ-123 and neuroprotection. Peptides 1994; 15:467-469.

81. Robertson CR, McCallum RM. Changing concepts in the pathophysiology of vasculitides. Curr Opinion Rheumatology 1994; 6:3-10.

82. Savage CO, Cooke SP. The role of the endothelium in systemic vasculitis. J Autoimmunity 1993; 6:237-249.

83. Mann DL, Young JB. Basic mechanisms in congestive heart failure. Recognizing the role of proinflammatory cytokines. Chest 1994; 105:897-904.

84. Hennig B, Diana JN, Toborek M et al. Influence of nutrients and cytokines on endothelial cell metabolism. J Am Coll Nutrition 1994; 13:224-231.

85. Douglas SA, Louden C, Vickery-Clark LM et al. A role for endogenous endothelin-1 in neointimal formation after rat carotid artery balloon angioplasty: protective effects of the novel nonpeptide endothelin receptor antagonist SB 209670. Circ Res 1994; 75:190-197.

86. Strieter RM, Kunkel SL, Bone RC. Role of tumor necrosis factor-α in disease states and inflammation. Crit Care Med 1993; 21(Suppl. 10):S447-S463.

87. Chen S-J, Chen Y-F, Meng QC et al. Endothelin-receptor antagonist bosentan prevents and reverses hypoxic pulmonary hypertension in rats. J Appl Physiol 1995; 79.

88. Gardiner SM, Kemp PA, March JE et al. Enhancement of the hypotensive and vasodilator effects of endotoxaemia in conscious rats by the endothelin antagonist, SB 209670. Br J Pharmacol 1995; 116:1718-1719.

89. Ruetten H, Thiemermann C, Vane JR. Effects of endothelin receptor antagonist, SB 209670, on circulatory failure and organ injury in endotoxic shock in the anaesthetized rat. Br J Pharmacol 1996; 118:198-204.

Endothelin Modulation of Renal Sodium and Water Transport

Donald E. Kohan

Introduction

Less than a decade ago, endothelin-1 (ET-1) was identified as the most potent vasoconstrictor known.[1] With this discovery, the scientific medical community, both academic and industrial, launched a massive attempt to implicate ET-1 in the pathogenesis of disorders associated with elevated vascular tone, including hypertension, atherosclerosis, myocardial infarction, pulmonary hypertension, and cerebrovascular accident. As a result, some clinical situations were identified in which ET-1 blockade may be therapeutic (e.g. acute renal failure), however, many other diseases (such as essential hypertension) did not appear to be primarily mediated by dysregulated ET-1 activity. During this rush to understand and treat vascular disorders associated with altered ET-1 effects, it became increasingly evident that the biology of endothelin was far more complex than originally suspected. Where we once thought ET-1 was primarily an endothelial cell-derived product (hence its name), we now know that ET-1 is actually a family of peptides that are produced by a tremendous variety of cell types. Furthermore, ET-1 has a broad spectrum of biological actions affecting multiple cell types, many of which are nonvascular. Our laboratory first became interested in one of these nonvascular sites in 1989 when we heard a report describing the distribution of immunoreactive ET-1 in the body.[2] This study determined that of all tissues in the body (using the pig as a model), the inner medulla of the kidney had by far the greatest concentration of ET-1. Why would this potent vasoconstrictor be expressed in such high amounts in a region virtually devoid of vascular smooth muscle? Part of the answer was suggested by another report about that time which described ET-1 inhibition of Na^+/K^+ ATPase activity in the inner medullary collecting duct (IMCD).[3] Could it be that ET-1 was produced by cells in the inner medulla and then acted locally to regulate collecting duct sodium transport? If so, could ET-1 regulation of renal tubule sodium reabsorption somehow be related to the vascular actions of the peptide? So began a large number of studies by our laboratory and many others into the role of ET-1 in modulating renal tubule transport processes. These investigations have revealed a highly complex system whereby ET-1 potently modulates solute and water transport systems throughout the nephron. This chapter will review much

Endothelin Receptors and Signaling Mechanisms, edited by David M. Pollock and Robert F. Highsmith. © 1998 Springer-Verlag and R.G. Landes Company.

of the work leading to our current understanding of ET-1 regulation of renal tubule water and electrolyte transport. In addition, the relevance of this system to fluid and electrolyte homeostasis will be discussed.

Endothelin Regulation of Water and Electrolyte Transport

Before discussing ET-1 regulation of nephron transport processes, it is useful to make a few general comments. First, as shall be described in more detail, ETs are produced by, bind to, and modulate fluid and electrolyte reabsorption in most regions of the nephron. Since the plasma concentrations of ET-1 is normally quite low, it is likely that ET-1 regulation of renal tubule function occurs locally in a paracrine and/or autocrine manner. Hence, as we review ET-1 modulation of fluid and electrolyte transport, it is important to bear in mind that ET-1 actions must be viewed in the context of where the peptide is produced and nearby regions on which it can act. The second point is that ET-1 appears to have variable effects on transport systems in different nephron segments. Thus, rather than trying to make general comments about ET-1 regulation of transport pathways, nephron segments will be considered individually. Finally, I will attempt to integrate individual nephron synthesis and effects of ETs into a global view of how these peptides are involved in normal and deranged regulation of renal tubule function.

Proximal Tubule

Initially, little attention was focused on ET-1 actions on the proximal tubule since this nephron segment did not appear to synthesize or bind appreciable amounts of the peptide.[4-6] Subsequent studies revealed, however, that proximal tubules did produce and respond to ET-1, albeit to a lesser degree than other nephron segments. Since this region of the nephron reabsorbs the bulk of filtered fluid and electrolytes, even modest effects on transport processes could have a profound impact on urinary excretion rates. It would seem important, therefore, to review how ET-1 might modulate transport systems in the proximal tubule.

While researchers believed that ET-1 modulated proximal tubule transport, there initially was little agreement as to what this effect might be. One group of investigators contended that ET-1 inhibited fluid, sodium, and bicarbonate reabsorption by the proximal tubule, while the other camp felt ET-1 had a stimulatory effect. Evidence in favor of an inhibitory action included the finding that ET-1 reduced fluid and bicarbonate absorption by the rat proximal straight tubule.[7] Infusion of low doses of ET-1 also increased end-proximal tubule fluid delivery as estimated by lithium clearance.[8,9] These findings were further supported by the observations that ET-1 reduced Na^+/K^+-ATPase activity in rat proximal tubules and increased cGMP levels in the LLC-PK$_1$ proximal tubule cell line.[7,10-12] These effects were felt due to activation of ET_B receptors since ET-1 and ET-3 were equipotent in reducing Na^+/K^+ ATPase activity and had a similar IC_{50} for [^{125}I]ET-1 binding to LLC-PK$_1$ cells.[10,11]

Evidence in favor of a stimulatory effect of ET-1 on proximal tubule solute transport was, however, equally convincing. ET-1 stimulated Na^+/H^+ exchange in proximal tubule-like OKP cells via a protein kinase C (PKC)-dependent pathway.[13] Similarly, ET-1 enhanced Na^+/H^+ exchange and Na^+/P_i cotransport by rat renal cortical slices by activation of both protein kinases A and C.[14] ET-1 also stimulated Na^+/H^+ antiporter and Na^+/HCO_3^- transporter activity in rabbit proximal tubule brush border membrane vesicles, although there was no effect on Na^+-glucose or Na^+-

succinate transporter activity.[15] The stimulatory effect of ET-1 on Na^+/H^+ exchange was ascribed to activation of ET_B receptors since antiporter activity increased in OKP cells transfected with the ET_B, but not the ET_A, receptor.[16,17] Interestingly, ET-1 increased proximal tubule cell calcium concentration (Ca^{2+}), inhibited adenylate cyclase activity and induced tyrosine phosphorylation of several proteins in OKP cells transfected with either ET_A or ET_B receptors, suggesting that ET-1 stimulation of Na^+/H^+ exchange in these cells was independent of Ca^{2+}, cAMP levels, or tyrosine kinase activity.[16] Finally, and to make matters worse, others found no effect of ET-1 on human proximal tubule sodium reabsorption (estimated by lithium clearance) or on rabbit proximal tubule Na^+/K^+-ATPase activity.[3,18] The problem clearly was how to reconcile these apparently disparate results.

A possible solution to this dilemma was provided by the observation that ET-1 has a biphasic effect on fluid absorption by the rat proximal straight tubule.[19] Low concentrations of the peptide (0.1 pM) increased fluid absorption by activating PKC, while high concentrations of ET-1 (1 nM) reduced fluid transport through PKC-, cyclooxygenase-, and lipoxygenase-dependent mechanisms. In support of these findings, others have noted that ET-1 inhibits proximal tubule fluid reabsorption through enhancing production of cyclooxygenase and lipoxygenase metabolites.[8,10] Taking all the above studies together, one can formulate a working hypothesis of how ET-1 regulates sodium and fluid transport in the proximal tubule. If the concentration of ET-1 is low enough to only activate PKC, proximal tubule sodium reabsorption is increased. At higher ET-1 concentrations, phospholipase A_2 metabolites accumulate and proximal tubule sodium and fluid reabsorption are reduced (Fig. 8.1).

The above considerations beg the question as to what is the concentration of ET-1 around the proximal tubule. As stated earlier, plasma concentrations of ET-1 are quite low, ranging between 1 and 10 pM. These concentrations lie above those shown to produce only a stimulatory effect on proximal tubule sodium transport, suggesting that any increase in local ET-1 levels would exert an inhibitory effect.[19] Such an increase in local ET-1 concentration would undoubtedly not arise from changes in plasma ET-1 concentration; rather, alterations in local ET-1 production (or even ET-1 receptor expression) would be much more likely to modulate proximal tubule transport processes. This ET-1 could derive from neighboring cortical collecting ducts (see below); however, it is also conceivable that the proximal tubule regulates its own reabsorptive pathways through an autocrine ET-1 system. Most studies have demonstrated relatively little,[4,20] or even no,[21,22] ET-1 production by proximal tubule cells derived from experimental animals. Significant ET-1 production by proximal tubules has been demonstrated, however, in humans as well as certain animal cell lines.[23-26] Furthermore, ET-1 synthesis by proximal tubules can be augmented by a variety of factors, including transforming growth factor-β, thrombin, interleukin-1, high density lipoproteins, hypoxia, cyclosporine A and FK-506.[24-28] Most importantly, proximal tubules can be induced to produce ET-1 in high enough amounts to function as an autocrine factor. This latter point is exemplified by the finding that blockade of ET_B receptors on cultured human proximal tubular cells reduced endogenous ET-1 stimulated cell proliferation (an effect of ET-1 discussed in later chapters).[26] Based on these observations, I propose that the proximal tubule is capable of regulating sodium and fluid transport by an autocrine ET-1-mediated pathway. Furthermore, I suggest that the effect of any increase in proximal tubule ET-1 production is to inhibit sodium transport by this

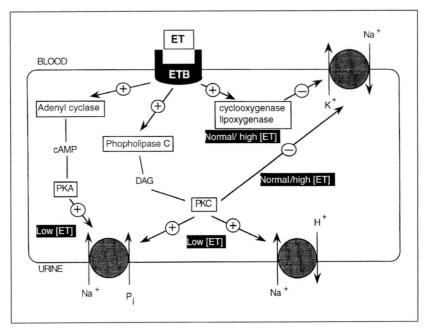

Fig. 8.1. Proposed schema of endothelin (ET)-regulated ion transport in the renal proximal tubule. PKA-protein kinase A; PKC-protein kinase C; DAG-diacylglycerol; ET_B-ET_B receptor; P_i-inorganic phosphate; cAMP-cyclic adenosine monophosphate.

region of the nephron. Unfortunately, no studies have, to my knowledge, clearly implicated ET-1 regulation of proximal tubule transport processes in the maintenance of body fluid and electrolyte homeostasis. This lack of information reflects the inherent difficulty in discriminating between local tubule effects of ET-1 agonists or antagonists as compared to more generalized renal and systemic actions. This remains a major stumbling block in advancing our knowledge of the role of ET-1 in modulating tubule fluid and electrolyte transport; new techniques under development, such as targeting gene disruption or overexpression to a particular tubule segment, may finally bring an answer.

Thick Ascending Limb of Henle's Loop

Endothelin biology in the thick ascending limb is poorly understood. This nephron segment secretes ET-1 and ET-3 and contains ET-1 mRNA.[4,20] A preliminary study indicates that these peptides, via activation of the ET_B receptor, inhibit chloride reabsorption in isolated mouse medullary and cortical thick ascending limbs via a cAMP-independent and Ca^{2+}-insensitive diacylglycerol-responsive PKC pathway.[29] As will be described below, this effect of ET-1 on signal transduction pathways involved in regulating ion transport in thick ascending limbs differs from that on cortical collecting tubules, again underscoring the importance of examining actions of ET-1 and ET-3 in the context of a particular cell type.

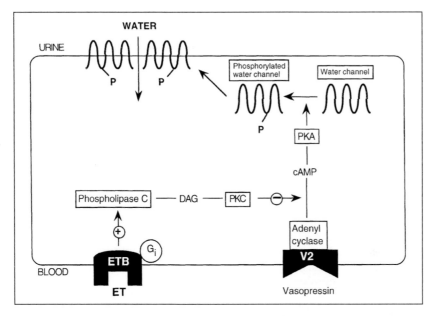

Fig. 8.2. Generalized schema of endothelin (ET)-regulated water reabsorption by the cortical and inner medullary collecting ducts. PKA-protein kinase A; PKC-protein kinase C; DAG-diacylglycerol; ET_B-ET_B receptor; cAMP-cyclic adenosine monophosphate; G_i-inhibitory G protein; V2-V2 vasopressin receptor.

Cortical Collecting Tubule

The collecting duct is the major nephron site of ET-1 production, binding, and action. Within this segment, ET-1 synthesis and receptor expression increases as one moves from the cortex to the inner medulla. This may reflect the relatively greater number of principal cells in the IMCD as compared to the cortical collecting tubule (CCT); however, intrinsic differences between the different regions of the collecting duct may also be responsible. A discussion of ET-1 biology in the CCT will be followed by examination of ET-1 in the IMCD in the next section.

The endothelin peptides have at least two major effects on the CCT: regulation of sodium and water reabsorption. With regard to water handling, ET-1 inhibits vasopressin (AVP)-stimulated water flux in the rat CCT via activation of the ET_B receptor.[6,30] This effect is due to a PKC-dependent reduction in cAMP accumulation and is independent of dihydropyridine-type Ca^{2+} channels and cyclooxygenase metabolites (Fig. 8.2).[31,32] ET-1 inhibition of water transport is dose-dependent, unlike its biphasic affect on proximal tubule fluid reabsorption.[30] Interestingly, AVP appears capable of desensitizing ET-1 binding to the CCT (by reducing K_d of the ET_B receptor) through a protein kinase A-dependent pathway.[33] Hence, ET-1 inhibition of AVP action may be limited, at least partially, due to AVP counter-regulation of ET-1 activity.

ET-1 also reduces mineralocorticoid and AVP-stimulated sodium and chloride reabsorption by the rat and rabbit CCT.[30,34,35] At least one mechanism by which this effect occurs is through inhibition of apical sodium and chloride entry.[34,35] Studies in A6 distal nephron cells indicate that basolateral ET-1 decreases amiloride-sensitive sodium channel activity by increasing channel mean closed time.[36] It is also possible that ET-1 inhibits Na^+/K^+-ATPase activity as it does in the IMCD (see below), however, this has not been studied in the CCT. The intracellular signaling pathways by which ET-1 modulates sodium reabsorption in the CCT are clearly different than those involved with water transport (Fig. 8.3). First, ET-1 may elicit a biphasic sodium channel response. ET-1 at 100 pM has been reported to inhibit sodium channel activity in A6 cells, while 10 nM ET-1 augmented channel activity.[36] Although not well studied, it has been suggested that the inhibitory effect is mediated by activation of ET_B receptors, while the stimulatory effect is due to ET_A receptor activity.[36] Similar to its effects on water transport, ET-1 regulation of CCT sodium transport is dependent upon PKC activation, however, it also involves an increase in intracellular Ca^{2+} concentration.[30,34,35] ET-1 elicits a biphasic increase in intracellular Ca^{2+} concentration in the CCT: the initial transient peak is caused by release of Ca^{2+} from cell stores, while the second sustained rise is primarily due to Ca^{2+} entry via dihydropyridine-sensitive channels.[37,38]

What then is the expected physiologic effect of ET-1 on sodium and water reabsorption by the CCT? I propose that ET-1 functions as an autocrine inhibitor of both sodium and water transport by this nephron segment. The CCT makes about five times as much ET-1 as does the proximal tubule and has relatively high endothelin receptor expression.[4,6] This degree of ET-1 production is, however, quite unlikely to rise much over 100 pM in the immediate vicinity of the CCT and would, therefore, probably mediate an inhibition of sodium and water reabsorption. Even if stimulated with factors shown to augment CCT cell line ET-1 production, such as transforming growth factor β,[39] it remains highly improbable that the CCT would release enough ET-1 to attain concentrations likely to stimulate apical sodium entry (10 nM).

Inner Medullary Collecting Duct

As mentioned in the beginning of this chapter, the renal inner medulla contains the highest concentration of ET-1 in the body.[2] In agreement with this initial finding, numerous studies have subsequently confirmed that the IMCD is the predominant nephron site of ET-1 production and receptor expression. Metabolic labeling, immunohistochemical and radioimmunoassay analysis of ET-1 levels, as well as reverse transcription-polymerase chain reaction quantitation (RT-PCR) of mRNA expression, have repeatedly shown that the IMCD produces up to ten times as much ET-1 as compared to any other cell type in the nephron.[4,20,22,40,41] Similarly, autoradiographic studies have found predominant ET-1 binding to renal medulla in several species[42-44] and localized this binding, at least in part, to collecting ducts.[45] More detailed binding studies, using microdissected rat nephron segments or in situ RT-PCR, have also shown that endothelin receptors are primarily expressed by the IMCD.[5,6,46] Taken together with the finding that IMCD cells secrete ET-1 from, and bind ET-1 on, the same (basolateral) side,[47] the above observations set the stage for ET-1 functioning as a powerful autocrine regulator of IMCD function.

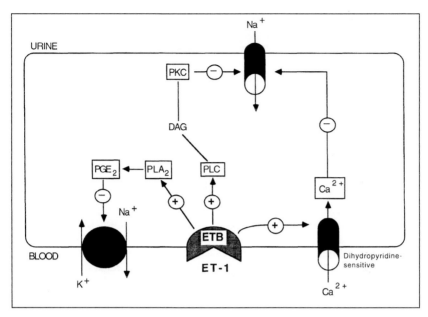

Fig. 8.3. Proposed generalized schema of endothelin (ET)-regulated sodium reabsorption by the cortical and inner medullary collecting ducts. PKC-protein kinase C; DAG-diacylglycerol; PLC-phospholipase C; PLA_2-phospholipase A_2; PGE_2-prostaglandin E_2; ET_B-ET_B receptor; cAMP-cyclic adenosine monophosphate; G_i-inhibitory G protein.

Like the CCT, ET-1 and possibly ET-3 appear to influence sodium and water transport by the IMCD. With regard to water reabsorption, ET-1 and ET-3 reduce AVP-stimulated cAMP accumulation[31,48,49] and osmotic water permeability[50,51] in the IMCD. This effect is primarily due to inhibiting adenyl cyclase activity since dibutyryl cAMP-stimulated osmotic water permeability in the IMCD is unaltered by ET-1.[50] Like the CCT, ET-1 regulation of water reabsorption is dose- and PKC-dependent and is unrelated to cyclooxygenase or dihydropyridine-sensitive Ca^{2+} channel activity.[31,48,51] ET-1 acts through an inhibitory G protein since pertussis toxin blocks the effects of ET-1 on AVP-stimulated cAMP accumulation and water flux in the IMCD.[48,51] Finally, several studies indicate that ET_B receptors mediate ET-1 inhibition of water reabsorption in the IMCD.[48,49,52] Hence, the IMCD appears to behave in all respects like the CCT in terms of ET-1 regulation of water transport (Fig. 8.2). Also like the CCT, AVP may downregulate ET-1 activity in the IMCD by reducing ET_B receptor expression, although the mechanism of this effect may differ somewhat between the two nephron segments.[53]

There is evidence that ET-1 also inhibits sodium transport by the IMCD; however, this is less well studied than in other regions of the nephron. Part of the difficulty in studying sodium transport systems in the IMCD is that relatively low ion fluxes are observed in this cell type under cultured or isolated perfused conditions. Nonetheless, one group has reported ET-1 inhibition of Na^+/K^+-ATPase activity in suspensions of rabbit IMCD.[3] This effect appears to be mediated by PGE_2

since it is inhibited by indomethacin and because ET-1 can augment PGE_2 production by the IMCD.[3,48] ET-1 does not alter basal or atrial natriuretic peptide(ANP)-induced cGMP accumulation, indicating that ANP-related pathways operate separately from those of ET-1.[54] There is no evidence that ET-1 modulates apical sodium entry into the IMCD in a manner analogous to that in the CCT. This should not be construed, however, to mean that ET-1 does not regulate the same sodium reabsorptive pathways in the two regions of the collecting duct. Rather, technical difficulties, as alluded to above, have prevented our detecting ET-1 modulation of IMCD apical sodium channel activity. Hence, until proven otherwise, I will illustrate ET-1 regulation of IMCD and CCT sodium transport processes in a similar manner (Fig. 8.3).

Having discussed ET regulation of transport systems in the IMCD, we must consider, as for the other regions of the nephron, the physiologic significance of ET-1 in the IMCD. As for the CCT, I think it highly likely that ET-1 (the major isoform produced by the IMCD[4]) regulates water, and possibly sodium, transport by the IMCD. Indeed, anti-ET-1 antibodies increase AVP-stimulated cAMP accumulation by IMCD cells, indicating that ET-1 is an autocrine inhibitor of AVP action in the terminal nephron.[55] How this system fits into the control of renal sodium and water excretion and maintenance of body fluid volume will be discussed in the following section.

Physiologic Significance of ET-1 Regulation of Nephron Transport

It is clear that endothelin peptides are capable of regulating renal tubule sodium and water reabsorption in in vitro preparations. Proving that ET-1 is a physiologic regulator of renal transport has, however, proven difficult: despite numerous attempts, there remains relatively little consensus on the role of ET-1 in controlling renal sodium and water excretion. This difficulty stems from the inability to selectively examine ET actions on the nephron (and individual nephron segments) as opposed to vascular effects of the peptides. For example, administration of endothelin agonists or antagonists almost invariably alters renal plasma flow (RPF) and glomerular filtration rate (GFR), effects which obviously impact urinary sodium and water excretion. In efforts to circumvent this problem, endothelin agonists or antagonists have been administered systemically in doses titrated to not detectably alter renal hemodynamics. Even under these circumstances, however, intravenously administered ET-1 or ET-3 have increased sodium excretion in some,[56,57] but not all,[58-61] studies. This inconsistency in ET-1 regulation of tubule sodium reabsorption may relate, at least in part, to activation of ET_A and ET_B receptors by the exogenous peptide. For example, when ET_A (but not ET_B) receptors are blocked, a natriuretic effect of ET-1 is uncovered.[62,63] In contrast to the studies on sodium excretion, there is much more agreement that systemically administered ET-1 increases water excretion, even in the absence of any increase in sodium excretion.[58-60] Taken together, these studies suggest that the nephron normally expresses endothelin receptors which can mediate inhibition of renal water, and possibly sodium, reabsorption.

Another means of assessing the role of ET-1 in regulating renal sodium and water excretion is to measure changes in renal ET-1 levels in response to maneuvers which alter body sodium and/or water balance. Using this approach, investigators have again failed to find consistent results with regard to sodium balance and renal ET-1 production. A high sodium diet has either increased[64] or not changed[65,66] urinary ET-1 excretion, a marker of renal tubule ET-1 production. Acute volume expansion has, in contrast, markedly lowered urinary ET-1 excretion.[67] Similarly, a low sodium diet has either reduced[68] or not changed[69] urinary ET-1 excretion. While these results do not support a role for endogenous renal ET-1 in modulating nephron sodium reabsorption, their interpretation is limited since urinary ET-1 excretion does not accurately reflect ET-1 production in any given nephron segment. As for the studies on administration of exogenous ET-1, renal ET-1 responses to altering water balance are much clearer than those for sodium. Water loading both humans and rats increases,[64,70-72] while water deprivation decreases[64,66] urinary ET-1 excretion. Thus renal ET-1 production does appear to be responsive to changes in water intake, although the nephron sites responsible for this cannot be determined from these studies.

An important observation that arises from the preceding studies is that water (and possibly sodium) loading increases, while water (and possibly sodium) deprivation reduces renal ET-1 production. Such a scenario makes sense when viewed from the perspective of maintaining extracellular fluid volume (ECFV) homeostasis. More specifically, one would predict that ECFV expansion would cause increased renal tubule ET-1 production and enhanced autocrine inhibition of fluid reabsorption. Conversely, ECFV depletion should lead to reduced nephron ET-1 synthesis and less autocrine inhibition of tubule fluid uptake. How would such a coupling between ECFV and renal tubule ET-1 production occur? To address this, we chose to examine the IMCD reasoning that: (1) this is the major source of ET-1 synthesis in the kidney; and (2) the IMCD is situated in a region which is highly responsive to changes in ECFV (e.g., medullary blood flow and tonicity). In the first series of studies, IMCD cells were exposed to media containing varying concentrations of NaCl, mannitol or urea.[64] Hypertonic NaCl or mannitol, but not urea, caused a marked reduction in ET-1 production by these cells, consistent with the prediction that dehydration, which increases medullary tonicity, causes a reduction in ET-1 autocrine inhibition of collecting duct water reabsorption. Similar observations have been made for CCT and MDCK cells.[73,74] To be fair, I should point out that others have not reported the same findings. One group found that IMCD cell ET-1 production was increased by hypertonic NaCl and raffinose, but not mannitol,[75] while another group observed increased IMCD ET-1 release in response to hypertonic NaCl or urea.[76] These latter findings are confusing, however, since one would expect impermeant solutes such as raffinose and mannitol to have the same effect, while urea, which is highly permeable, should not behave like NaCl. I propose that, at least as a working hypothesis, we consider that IMCD ET-1 production is regulated by medullary tonicity. When inner medullary tonicity rises, such as dehydration, IMCD ET-1 production is inhibited and water retention is favored. When inner medullary tonicity falls, such as water loading, ET-1 autocrine inhibition of IMCD fluid reabsorption is enhanced. In vivo confirmation of this hypothesis awaits studies designed to specifically modulate collecting duct ET-1 production.

Conclusion

It stands to reason that ET-1 should be involved in the regulation of renal transport processes since the nephron produces large amounts of the peptide, has high receptor expression, and responds to extremely low ET-1 concentrations. Conclusive proof of this, under either physiologic or pathophysiologic conditions, is unfortunately still lacking. Nonetheless, I believe that future studies will reveal a role for ET-1 in modulating renal fluid and electrolyte excretion. Since ET-1 is primarily regulated at the level of gene transcription and has long-lasting effects, it may be that these peptides are more important in states associated with prolonged alterations ET-1 activity. These might include normal adaptive renal responses to chronic changes in salt or water intake as well as abnormal renal responses associated with deranged kidney function. The factors responsible for controlling ET-1 production, receptor expression, and signaling mechanisms remain to be determined for most tubule segments. As stated previously, new gene targeting strategies may help us to define the role of ET-1 in regulating transport processes within specific regions of the nephron.

Direct inhibition of nephron fluid reabsorption by ET-1 contrasts with its vasoconstrictive, fluid-retaining effects. While this does not strictly correlate with receptor subtype expression, it is generally true that ET_B receptors mediate tubular effects while ET_A receptors mediate vascular actions. Thus, when designing endothelin receptor antagonists to reduce pathologic vasoconstriction, it may prove most beneficial to use selective ET_A receptor blockers. Not only would these agents reduce ET-1-mediated vasoconstriction, but they may uncover, as in experimental models discussed earlier, a natriuretic and/or diuretic effect of elevated endogenous ET-1. Clearly such a response would be ideal for a disorder such as congestive heart failure associated with elevated ET-1 levels, excessive vasoconstriction, and renal fluid retention. It now remains to be seen how trials using endothelin receptor antagonists in human disease will live up to our hopes for a new and important class of therapeutic agents.

References

1. Yanagisawa M, Kurihara H, Kimura S et al. A novel potent vasoconstrictor peptide produced by vascular endothelial cells. Nature 1988; 332:411-415.
2. Kitamura K, Tanaka T, Kato J et al. Regional distribution of immunoreactive endothelin in porcine tissue: abundance in inner medulla of kidney. Biochem Biophys Res Commun 1989; 161:348-352.
3. Zeidel ML, Brady HR, Kone BC et al. Endothelin, a peptide inhibitor of Na^+-K^+-ATPase in intact tubular epithelial cells. Am J Physiol 1989; 257:C1101-C1107.
4. Kohan DE. Endothelin synthesis by rabbit renal tubule cells. Am J Physiol 1991; 261:F221-F226.
5. Terada Y, Tomita K, Nonoguchi H et al. Different localization of two types of endothelin receptor mRNA in microdissected rat nephron segments using reverse transcription and polymerase chain reaction assay. J Clin Invest 1992; 90:107-112.
6. Takemoto F, Uchida S, Ogata E et al. Endothelin-1 and endothelin-3 binding to rat nephrons. Am J Physiol 1993; 264:F827-F832.
7. Garvin J, Sanders K. Endothelin inhibits fluid and bicarbonate transport in part by reducing Na^+/K^+ ATPase activity in the rat proximal straight tubule. J Am Soc Nephrol 1991; 2:976-982.
8. Perico N, Cornejo RP, Benigni A et al. Endothelin induces diuresis and natriuresis in the rat by acting on proximal tubular cells through a mechanism mediated by lipoxygenase products. J Am Soc Nephrol 1991; 2:57-69.
9. Harris PJ, Zhuo J, Mendelsohn FAO et al. Haemodynamic and renal tubular effects of low doses of endothelin in anesthetized rats. J Physiol 1991; 433:25-39.
10. Ominato M, Satoh T, Katz AI. Endothelins inhibit Na-K-ATPase activity in proximal tubules: studies of mechanisms. J Am Soc Nephrol 1994; 5:588.

11. Ozaki S, Ihara M, Saeki T et al. Endothelin ET_B receptors couple to two distinct signaling pathways in porcine kidney epithelial LLC-PK$_1$ cells. J Pharmacol Exp Ther 1994; 270:1035-1040.

12. Ishii K, Warner TD, Sheng H et al. Endothelin increases cyclic GMP levels in LLC-PK$_1$ porcine kidney epithelial cells via formation of an endothelium-derived relaxing factor-like substance. J Pharmacol Exp Ther 1991; 259:1102-1108.

13. Walter R, Helmle-Kolb C, Forgo J et al. Stimulation of Na^+/H^+ exchange activity by endothelin in opossum kidney cells. Pflügers Arch-Eur J Physiol 1995; 430:137-144.

14. Guntupalli J, DuBose TDJ. Effects of endothelin on rat renal proximal tubule Na^+-P$_i$ cotransport and Na^+/H^+ exchange. Am J Physiol 1994; 266:F658-F666.

15. Eiam-Ong S, Hilden SA, King AJ et al. Endothelin-1 stimulates the Na^+/H^+ and Na^+/HCO_3^- transporters in rabbit renal cortex. Kidney Int 1992; 42:18-24.

16. Peng Y, Cano A, Yanagisawa M et al. ET_A and ET_B receptors activate similar signaling pathways, but only ET_B receptors activate NHE-3. J Am Soc Nephrol 1996; 7:1683.

17. Chu TS, Cano A, Yanagisawa M et al. Endothelin activates the Na/H antiporter in OKP cells stably transfected with the ET_B receptor cDNA. J Am Soc Nephrol 1993; 4:250.

18. Sørensen SS, Madsen JK, Pedersen EB. Systemic and renal effect of intravenous infusion of endothelin-1 in healthy human volunteers. Am J Physiol 1994; 266:F411-F418.

19. Garcia NH, Garvin JL. Endothelin's biphasic effect on fluid absorption in the proximal straight tubule. J Clin Invest 1994; 93:2572-2577.

20. Chen M, Todd-Turla K, Wang W et al. Endothelin-1 mRNA in glomerular and epithelial cells of kidney. Am J Physiol 1993; 265:F542-F550.

21. Uchida K, Ballermann B. Sustained activation of PGE$_2$ synthesis in mesangial cells cocultured with glomerular endothelial cells. Am J Physiol 1992; 263:C200-C209.

22. Ujiie K, Terada Y, Nonoguchi H et al. Messenger RNA expression and synthesis of endothelin-1 along rat nephron segments. J Clin Invest 1992; 90:1043-1048.

23. Zoja C, Morigi M, Figliuzzi M et al. Proximal tubular cell synthesis and secretion of endothelin-1 on challenge with albumin and other proteins. Am J Kid Dis 1995; 26(6):934-941.

24. Ohta K, Hirata Y, Imai T et al. Cytokine-induced release of endothelin-1 from porcine renal epithelial cell line. Biochem Biophys Res Commun 1990; 169:578-584.

25. Ong ACM, Jowett TP, Moorhead JF et al. Human high density lipoproteins stimulate endothelin-1 release by cultured human renal proximal tubular cells. Kidney Int 1994; 46:1315-1321.

26. Ong ACM, Jowett TP, Firth JD et al. An endothelin-1 mediated autocrine growth loop involved in human renal tubular regeneration. Kidney Int 1995; 48:390-401.

27. Nakahama H. Stimulatory effect of cyclosporine A on endothelin secretion by a cultured renal epithelial cell line, LLC-PK1 cells. Eur J Pharmacol 1990; 180:191-192.

28. Moutabarrik A, Ishibashi M, Fukunaga M et al. FK 506 Mechanism of nephrotoxicity: stimulatory effect on endothelin secetion by cultured kidney cells and tubular cell toxicity in vitro. Transplant Proc 1991; 23:3133-3136.

29. Bailly C, Ferreira MC. Endothelin inhibits Cl reabsorption in TAL via a Ca-dependent pathway. J Am Soc Nephrol 1996; 7:1627.

30. Tomita K, Nonoguchi H, Terada Y et al. Effects of ET-1 on water and chloride transport in cortical collecting ducts of the rat. Am J Physiol 1993; 264:F690-F696.

31. Tomita K, Nonoguchi H, Marumo F. Effects of endothelin on peptide-dependent cyclic adenosine monophosphate accumulation along the nephron segments of the rat. J Clin Invest 1990; 85:2014-2018.

32. Tomita K, Nonoguchi H, Marumo F. Inhibition of fluid transport by endothelin through protein kinase C in collecting duct of rats. Basel: Karger, 1991:207-215. (Koide H, Endou H, Kurokawa K, ed. Contrib Nephrol; vol 95).

33. Takemoto F, Uchida S, Katagirl H et al. Desensitization of endothelin-1 binding by vasopressin via a cAMP-mediated pathway in rat CCD. Am J Physiol 1995; 268((Renal Fluid Electrolyte Physiol 37)):F385-F390.

34. Ling BN. Luminal endothelin-1 inhibits apical Na^+ and Cl^- channels in cultured rabbit CCT principal cells. J Am Soc Nephrol 1994; 5:292.

35. Kurokawa K, Yoshitomi K, Ikeda M et al. Regulation of cortical collecting duct function: effect of endothelin. Am Heart J 1993; 125:582-588.

36. Gallego MS, Ling BN. Regulation of amiloride-sensitive sodium channels by endothelin-1 in distal nephron cells. Am J Physiol (in press).

37. Naruse M, Uchida S, Ogata E et al. Endothelin-1 increases cell calcium in mouse collecting tubule cells. Am J Physiol 1991; 261:F720-F725.

38. Korbmacher C, Boulpaep EL, Giebisch G et al. Endothelin increases $[Ca^{2+}]_i$ in M-1 mouse cortical collecting duct cells by a dual mechanism. Am J Physiol 1993; 265:C349-C357.

39. Schnermann JB, Zhu XL, Shu X et al. Regulation of endothelin production and secretion in cultured collecting duct cells by endogenous transforming growth factor-β. Endocrinol 1996; 137:5000-5008.

40. Pupilli C, Brunori M, Misciglia N et al. Presence and distribution of endothelin-1 gene expression in human kidney. Am J Physiol 1994; 267:F679-F687.

41. Uchida S, Takemoto F, Ogata E et al. Detection of endothelin-1 mRNA by RT-PCR in isolated rat renal tubules. Biochem Biophys Res Commun 1992; 188:108-113.

42. Waeber C, Hoyer D, Palacios J-M. Similar distribution of [125I]sarafotoxin-6b and [125I]endothelin-1,-2,-3 binding sites in the human kidney. Eur J Pharmacol 1990; 176:233-236.

43. Wilkes BM, Susin M, Mento PF et al. Localization of endothelin-like immunoreactivity in rat kidneys. Am J Physiol 1991; 260:F913-F920.

44. Davenport AP, Nunez DJ, Brown MJ. Binding sites for ^{125}I-labelled endothelin-1 in the kidneys: differential distribution in rat, pig and man demonstrated by using quantitative autoradiography. Clin Sci 1989; 77:129-131.

45. Yukimura T, Notoya M, Mizojiri K et al. High resolution localization of endothelin receptors in rat renal medulla. Kidney Int 1996; 50:135-147.

46. Chow LH, Subramanian S, Nuovo GJ et al. Endothelin receptor mRNA expression in renal medulla identified by in situ PT-PCR. Am J Physiol 1995; 269((Renal Fluid Electrolyte Physiol 38)):F449-F457.

47. Kohan DE, Padilla E. Endothelin-1 is an autocrine factor in rat inner medullary collecting ducts. Am J Physiol 1992; 263:F607-F612.

48. Kohan DE, Padilla E, Hughes AK. Endothelin B receptor mediates ET-1 effects on cAMP and PGE2 accumulation in rat IMCD. Am J Physiol 1993; 265:F670-F676.

49. Edwards RM, Stack EJ, Pullen M et al. Endothelin inhibits vasopressin action in rat inner medullary collecting duct via the ET_B receptor. J Pharmacol Exp Ther 1993; 267:1028-1033.

50. Oishi R, Nonoguchi H, Tomita K et al. Endothelin-1 inhibits AVP-stimulated osmotic water permeability in rat medullary collecting duct. Am J Physiol 1991; 261:F951-F956.

51. Nadler SP, Zimplemann JA, Hebert RL. Endothelin inhibits vasopressin-stimulated water permeability in rat terminal inner medullary collecting duct. J Clin Invest 1992; 90:1458-1466.

52. Woodcock EA, Land SL. Functional endothelin ET_B receptors on renal papillary tubules. Eur J Pharmacol 1993; 247:93-95.

53. Wong BPH, Wong NLM. Mechanisms of vasopressin-induced endothelin receptor desensitization. J Am Soc Nephrol 1996; 7:1575.

54. Cernacek P, Legault L, Stewart DJ et al. Specific endothelin binding sites in renal medullary collecting duct cells: lack of interaction with ANP binding and cGMP signalling. Can J Physiol Pharmacol 1992; 70:1167-1174.

55. Kohan DE, Hughes AK. Autocrine role of endothelin in rat IMCD: inhibition of AVP-induced cAMP accumulation. Am J Physiol 1993; 265:F126-F129.

56. Denton KM, Anderson WP. Vascular actions of endothelin in the rabbit kidney. Clin Exp Pharmacol Physiol 1990; 17:861-872.

57. Schramek H, Willinger CC, Gstraunthaler G et al. Endothelin-3 modulates glomerular filtration rate in the isolated perfused rat kidney. Renal Physiol Biochem 1992; 15:325-333.

58. Schnermann J, Lorenz JN, Briggs JP et al. Induction of water diuresis by endothelin in rats. Am J Physiol 1992; 263:F516-F526.

59. Kamphuis C, Yates NA, McDaugall JG. Differential blockade of the renal vasoconstrictor and diuretic responses to endothelin-1 by endothelin antagonist. Clin Exper Pharmacol Physiol 1994; 21:329-333.

60. Goetz K, Wang BC, Leadley RJ et al. Endothelin and sarafotoxin produce dissimilar effects on renal blood flow, but both block the antidiuretic effects of vasopressin. Proc Soc Exp Biol Med 1989; 191:425-427.

61. Freed MI, Thompson KA, Wilson DE et al. Endothelin receptor antagonism does not alter renal hemodynamic responses or urinary sodium excretion in healthy humans. J Am Soc Nephrol 1996; 7:1580.

62. Brooks DP, De Palma PD, Pullen M et al. Identification and function of putative ET_B receptor subtypes in the dog kidney. J of Cardiovasc Pharmacol 1995; 26(3):S322-S325.

63. Clavell AL, Stingo AJ, Margulies KB et al. Role of endothelin receptor subtypes in the in vivo regulation of renal function. Am J Physiol 1995; 268(Renal Fluid Electrolyte Physiol 37):F455-F460.

64. Kohan DE, Padilla E. Osmolar regulation of endothelin-1 production by rat inner medullary collecting duct. J Clin Invest 1993; 91:1235-1240.

65. Ohta K, Hirata Y, Schichiri M et al. Urinary excretion of endothelin-1 in normal subjects and patients with renal disease. Kidney Int 1991; 39:307-311.

66. Michel H, Bäcker A, Meyer-Lehnert H et al. Rat renal, aortic and pulmonary endothelin-1 receptors: effects of changes in sodium and water intake. Clin Sci 1993; 85:593-597.

67. Serneri GGN, Modesti PA, Cecioni I et al. Plasma endothelin and renal endothelin are two distinct systems involved in volume homeostasis. Am J Physiol 1995; 268((Heart Circ Physiol 37)):H1829-H1837.

68. Rascher W, Gyodi G, Worgall S et al. Effect of sodium chloride supplementation on urinary sodium excretion in premature infants. J Pediatr 1994; 125:793-797.

69. Firth JD, Schricker K, Ratcliffe PJ et al. Expression of endothelins 1 and 3 in the rat kidney. Am J Physiol 1995; 269:F522-F528.

70. Mattyus I, Zimmerhack LB, Schwarz A et al. Renal excretion of endothelin in children is influenced by age and diuresis. Acta Pædiatr 1994; 83:468-472.

71. Worgall S, Manz F, Kleschin K et al. Elevated urinary excretion of endothelin-like immunoreactivity in children with renal disease is related to urine flow rate. Clin Nephrol 1994; 41:331-337.

72. Zeiler M, Löffler B-M, Bock HA et al. Water diuresis increases endothelin-1 excretion in humans. J Am Soc Nephrol 1995; 6:751.

73. Schramek H, Gstraunthaler G, Willinger CC et al. Hyperosmolality regulates endothelin release by Madin-Darby canine kidney cells. J Am Soc Nephrol 1993; 4:206-213.

74. Todd-Turla KM, Zhu XL, Shu X et al. Synthesis and secretion of endothelin in a cortical collecting duct cell line. Am J Physiol 1996; 271:F330-F339.

75. Yang T, Terada Y, Nonoguchi H et al. Effect of hyperosmolality on production and mRNA expression of endothelin-1 in inner medullary collecting duct. Am J Physiol 1993; 264:F684-F689.

76. Migas I, Bäcker A, Meyer-Lehnert H et al. Endothelin synthesis by porcine inner medullary collecting duct cells. Am J Hyperten 1995; 8:748-752.

CHAPTER 9

Endothelin and Ion Channels

Tracy L. Keith and Robert F. Highsmith

Endothelin was initially characterized as a peptidergic vasoconstricting factor in 1985,[1] and its amino acid sequence was reported in 1988.[2] Three isoforms have since been identified and specific receptors localized in various organ systems (for reviews see refs. 3-5). In the vasculature ET-1 is known to interact with at least two receptors, ET_A on smooth muscle mediating constriction, and ET_B on endothelial cells mediating vasodilation.[6] These actions have been noted in the coronary and cerebral circulations as well as in the pulmonary, renal and mesenteric vascular beds.[3,4] In addition to its vasoactive properties ET has been reported to contract nonvascular smooth muscle in bronchial airways, the gastrointestinal tract and in the uterus.[7] It has also been shown to regulate secretion from neuroendocrine cells[8] and it is known to be a growth factor, stimulating proliferation of smooth muscle cells.[9] While numerous reports have described these actions of ET-1, the signaling mechanisms in these respective cell types are incompletely resolved. Some common themes involve receptor activation and mobilization of calcium as well as recruitment of a variety of second messengers. The focus of this chapter is on the role of ion channels in the signaling processes triggered by ET-1 in a variety of cell types.

Ca^{2+} Mobilization: A Common Theme in ET-Signaling

While the biological actions of ET vary among tissues, the key event in cellular signaling induced by ET is a biphasic increase in intracellular calcium. An initial transient peak is attributed to release of intracellular Ca^{2+} from IP_3-sensitive stores, while the second, sustained phase is initiated by influx of Ca^{2+} from the extracellular space.[10] In vascular smooth muscle, the sustained elevation in $[Ca^{2+}]_i$ that results from Ca^{2+} influx is thought to be responsible for the prolonged constricting actions of ET. The routes of Ca^{2+} influx as well as the mechanisms by which those pathways are activated are still subjects of great debate. The early hypothesis that ET was a direct activator of voltage-dependent Ca^{2+} channels was quickly disproved when it was demonstrated that ET could not displace specific L-type Ca^{2+} channel agonists. Distinct receptors for ET were then identified[11] and it was believed that the opening of these channels and subsequent influx of Ca^{2+} occurred indirectly, following the formation of the ligand-receptor complex.[12,13] The precise mechanism of Ca^{2+} channel activation in response to ET remains incompletely resolved, although it has been speculated that voltage-gated Ca^{2+} channels are activated secondarily to membrane depolarization induced by the movement of ions through

Endothelin Receptors and Signaling Mechanisms, edited by David M. Pollock and Robert F. Highsmith. © 1998 Springer-Verlag and R.G. Landes Company.

other ion channels.[14] Alternatively, L-type channels may be opened by the actions of second messengers such as protein kinase C (PKC) or by Ca^{2+} released from intracellular stores. Involvement of receptor operated as well as nonselective cation channels in the ET-stimulated mobilization of extracellular Ca^{2+} has also been reported, although the relative contributions of the individual currents to the total Ca^{2+} influx is difficult to assess.

Most studies of ET-induced signal transduction have utilized vascular smooth muscle in which the ET_A receptor is the predominant subtype. Although, the ET_B receptor has also been shown to mediate constriction in certain vascular beds,[15,16] the actions of ET-1 in vascular smooth muscle are typically described by the events that follow activation of the ET_A receptor. In endothelial cells, as well as in renal mesangial cells, the ET_B receptor is abundant and is activated by ET-1 as well as ET-3.[17] However, much less is known about ET_B-mediated signal transduction pathways. The following discussion of the activation of Ca^{2+} channels by ET will first focus on vascular smooth muscle subsequent to binding of ET-1 to the ET_A receptor. Next, the role of K^+ and Cl^- channel activation in ET-induced membrane depolarization will be described. Finally, the involvement of ion channels in ET-signaling in other cell types will be discussed.

Actions of ET on Vascular Smooth Muscle

Activation of L-Type Ca^{2+} Channels by ET-1

Following the first report that an endothelium derived constricting factor was produced by endothelial cells in culture, a flurry of research activity was initiated to further characterize the substance and to elucidate its mechanism of action. Hickey and coworkers (1985) demonstrated that the peptide, later named endothelin (ET), induced a sustained constriction of porcine, as well as bovine and canine coronary arteries.[1] The constriction could be abolished by removal of Ca^{2+} from the incubation medium, and it was markedly inhibited by pretreatment with the Ca^{2+} channel antagonist verapamil.[1] These observations suggested that the vasoconstricting actions of ET were absolutely dependent on extracellular Ca^{2+} and that Ca^{2+} influx occurred through voltage-dependent Ca^{2+} channels. In 1988, elucidation of the amino acid sequence for ET by Yanagisawa et al revealed a high degree of homology with a class of snake venom toxins (sarafotoxins) known to bind to voltage gated Na^+ channels.[2] It was theorized that ET might be an endogenous ligand of voltage operated Ca^{2+} channels, activating them directly. The first challenge in studying ET-induced Ca^{2+} mobilization was to determine the identity of the channels that facilitated Ca^{2+} entry; the second was to elucidate the sequence of events leading to their activation. In 1989, Goto et al presented evidence that substantiated the claim that the ET binding site was in fact a voltage-dependent Ca^{2+} channel. Porcine coronary artery strips were treated with nicardipine, a specific antagonist of dihydropyridine (DHP)-sensitive Ca^{2+} channels. The contractile response to ET-1 was inhibited, indicating that Ca^{2+} influx essential for mediating constriction occurred via DHP-sensitive channels.[18] Voltage clamp studies in single coronary smooth muscle cells confirmed that this route of Ca^{2+} entry was activated by ET.[18] A later study by Kasuya also demonstrated that the effects of ET could be antagonized by nicardipine; however diltiazem and verapamil, antagonists of a different class of Ca^{2+} channels produced the same results.[19] Binding assays showed that ET did not displace the binding of [^{125}I]imodipine, a specific agonist of the DHP-sensitive Ca^{2+} channel,[19] indicating that the binding site for ET must be independent of the DHP-sensitive channel. Thus it was apparent that the activation of voltage-dependent Ca^{2+} channels occurred secondarily to the binding of ET at the smooth muscle cell membrane.

These studies in porcine coronary arteries were the first to claim that the sustained phase of the Ca^{2+} response and the ability to generate force in response to ET were dependent on the activation of specific L-type Ca^{2+} channels. Subsequent observations in other vessels corroborated this finding. DHP-sensitive Ca^{2+} channels were shown to be involved in mediating the contractile response to ET in bovine cerebral arteries,[20] human forearm circulation,[21] rabbit mesenteric resistance arteries,[22] renal arteries of the rat vasculature[23] and human intracranial arteries.[24] Furthermore, studies in cultured and freshly dispersed vascular smooth muscle cells of various origin also confirmed that ET-stimulated Ca^{2+} influx occurred via DHP-sensitive Ca^{2+} channels.[25-27]

While the role of voltage-dependent DHP-sensitive Ca^{2+} channels in ET-signaling appeared to be well established, a number of subsequent reports could not confirm this finding. Evidence for the involvement of other channels in facilitating Ca^{2+} influx began to accumulate. D'Orleans-Juste and colleagues compared the effects of nicardipine on the actions of ET and Bay K 8644, an agonist of the DHP-sensitive channel. Contractile responses elicited by ET-1 in rabbit and canine vessels were unaffected by nicardipine, whereas Bay K 8644-induced contractions were markedly inhibited in the presence of the antagonist.[28] These observations suggested that DHP-sensitive Ca^{2+} channels were not involved in mediating Ca^{2+} influx in any of these preparations. Similar findings were reported in isolated rat aorta in which nifedipine had no effect on ET-induced constriction.[29] In additional experiments nifedipine or diltiazem had no effect on ET-1-stimulated Ca^{2+} influx in vascular smooth muscle cells cultured from rat thoracic aortae,[29] suggesting that, in this tissue, voltage-dependent Ca^{2+} channels were not involved in facilitating influx of extracellular Ca^{2+}. In 1992, Bodellson et al studied the effects of Ca^{2+} channel blockade on ET-1-induced constriction of human uterine arteries. Nicardipine had no effect on the contractile response to ET indicating that DHP-sensitive channels were not activated.[30] However inhibition of ET-induced contraction was noted in the presence of verapamil. It was concluded that ET-stimulated Ca^{2+} entry in uterine arteries occurred through DHP-resistant Ca^{2+} channels.[30] Collectively these data imply that the mechanism by which ET induces vasoconstriction involves more than one Ca^{2+} influx pathway. Furthermore Ca^{2+} influx pathways appear to be highly variable among species and among vascular beds. Activation of DHP-sensitive voltage-dependent Ca^{2+} channels is crucial in some preparations, while it appears to play a minimal role in others.

Activation of T-type Ca^{2+} Channels by ET-1

Many of the original reports describing the vasoconstricting actions of ET-1 were conducted in the porcine coronary artery providing a great deal of data regarding ET-1-signaling in this vessel. One pivotal observation in the quest to elucidate ET-1's mechanism of action was that Ca^{2+} influx occurred exclusively through DHP-sensitive Ca^{2+} channels. A more recent study conducted in the same model system provided additional information and offered new insights into the complexity of ET-1-signaling. Blackburn and Highsmith demonstrated that the contraction elicited by ET-1 in porcine coronary artery could be partially inhibited by

Ni^{2+}.[31] Ni^{2+} is a heavy metal with demonstrated Ca^{2+} channel blocking properties, although at the concentration used in this study (360µM), Ni^{2+} has a very low affinity for L-type channels. Thus in addition to influx through DHP-sensitive pathways, Ni^{2+}-sensitive Ca^{2+} channels were shown to play a role in ET-induced Ca^{2+} mobilization.[31] It was speculated that these channels were members of a class of voltage-dependent Ca^{2+} channels distinct from L-type channels, known as T-type Ca^{2+} channels. These channels have been previously identified in vascular smooth muscle and are known to be sensitive to the actions of Ni^{2+}.[32,33] The possibility that T-type channels are activated in response to ET was reinforced by Chen and Wagoner in 1991 who identified a Ni^{2+}-sensitive Ca^{2+} influx pathway in rat aortic smooth muscle cells stimulated with ET.[34] Together these findings support the idea that Ca^{2+} entry in vascular smooth muscle occurs via numerous pathways. Furthermore it is likely that the precise coordination of signaling events leads to activation of several distinct classes of Ca^{2+} channels, each contributing to the total Ca^{2+} influx.

Activation of Nonselective Cation Channels by ET-1

While the importance of voltage-dependent Ca^{2+} channels in ET-signaling has been established, another hypothesis that has received attention is that receptor operated or nonselective cation channels may also be stimulated by ET. In 1988, van Rhenterghem et al identified a nonselective cation channel in rat aortic smooth muscle cells that could be activated by vasopressin.[35] The channel was shown to facilitate a long-lasting Ca^{2+} current that was not dependent on membrane depolarization. Further investigation revealed that ET was also capable of stimulating a current through this channel. In rat aortic strips contracted with ET-1, the L-type Ca^{2+} channel antagonist, PN200-110, caused a marked reduction in force.[36] Because the contractile response was not completely blocked, it was theorized that Ca^{2+} influx may be mediated by an additional pathway. Using A7r5 cells, a smooth muscle-like cell line, voltage clamp experiments indicated that ET-1 did not activate the DHP-sensitive Ca^{2+} current directly, but instead a nonspecific cation current was evoked.[36] In 1991 Chen and Wagoner conducted a series of voltage clamp experiments in rat thoracic aorta using the whole cell patch technique. Consistent with the findings of van Rhenterghem, they concluded that ET activated a nonselective cation current.[34] Furthermore they determined that the activity of this channel was dependent on the presence of Ca^{2+}. Ion substitution experiments revealed that the channel was permeable to Na$^+$, Li$^+$, K$^+$ and Cs$^+$ as well as Ca^{2+}, suggesting that the current is carried by a nonvoltage dependent ion channel.[34] In a similar series of experiments conducted in rabbit aorta, Watanabe observed that nifedipine could not completely reverse a contraction induced by ET. It was hypothesized that a receptor operated channel may be involved.[37] Support for this theory was provided by Enoki et al who recorded an ET-stimulated cation current in freshly dispersed cells of rabbit aorta.[38] The channel that was activated was independent of membrane potential and was permeable to Ca^{2+} as well as to monovalent cations.[38] Furthermore the current was linked to activation of the ET$_A$ receptor by ET-1. This finding was confirmed in whole cell recordings from mouse fibroblasts transfected with recombinant ET$_A$ receptors.[38] Thus in aortic preparations from rat and rabbit, there is evidence that ET-stimulated Ca^{2+} influx occurs, at least in part, through nonselective cation channels. The relative contribution of this current to the total Ca^{2+} influx is uncertain, but it does appear to be involved in mediating vasoconstriction.

ET-1-induced Mobilization of Intracellular Ca²⁺

In contrast to early reports that ET-induced constriction was absolutely dependent on the influx of extracellular Ca^{2+}, later studies demonstrated that it was possible to elicit a contraction in Ca^{2+} free media.[39,40] Increases in intracellular Ca^{2+} were observed upon stimulation by ET suggesting that Ca^{2+} released from intracellular stores could be coupled to contraction. Several studies noted a marked increase in intracellular IP_3 following the binding of ET to its receptor at the cell surface.[41-43] It was concluded that increases in cytosolic Ca^{2+} were attributed to the release of intracellular Ca^{2+} from IP_3-sensitive stores. The IP_3 receptor, classified as a receptor operated Ca^{2+} channel, facilitates efflux of Ca^{2+} from the sarcoplasmic reticulum that results in an elevated cytosolic Ca^{2+} concentration. Though these channels are activated by ET indirectly, this finding further extends our understanding of the involvement of ion channels in ET-signaling.

Gardner et al demonstrated that ET induced a biphasic Ca^{2+} response in umbilical artery smooth muscle cells in which a portion of the Ca^{2+} was mobilized from intracellular stores and a second phase was attributed to influx from the extracellular space.[44] Unlike other smooth muscle cell types, however, the presence of the L-type channel antagonists, verapamil and nifedipine, caused a marked reduction in both phases of the response.[44] These findings support the view that DHP-sensitive Ca^{2+} channels are involved in ET-induced Ca^{2+} mobilization and suggest that there may be an interaction between regulation of Ca^{2+} influx and its release from intracellular pools. It has been proposed that the release of Ca^{2+} from intracellular stores may exert a regulatory influence on Ca^{2+} influx pathways. This is an intriguing feature of ET-induced Ca^{2+} signaling that is discussed in more detail by Brock in chapter 10.

In summary, it is obvious that there are several types of Ca^{2+} channels activated in vascular smooth muscle in the course of ET-signaling. Some variability may be accounted for by differences in the distribution of respective Ca^{2+} channels among species and among vascular beds. However, it appears that a more important factor in interpreting experimental data regarding ET-signaling in diverse model systems is the concentration of the agonist used. At low doses of ET, influx of Ca^{2+} from the extracellular space is the pivotal event in mediating the contractile response and Ca^{2+} channel activation is critical. At higher concentrations of ET, the contribution of Ca^{2+} released from IP_3-sensitive stores may be significant. While many of these details remain unresolved, it is indisputable that the interaction of ion currents and the involvement of numerous ion channels in the signal transduction pathways triggered by ET represent a complex cellular event.

Activation of K⁺ and Cl⁻ Channels by ET

The hallmark of ET-1-signaling in vascular smooth muscle is the increase in intracellular $[Ca^{2+}]$ that is essential for eliciting a contraction. Consequently the focus of any discussion on ET-1-signaling tends to be centered on the events that mediate Ca^{2+} mobilization, in particular influx from the extracellular space. While the role of Ca^{2+} channels in facilitating Ca^{2+} entry has been discussed extensively, it is also important to consider the mechanism by which these influx pathways are activated. Following the identification of specific receptors for ET-1, the hypothesis that voltage-dependent Ca^{2+} channels were opened directly by ET-1 was rejected. At that time it was generally believed that depolarization of the cell membrane must precede activation of voltage-dependent channels. Because ET-1 itself had no significant effect on membrane potential, it was concluded that additional mechanisms must be involved. It was proposed that other ion channels were responsible for inducing membrane depolarization which in turn led to activation of L-type channels. In a series of voltage clamp experiments in A7r5 cells in 1988, van Rhenterghem et al observed an initial transient outward Ca^{2+}-activated K^+ current in response to ET-1.[36] This current caused a hyperpolarization response that in turn activated a Ca^{2+} permeable nonselective cation channel.[36] Entry of Ca^{2+} into the cell through this channel was thought to induce depolarization, which in turn caused activation of L-type channels and facilitated additional Ca^{2+} entry. The contribution of the nonselective Ca^{2+} current to the total Ca^{2+} influx appeared to be minor. However, its role in the activation of L-type Ca^{2+} channels was significant in terms of ET-signaling. In subsequent studies, van Rhenterghem and Lazdunski also identified an ET-stimulated chloride current in rat aortic smooth muscle cells.[45] Activation of this current was shown to contribute to sustained depolarization of the cell membrane. Single channel recordings revealed two distinct low conductance Cl^- channels, one of which was dependent on Ca^{2+}.[45] This Ca^{2+}-dependent Cl^- channel is presumably linked to the biphasic increase in Ca^{2+} which is known to be characteristic of ET signaling.

An alternative scheme presented by Klockner and Isenberg in 1991 was also based on the premise that L-type Ca^{2+} channels are activated secondarily to membrane depolarization. The effect of ET on ion currents was examined in myocytes from the porcine coronary and human mesenteric arteries using the voltage clamp technique. Their results indicated that the initial depolarization response to ET-1 was induced by a Ca^{2+} activated Cl^- current ($I_{Cl(Ca)}$)[46] rather than a nonselective cation current as van Rhenterghem had described.[36] Klockner and Isenberg suggested that Ca^{2+} released in response to IP_3 was responsible for activating the Ca^{2+} activated Cl^- channel as well as a Ca^{2+} activated K^+ channel.[46] Both of these currents were purported to be involved in the sustained depolarization of the membrane necessary to activate L-type Ca^{2+} channels. They further speculated that depletion of the IP_3-sensitive Ca^{2+} store would result in inactivation of these channels, leading to repolarization and subsequent closing of L-type channels. Support for the role of a Cl^- channel in mediating the initial depolarization response to ET-1 was provided by a series of reports in cultured aortic smooth muscle cells and in renal mesangial cells.[47-49] In line with the previous studies, release of intracellular Ca^{2+} from an IP_3-sensitive store was thought to be the primary signal leading to activation of Ca^{2+}-dependent Cl^- channels. Cl^- efflux in response to ET-1 was blocked by the Cl^- channel antagonist, indanyloxyacetic acid (IAA-94), and membrane depolarization was inhibited.[47] Consequently Ca^{2+} channels were not activated and there was no increase in $[Ca^{2+}]_i$ observed. These findings were confirmed and extended in studies conducted in the renal microvasculature in which ET-induced vasoconstriction was blocked by treatment with IAA-94, the Cl^- channel blocker.[48]

The role of the K_{Ca} channel in the depolarization response induced by ET-1 was further characterized in smooth muscle cells of the rat renal microvasculature.[49] Consistent with the previous reports, a Ca^{2+} dependent Cl^- channel was shown to be responsible for the initial depolarization of the cell membrane that preceded Ca^{2+} influx. However, depolarization also lead to activation of a transient outward Ca^{2+} dependent K^+ channel that induced an opposing hyperpolarization.[49] This current was thought to be involved in terminating the signaling events initiated by ET-1.

Clearly the sequence of events leading to the depolarization of the smooth muscle cell membrane and the activation of L-type Ca^{2+} channels is incompletely resolved. The precise roles of K^+ and Cl^- channels in activating and inactivating voltage-dependent Ca^{2+} channels continue to be investigated. Furthermore it is likely that cellular mediators such as IP_3 participate, either directly or indirectly, in the activation of these currents. Activation of L-type Ca^{2+} channels by second messengers, independent of membrane depolarization, has also been described.[50] PKC, for example, is thought to be involved in a distinct pathway by which ET-1- stimulated Ca^{2+} influx occurs. The role of second messengers in the activation of ion channels is discussed elsewhere.

Activation of K^+ channels in vascular smooth muscle in response to ET-1 has also been proposed to mediate relaxation. Although it has not been coupled to a specific receptor, direct activation of ATP-sensitive K^+ channels by ET-1 resulted in vasodilation of the cat pulmonary artery.[51] It was theorized that this current would hyperpolarize the cell membrane and prevent the influx of Ca^{2+} necessary to sustain constriction. In a similar study in the porcine coronary artery, Ca^{2+}-activated K^+ channels were shown to be activated in response to ET-1 and vasodilation ensued.[52]

Actions of ET in Endothelial Cells

While the physiologic response to ET-1 in vascular smooth muscle is characterized by vasoconstriction and is mediated by the ET_A receptor, activation of ET_B receptors on endothelial cells almost always results in vasodilation. In endothelial cells, the signaling events that follow the binding of either ET-1 or ET-3 are not as well known as those associated with the ET_A receptor. However, an increase in $[Ca^{2+}]_i$ is thought to be instrumental in mediating the effects of ET. As in vascular smooth muscle, activation of phospholipase C (PLC) and production of IP_3 leads to release of Ca^{2+} from intracellular stores which is followed by influx of extracellular Ca^{2+}. Increasing $[Ca^{2+}]_i$ in ECs leads to the production of endothelium derived relaxing factors (EDRF) that mediate vasodilation. Unlike the ET_A mediated signaling pathway in vascular smooth muscle, however, Ca^{2+} influx in ECs is not dependent on membrane depolarization. Rather, hyperpolarization of the endothelial cell membrane often precedes Ca^{2+} entry. While the identity of Ca^{2+} influx pathways in ECs remains uncertain, it is clear that voltage dependent Ca^{2+} channels are not involved. It has been speculated that receptor-operated or nonselective cation channels are activated in the course of ET-signaling; however, little is known about the electrophysiological profile of ECs particularly in the presence of agonists. Luckhoff and Busse conducted a study in bovine aortic endothelial cells to examine the role of membrane potential in agonist-induced Ca^{2+} mobilization.[53] In response to bradykinin, an initial transient increase in Ca^{2+} was attributed to Ca^{2+} release from intracellular stores and a sustained elevation was due to influx of Ca^{2+} from the extracellular space. Treating cells with a hyperpolarizing drug, BRL 34915, increased the magnitude of the Ca^{2+} response during the sustained phase, whereas depolarization of the cell membrane with a high K^+ solution actually inhibited Ca^{2+} influx.[53] Based on these findings, it was hypothesized that certain K^+ channels may be activated in the course of bradykinin signaling, inducing membrane hyperpolarization and subsequently triggering Ca^{2+} entry. This concept was substantiated by Kukovetz et al when a Ca^{2+}-sensitive K^+ channel was found to induce hyperpolarization in porcine aortic endothelial cells.[54] Simultaneous measurement of Ca^{2+}

and membrane potential revealed that intracellular Ca^{2+} increased with membrane potential in response to bradykinin. Furthermore, it was determined that voltage-dependent L-type Ca^{2+} channels were not involved in agonist-induced Ca^{2+} entry in endothelial cells.[54] These studies were conducted with bradykinin, but similar results were obtained in a later study using ET-1.

In rat brain ECs, the DHP-sensitive Ca^{2+} channel blockers, nifedipine and nitrendipine, had no effect on Ca^{2+} uptake in response to ET-1.[55] Thus it was hypothesized that ET-1-induced Ca^{2+} influx in these cells occurred via receptor operated channels. In 1994 Zhang et al conducted a series of experiments using the voltage clamp technique to directly measure ion currents activated in response to ET-1. In bovine pulmonary artery endothelial cells (BPAECs), ET-1 was shown to inhibit the inward rectifying K^+ channel, which is thought to be involved in maintaining the resting membrane potential.[56] Inhibiting the current would result in hyperpolarization, which has previously been demonstrated to stimulate Ca^{2+} influx in ECs.[53,54] Additional experiments in human umbilical vein endothelial cells (HUVECs) revealed that ET-1 activated a nonselective Ca^{2+} permeable cation channel.[56] This finding was extended in a subsequent study in which the current could be partially blocked by SK&F 96365, an inhibitor of receptor operated Ca^{2+} channels.[57] Thus it appears that ET-stimulated Ca^{2+} influx in ECs occurs by direct activation of a receptor operated, nonselective cation channel. Furthermore, Ca^{2+} influx may be stimulated indirectly as a consequence of ET-induced hyperpolarization. The effects of ET reported here are presumably mediated by the ET_B receptor, although there has been no evidence presented suggesting that ET_B is coupled to ion channels in ECs. Thus the role of ion currents in mediating cellular responses to ET in ECs, particularly regarding the synthesis and release of EDRF, appears to be at the level of facilitating Ca^{2+} influx. Elevating intracellular Ca^{2+} activates nitric oxide synthase and leads to the generation of NO, which induces vasodilation by relaxing underlying smooth muscle cells.

Actions of ET in the Kidney

In addition to its effects on ion channels in the renal vasculature, which were discussed previously, the actions of ET have also been documented in other cell types within the kidney. For example, renal mesangial cells, located within and between the glomerular capillaries, exhibit a contractile phenotype that is similar to vascular smooth muscle cells and contract in the presence of ET. In this regard ET is thought to be an important regulator of glomerular function and has been implicated in a number of renal pathologies. The signaling events that follow activation of the ET receptor in mesangial cells parallel those described in vascular smooth muscle. A biphasic increase in intracellular Ca^{2+} is characterized by release of Ca^{2+} from intracellular stores as well as influx of Ca^{2+} from the extracellular space.[58] The latter occurs through voltage-dependent Ca^{2+} channels, which are thought to be activated secondarily to depolarization of the cell membrane.[58] To investigate the events that precede activation of these channels, Hu et al studied the effect of ET on ion currents in primary cultures of rat mesangial cells. In whole cell recordings designed to characterize ion conductances at the single channel level, ET-1 was shown to augment a Cl^- current while simultaneously inhibiting an outward K^+ channel.[59] This resulted in depolarization of the cell membrane and activation of voltage-gated Ca^{2+} channels. In additional experiments the effect of ET on both ion channels could be blocked by the ET_B receptor antagonist IRL 1038, whereas the ET_A receptor antagonist BQ 123 had no effect. These results indicate that ET binding to the ET_B receptor modulates a Cl^- current as well as a K^+ current in renal mesangial cells. Ling and coworkers[60] confirmed the findings of Hu et al. Using the patch clamp technique, the electrophysiological profile of mesangial cells was characterized and several ion channels were identified, including a Ca^{2+}-dependent Cl^- channel, a nonselective cation channel that was insensitive to changes in membrane polarity and a receptor operated Ca^{2+} channel. In the presence of 10 nM ET, the Cl^- conductance was markedly enhanced, and the activity of the nonselective cation channel was increased tenfold.[60] This observation was similar to that reported by van Rhenterghem in vascular smooth muscle in which ET stimulated both a Ca^{2+}-dependent Cl^- channel as well as a nonselective cation channel. Thus ET activates Ca^{2+} channels in renal mesangial cells indirectly following modulation of other ion conductances that results in membrane depolarization. These findings suggest a common scheme in ET signaling, although the particular receptors mediating these events may vary among cell types.

Actions of ET in Cardiac Muscle

In addition to potent constricting actions in the vasculature and in the kidney, ET exerts a profound influence on cardiac muscle. Ishikawa et al documented that ET was a positive inotropic and chronotropic agent in guinea pig hearts.[61] These observations suggested that ET altered the electrical activity of cardiac myocytes, although there are several conflicting reports regarding the mechanism by which these actions of ET are mediated. Regulation of ion channels by ET in cardiac muscle remains incompletely understood. Voltage clamp studies in isolated rat atrial cells were conducted by Kim who demonstrated that ET activates an inwardly rectifying K^+ channel.[62] Evoking this current results in a cessation of spontaneous beating, a finding that was later corroborated in myocytes of rabbit and guinea pig atria by Ono et al.[63] In additional experiments using the same preparations, Ono demonstrated that ET inhibited the L-type Ca^{2+} current. Together these data imply that ET has *negative* chronotropic effects in cardiac myocytes and conflict with earlier reports that ET increases heart rate. While the significance of these observations is unclear, it is possible that the effects of ET on the heart are mediated by specific receptors in different cell types. James and coworkers studied ion currents in guinea pig ventricular myocytes and showed that ET inhibited a PKA-dependent Cl^- current.[64] Under normal conditions, this Cl^- conductance opposes the L-type Ca^{2+} current and shortens the action potential duration. In the presence of ET-1, inhibition of the Cl^- channel resulted in prolongation of the action potential duration. Sustained activation of the Ca^{2+} current subsequently potentiates the positive inotropic effect of ET. The actions of ET on this particular Cl^- channel are mediated indirectly by inhibiting adenylyl cyclase, which in turn, prevents the accumulation of cyclic AMP (cAMP) and the activation of protein kinase A (PKA). Consequently the Cl^- channel is not activated and the L-type Ca^{2+} current is sustained. Thus the signaling processes involved in ET-induced activation of ion channels in cardiac muscle are very complex and remain incompletely understood. In addition, variability of responses among cell types and in ET receptor distribution may also account for disparate results.

Actions of ET in Nonvascular Smooth Muscle

The effects of ET on nonvascular smooth muscle include generation of prolonged and sustained contractions of tracheal and bronchial smooth muscle as well as graded contractions in the stomach and intestinal tract.[7] In addition ET is a potent constrictor of the uterus.[7] While the distribution of ET receptor subtypes varies widely among these tissues,[65,66] the signaling cascade leading to contraction is similar to that mediated by the ET_A receptor in vascular smooth muscle. Thus the primary mediator of ET's action in these tissues is Ca^{2+}, initially released from intracellular stores in response to IP_3 formation. Secondarily, and perhaps more importantly, Ca^{2+} enters from the extracellular space, leading to a sustained increase in cytosolic $[Ca^{2+}]$ and activation of contractile proteins. As in vascular smooth muscle, there is variability among species and among cell types as to the route of Ca^{2+} entry. DHP-sensitive Ca^{2+} channels are known to be activated in some cell types,[67,70] yet seem to be unimportant in others.[68-71] In spite of the diversity in ion channel activation, the onset of the contraction induced by ET is similar in a variety of smooth muscle cell types studied. One notable exception was reported in intestinal smooth muscle by Lin and Lee in which a transient relaxation in response to ET was observed prior to initiation of the contractile response.[72] The relaxation could not be attributed to the release of vasodilating substances from the endothelium, rather it was due to the activation of a Ca^{2+} dependent K^+ channel. This ET-stimulated current hyperpolarized the smooth muscle cell membrane and induced the observed relaxation.

Actions of ET in Other Cell Types

While the actions of ET-1 are largely described in terms of its ability to induce a contraction, its actions are not limited to cell types exhibiting a contractile phenotype. The peptide also stimulates secretion in neuroendocrine cells[73] and it is a mitogen, regulating DNA synthesis in fibroblasts, osteoblasts and melanocytes, as well as in vascular smooth muscle and cardiac myocytes.[9] Increases in intracellular Ca^{2+} concentration, specifically via Ca^{2+} influx, are critical in mediating ET's effects in these tissues. With regard to secretory function, ET has been demonstrated to stimulate the release of LH, FSH, GH and TSH from the anterior pituitary while inhibiting the release of prolactin.[74] These actions were blocked in the presence of the DHP-sensitive Ca^{2+} channel antagonist, nifedipine. In separate studies aldosterone[75] and vasopressin[76] secretion were also shown to be induced by ET and were described as Ca^{2+} dependent phenomena. Collectively these findings suggest that ET-stimulated hormone secretion is dependent on Ca^{2+} influx via voltage-dependent Ca^{2+} channels. In the case of ET-stimulated vasopressin secretion from neurohypophysial explants, ET was also shown to activate a charybdotoxin-sensitive K^+ channel.[76] This particular class of K^+ channels is activated in response to an increase in cytosolic Ca^{2+}, which is achieved by influx of extracellular Ca^{2+} through voltage operated Ca^{2+} channels. Induction of a current through this channel repolarizes the cell and essentially turns off the secretory function.

The mitogenic effects of ET are also dependent on the mobilization of extracellular Ca^{2+}.[9] The stimulatory effect of ET on DNA synthesis was inhibited by nifedipine in vascular smooth muscle cells, indicating that DHP-sensitive Ca^{2+} channels are involved in this process as well. Other classes of Ca^{2+} channel antagonists were partially effective in inhibiting the mitogenic effects of ET, although there is some variability among cell types.[77-79] Thus, while numerous details regarding the signal transduction cascades leading to these respective responses remain to be elucidated, it is apparent that influx of extracellular Ca^{2+} via voltage-dependent Ca^{2+} channels is a critical feature. The roles of other ion channels in mediating these responses to ET are incompletely understood.

Conclusions

Almost without exception the actions of ET are dependent on the influx of extracellular Ca^{2+}. The predominant route of entry is the L-type DHP-sensitive Ca^{2+} channel although additional classes of voltage-dependent Ca^{2+} channels may also be involved. In spite of considerable variability among cell types and among species, the importance of Ca^{2+} influx is paramount. Nonselective cation channels and receptor operated channels also appear to play a role in mediating these effects, although their functions are not clearly defined. The mechanism of activation of voltage-dependent Ca^{2+} channels often requires an initial depolarization that is accomplished by modulating other ion conductances. Ca^{2+}-dependent Cl^- channels as well as certain classes of K^+ channels may be activated in the initiating events following ET receptor activation.

References

1. Hickey KA, Rubanyi G, Paul RJ et al. Characterization of a coronary vasoconstrictor produced by cultured endothelial cells. Am J Physiol 1985; 248 (Cell Physiol 17):C550-C556.
2. Yanagisawa M, Kurihara H, Kimura S et al. A novel potent vasoconstrictor peptide produced by vascular endothelial cells. Nature 1988; 332:411-415.
3. Rubanyi GM, Polokoff MA. Endothelins: Molecular biology, biochemistry, pharmacology, physiology and pathophysiology. Pharmacological Reviews 1994; 46:325-415.
4. Masaki T, Yanagisawa M, Goto, K. Physiology and pharmacology of endothelins. Medicinal Research Reviews 1992; 12(4):391-421.
5. Masaki T, Kimura S, Yanagisawa M et al. Molecular and cellular mechanism of endothelin regulation: Implications for vascular function. Circulation 1991; 84(4):1457-1468.
6. Huggins JP, Pelton JT, Miller RC. The structure and specificity of endothelin receptors: Their importance in physiology and medicine. Pharmac Ther 1993; 59:55-123.
7. Rae GA, Calixto JB, D'Orleans-Juste P. Effects and mechanisms of action of endothelins on nonvascular smooth muscle of the respiratory, gastrointestinal and urogenital tracts. Regulatory Peptides 1995; 55:1-46.
8. Naruse M, Naruse K, Demura H. Recent advances in endothelin research on cardiovascular and endocrine systems. Endocrine Journal 1994; 41(5):491-507.
9. Battistini B, Chailler P, D'Orleans-Juste P et al. Growth regulatory properties of endothelins. Peptides 1993; 14:385-399.
10. Marsden PA, Danthuluri NR, Brenner BM et al. Endothelin action on vascular smooth muscle involves inositol trisphosphate and calcium mobilization. Biochem Biophys Res Comm 1989; 158:86-93.
11. Ihara M, Saeki T, Funabashi K et al. Two endothelin receptor subtypes in porcine coronary arteries. J Cardiovasc Pharmacol 1991; 17(Supp 7):S119-121.
12. Chabrier PE, Auguet M, Roubert P et al. Vascular mechanisms of action of endothelin-1: Effect of Ca^{2+} antagonists. J Cardiovasc Pharmacol 1989; 13(Supp 5):S32-S35.

13. Sakata K, Ozaki H, Kwon SC et al. Effects of endothelin on the mechanical activity and cytosolic calcium levels of various types of smooth muscle. Br J Pharmacol 1989; 98:483-492.

14. Gordan, J. Vascular biology: Put out to contract. Nature 1988; 332:395-396.

15. Pollock DM, Opgenorth TJ. Evidence for endothelin-induced renal vasoconstriction independent of ET_A receptor activation. Am J Physiol 1993; 264(Regulatory Integrative Comp Physiol 33):R222-226.

16. Clozel M, Gray GA, Breu V et al. The endothelin ET_B receptor mediates both vasodilation and vasoconstriction in vivo. Biochem Biophys Res Comm 1992; 186:867-873.

17. Inoue A, Yanagisawa M, Kimura S et al. The human endothelin family: three structurally and pharmacologically distinct isopeptides predicted by three separate genes. Proc Natl Acad Sci USA 1989; 86:2863-2867.

18. Goto K, Kasuya Y, Matsuki N et al. Endothelin activates the dihydropyridine-sensitive, voltage-dependent Ca^{2+} channel in vascular smooth muscle. Proc Natl Acad Sci USA 1989; 86:3915-3918.

19. Kasuya Y, Ishikawa T, Yanagisawa M et al. Mechanism of contraction to endothelin in isolated porcine coronary artery. Am J Physiol 1989; 257 (Heart Circ Physiol 26):H1828-1835.

20. Encabo A, Ferrer M, Marin J et al. Vasoconstrictive responses elicited by endothelin in bovine cerebral arteries. Gen Pharmac 1992; 23:263-267.

21. Luscher TF. Endothelin: systemic arterial and pulmonary effects of a new peptide with potent biologic properties. Am Rev Respir Dis 1992; 146:S56-S60.

22. Yoshida M, Suzuki A, Itoh T. Mechanisms of vasoconstriction induced by endothelin-1 in smooth muscle of rabbit mesenteric artery. J Physiol 1994; 477:253-265.

23. Godfraind T, Mennig D, Morel N et al. Effect of endothelin-1 on calcium channel gating by agonists in vascular smooth muscle. J Cardiovasc Pharmacol 1989; 13 (Supp 5):S112-S117.

24. Hardebo JE, Kahrstrom J, Owman C et al. Endothelin is a potent constrictor of human intracranial arteries and veins. Blood Vessels 1989; 26:249-253.

25. Inoue Y, Oike M, Nakao K et al. Endothelin augments unitary calcium channel currents on the smooth muscle cell membrane of guinea pig portal vein. J Physiol 1990; 423:171-191.

26. Xuan YT, Whorton AR, Watkins WD. Inhibition by nicardipine of endothelin mediated inositol phosphate formation and Ca^{2+} mobilization in smooth muscle cells. Biochem Biophys Res Commun 1989; 160:758-764.

27. Silberberg SD, Poder TC, Lacerda AE. Endothelin increases single channel calcium currents in coronary arterial smooth muscle cells. FEBS Letters 1989; 247:68-72.

28. D'Orleans-Juste P, De Nucci G, Vane J. Endothelin-1 contracts isolated vessels independently of dihydropyridine-sensitive Ca^{2+} channel activation. Eur J Pharmacol 1989; 165:289-295.

29. Chabrier PE, Auguet M, Roubert P et al. Vascular mechanism of action of endothelin-1: effect of Ca^{2+} antagonists. J Cardiovasc Pharmacol 1989; 13 (Supp 5):32-35.

30. Bodelsson G, Sjoberg NO, Stjernquist M. Contractile effect of endothelin in the human uterine artery and autoradiographic localization of its binding sites. Am J Obstet Gynecol 1992; 167:745-750.

31. Blackburn K, Highsmith, RF. Nickel inhibits endothelin-induced contractions of vascular smooth muscle. Am J Physiol 1990; 258 (Cell Physiol 27):C1025-C1030.

32. Benham CD, Hess P, Tsien RW. Two types of calcium channels in single smooth muscle cells from rabbit ear artery studied in whole-cell and single-channel recordings. Circ Res 1987; 61:111-116.

33. Friedman ME, Suarez-Kurtz G, Kaczorowski GJ. Two calcium currents in a smooth muscle cell line. Am J Physiol 1986; 250:H699-H703.

34. Chen, C and Wagoner, PK. Endothelin induces a nonselective cation current in vascular smooth muscle cells. Circ Res 1991; 69:447-454.

35. Van Rhenterghem C, Romey G, Lazdunski M. Vasopressin modulates the spontaneous electrical activity in aortic cells (line A7r5) by acting on three different types of ionic channels. Proc Natl Acad Sci USA 1988; 85:9365-9369.

36. Van Rhenterghem C, Vigne P, Barhanin, J et al. Molecular mechanism of action of the vasoconstrictor peptide endothelin. Biochem Biophys Res Commun 1988; 157:977-985.

37. Watanabe T, Kusumoto K, Kitayoshi T. Positive inotropic and vasoconstrictive effects of endothelin-1 in vivo and in vitro experiments: characteristics and the role of L type calcium channels. J Cardiovasc Pharmacol 1989; 13 (Supp 5):S108-S111.

38. Enoki T, Miwa S, Sakamoto A et al. Long-lasting activation of cation current by low concentration of endothelin-1 in mouse fibroblasts and smooth muscle cells of rabbit aorta. Br J Pharmacol 1995; 115:479-485.

39. Kasuya Y, Takuwa Y, Yanagisawa M et al. Endothelin-1-induces vasoconstriction through two functionally distinct pathways in porcine coronary artery: Contribution of phosphoinositide turnover. Biochem Biophys Res Comm 1989; 161: 1049-1055.

40. Resink T, Scott-Burden T, Buhler F. Activation of multiple signal transduction pathways by endothelin in cultured human vascular smooth muscle cells. Eur J Biochem 1990; 189:415-421.

41. Mitsuhashi T, Morris RC, Ives HE. Endothelin-induced increases in vascular smooth muscle Ca^{2+} do not depend on dihydropyridine-sensitive Ca^{2+} channels. J Clin Invest 1989; 84:635-639.

42. Kai H, Kanaide H, Nakamura M. Endothelin-sensitive intracellular Ca^{2+} store overlaps with caffeine-sensitive one in rat aortic smooth muscle cells in primary culture. Biochem Biophys Res Commun 1989; 158:235-243.

43. Muldoon LL, Anslen H, Rodland KD et al. Stimulation of Ca^{2+} influx by endothelin-1 is subject to negative feedback by elevated intracellular Ca^{2+}. Am J Physiol 1991; 260:C1273-C1281.

44. Gardner J, Tokudome P, Tomonari H et al. Endothelin-induced calcium responses in human vascular smooth muscle cells. Am J Physiol 1992; 262:C148-C155.

45. Van Rhenterghem C, Lazdunski M. Endothelin and vasopressin activate low conductance chloride channels in aortic smooth muscle cells. Pflugers Arch 1993; 425:156-163.

46. Klockner U, Isenberg G. Endothelin depolarizes myocytes from porcine coronary and human mesenteric arteries through a Ca-activated chloride current. Pflugers Arch 1991; 418:168-175.

47. Iijima K, Lan L, Nasjletti A et al. Intracellular signaling pathway of endothelin-1. J Cardiovasc Pharmacol 1991; 17(Supp 7):S146-S149.

48. Takenaka T, Epstein M, Forster H et al. Attenuation of endothelin effects by a chloride channel inhibitor, indanyloxyacetic acid. Am J Physiol 1992; 262: F799-F806.

49. Gordienko DV, Clausen C, Goligorsky MS. Ionic currents and endothelin signaling in smooth muscle cells from rat renal resistance arteries. Am J Physiol 1994; 266:F325-F341.

50. Xuan YT, Wang OL, Whorton AR. Regulation of endothelin-induced Ca^{2+} mobilization in smooth muscle cells by protein kinase C. Am J Physiol 1994; 266:C1560-C1567.

51. Luckhoff A, Busse R. Calcium influx into endothelial cells and formation of endothelium-derived relaxing factor is controlled by the membrane potential. Pflugers Arch 1990; 416:305-311.

52. Lippton HL, Cohen GA, McMurtry IF. Pulmonary vasodilation to endothelin isopeptides in vivo is mediated by potassium channel activation. J Appl Physiol 1991; 70:947-952.

53. Hu S, Kim HS, Jeng AJ. Dual action of endothelin-1 on the Ca^{2+}-activated K^+ channel in smooth muscle cells of porcine coronary artery. Eur J Pharmacol 1991; 194:31-37.

54. Kukovetz WR, Graier WF, Groschner K. Contribution of agonist-induced hyperpolarization to Ca^{2+} influx and formation of EDRF in vascular endothelial cells. Jpn J Pharmacol 1992; 58:213-219.

55. Stanimirovic DB, Nikodijevic B, Nikodijevic-Kedeva D et al. Signal transduction and Ca^{2+} uptake activated by endothelins in rat brain endothelial cells. Eur J Pharmacol 1994; 288:1-8.

56. Zhang H, Inazu M, Weir B et al. Endothelin-1 inhibits inward rectifier potassium channels and activates nonspecific cation channels in cultured endothelial cells. Pharmacology 1994; 49:11-22.

57. Inazu M, Zhang H, Daniel EE. Different mechanisms can activate Ca^{2+} entrance via cation currents in endothelial cells. Life Sciences 1995; 56:11-17.

58. Simonson MS, Dunn MJ. Endothelin peptides and the kidney. Annu Rev Physiol 1993; 55:249-265.

59. Hu S, Kim HS, Lappe RW et al. Coupling of endothelin receptors to ion channels in rat glomerular mesangial cells. J Cardiovasc Pharmacol 1993; 22(Supp 8):S149-S153.

60. Ling BN, Matsunaga H, Ma H et al. Role of growth factors in mesangial cell ion channel regulation. Kidney International 1995; 48:1158-1166.

61. Ishikawa T, Yanagisawa M, Kimura S. Positive inotropic action of novel vasoconstrictor peptide endothelin on guinea pig atria. Am J Physiol 1988; 255:H970-H973.

62. Kim D. Endothelin activation of an inwardly rectifying K^+ current in atrial cells. Circ Res 1991; 69:250-255.

63. Ono M, Tsujimoto G, Sakamoto A et al. Endothelin A receptor mediates cardiac inhibition by regulating calcium and potassium currents. Nature 1994; 370:301-304.

64. James AF, Xie LH, Fujitani Y. Inhibition of the cardiac protein kinase A-dependent chloride conductance by endothelin-1. Nature 1994; 370:297-300.

65. Henry PJ. Endothelin-1-induced contraction in rat isolated trachea—involvement of ET_A and ET_B receptors and multiple signal transduction systems. Br J Pharmacol 1993; 110:435-441.

66. Noguchi K, Noguchi Y, Hirose H et al. Role of endothelin ET_B receptors in bronchoconstrictor and vasoconstrictor responses in guinea pigs. Eur J Pharmacol 1993; 233:47-51.

67. Uchida Y, Ninomiya H, Saotome M et al. Endothelin, a novel vasoconstrictor peptide, as potent bronchoconstrictor. Eur J Pharmacol 1988; 154:227-228.

68. Turner NC, Power RF, Polak JM et al. Contraction of rat tracheal smooth muscle by endothelin. Br J Pharmacol 1989; 96:103-107.

69. Grunstein MM, Rosenberg SM, Schramm CM et al. Mechanism of action of endothelin-1 in maturing rabbit airway smooth muscle. Am J Physiol 1991; 260:L434-L443.

70. Kozuka M, Ito T, Hirose S et al. Endothelin induces two types of contractions of rat uterus: phasic contractions by way of voltage-dependent calcium channels and developing contractions through a second type of calcium channel. Biochem Biophys Res Commun 1989; 159:317-323.

71. Molnar M, Hertelendy F. Signal transduction in rat myometrial cells: comparison of the actions of endothelin-1, oxytocin and prostaglandin $F_{2\alpha}$. Eur J Pharmacol 1995; 133:467-474.

72. Lin WW and Lee CY. Intestinal relaxation by endothelin isopeptides: involvement of Ca^{2+}-activated K^+ channels. Eur J Pharmacol 1992; 219:355-360.

73. Kennedy RL, Haynes WG, Webb DJ. Endothelins as regulators of growth and function in endocrine tissues. Clinical Endocrinology 1993; 39:259-265.

74. Samson WK, Skala KD, Alexander B et al. Possible neuroendocrine actions of endothelin-3. Endocrinology 1991; 128:1465-1473.

75. Zeng ZP, Naruse M, Guan BJ et al. Endothelin stimulates aldosterone secretion in vitro from normal adrenocortical tissue, but not adenoma tissue in primary aldosteronism. J Clin Endocrinol Metab 1992; 74:874-878.

76. Rossi NF. Cation channel mechanisms in ET-3-induced vasopressin secretion by rat hypothalamo-neurohypophysial explants. Am J Physiol 1995; 268:E467-E475.

77. Hirata Y, Takagi Y, Fukuda Y et al. ET is a potent mitogen for rat vascular smooth muscle cells. Atherosclerosis 1989; 78:225-228.

78. Nakaki T, Nakayama M, Yamamoto S et al. ET-mediated stimulation of DNA synthesis in vascular smooth muscle cells. Biochem Biophys Res Commun 1989; 158:880-883.

79. Supattapone S, Simpson AWM, Ashley CC. Free calcium rise and mitogenesis in glial cells caused by ET. Biochem Biophys Res Commun 1989; 165:1115-1122.

CHAPTER 10

Endothelin and Calcium Signaling

E. Radford Decker and Tommy A. Brock

Introduction

Intensive investigation in recent years has led to the clarification of mechanisms involved in the synthesis, degradation and intracellular signaling pathways related to endothelin (ET), a family of highly potent, autocrine and paracrine polypeptides, that exert their pleiotropic actions in diverse cell types. Hickey et al[1] were the first to describe the biological activity of this potent endothelium-derived constricting factor in 1985, but the identity of this factor remained elusive until 1988 when Yanagisawa and coworkers[2] reported the successful purification and cloning of ET. Subsequent molecular cloning studies documented the existence of three distinct genes (ET-1, ET-2 and ET-3)[3-6] within the human genome. One of the initial observations made concerning the ET family was that it shared striking structural similarity to sarafotoxins, a family of four polypeptides found in the venom of *Atractaspis engaddenas*[7] which appear to act via common receptors.

This chapter will summarize what is currently known about the initial signaling events involved in ET receptor signal transduction with special emphasis on vascular smooth muscle and endothelial cells. Since intracellular free calcium concentration ($[Ca^{2+}]_i$) is a key mechanism by which ET appears to mediate many of its cellular actions, our principal focus will be to review the molecular mechanisms underlying calcium signaling dynamics. Other signal transduction pathways which potentially initiate or modulate ET-induced $[Ca^{2+}]_i$ transients and are activated in response to changes in $[Ca^{2+}]_i$ will also be discussed. References will be made to other cell types when substantial differences or gaps in our knowledge exist about vascular cells.

Signal Transduction Mechanisms

In the past five years, there has also been a rapid development in our understanding of ET receptor subtypes. A general scheme of the intracellular signaling cascade resulting from ET binding to its cell surface receptor on different cell types is illustrated in Figure 10.1. The interaction between ET receptors and a specific guanine nucleotide regulatory protein (G-protein) leads to the activation of multiple effector proteins. In particular, phospholipase C activation results in the rapid hydrolysis of the membrane inositol phospholipid, phosphatidylinositol 4,5-bisphosphate, yielding inositol 1,4,5-triphosphate (IP$_3$) and *sn*1,2-diacylglycerol (DAG). A common event elicited by ET receptor activation in many cell types is a

Endothelin Receptors and Signaling Mechanisms, edited by David M. Pollock and Robert F. Highsmith. © 1998 Springer-Verlag and R.G. Landes Company.

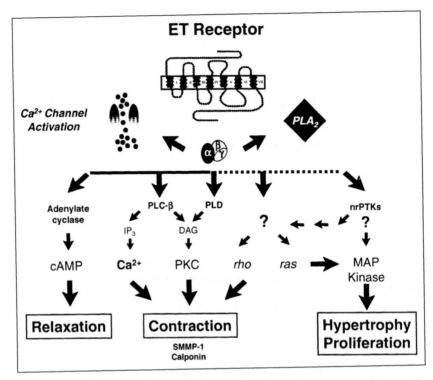

Fig. 10.1. General scheme of ET-1 signal transduction pathways in vascular smooth muscle. ET, Endothelin-1; G-protein, guanine nucleotide binding protein; PLC, phospholipase C; IP$_3$, inositol 1,4,5 trisphosphate; Ca$_i$, intracellular Ca^{2+} concentration; DAG, diacylglycerol; PKC, protein kinase C; PLD, phospholipase D; PA, phosphatidic acid; PLA$_2$, phospholipase A$_2$; AA, arachidonic acid; PG, prostaglandin; nrPTKs, nonreceptor protein tyrosine kinases; MAP-kinase, mitogen activated protein kinase; SMMP-1, smooth muscle myosin phosphatase-1.

biphasic increase in [Ca^{2+}]$_i$. The rapid rise in IP$_3$ initiates calcium (Ca^{2+}) release from intracellular storage sites. This burst of IP$_3$-dependent Ca^{2+} release activates other Ca^{2+}-dependent processes which lead to cell membrane depolarization and sustained extracellular Ca^{2+} influx. Diacylglycerol is believed to be the physiological stimulus for protein kinase C activation (PKC). Several cellular actions of ET-1 can be mimicked by phorbol esters, potent and persistent activators of PKC. In VSMCs, the two primary biological consequences of ET stimulation are contraction and mitogenesis.

Endothelin Receptor-G-protein coupling

Receptor Subtypes

To date, cDNAs of three distinct high-affinity ET receptor subtypes, designated ET$_A$, ET$_B$, and ET$_C$, have been identified in different species.[8] Endothelin receptors have been shown to be members of the rhodopsin superfamily of G protein-coupled receptors which contain seven transmembrane spanning domains.[9,10] The human cDNAs for ET_A and ET_B receptors are approximately 55% homologous. An ET_C has been cloned from *Xenopus laevis* melanocytes which shows increased specificity for ET-3[11] and exhibits 50% amino acid sequence homology with ET_A and ET_B receptors. A homolog of this receptor subtype has not been identified in mammals. Recently, pharmacological evidence has been presented suggesting that subtypes of ET_A[12-14] and ET_B[14-16,22] exist. Within blood vessels, ET_A receptors are found on vascular smooth muscle cells (VSMCs) and are associated with vasoconstriction, while ET_B receptors are found predominantly on endothelial cells and are associated with the release of prostacyclin and nitric oxide. However, there are vascular beds, including rat renal vasculature, rabbit pulmonary artery and saphenous vein, and human internal mammary artery in which VSMC responses to ET are mediated via a mixture of ET_A and ET_B receptors. Numerous clinical studies using mixed ET_A/ET_B and ET_A selective antagonists are currently in progress for several disease indications, including congestive heart failure, cerebrovascular vasospasm, pulmonary hypertension, and acute renal failure.[17]

G-Protein-Effector Protein Interactions

In all eukaryotic organisms, heterotrimeric G-proteins play an important role in the transduction of extracellular signals to the cell interior. It has been estimated that at least a third of all signal transduction processes involve heterotrimeric G-proteins. These protein complexes consist of α (molecular mass = 39 - 46 kDa), β (37 kDa), and γ (8 kDa) subunits.[18,19] The G_α subunit binds guanine nucleotides (GDP and GTP) and contains intrinsic GTPase activity. The GDP bound form of the G_α subunit binds tightly to the $G_{\beta\gamma}$ subunits. When GTP replaces the GDP bound to the G_α subunit, the G_α subunit dissociates from the $G_{\beta\gamma}$ subunits. The $G_{\beta\gamma}$ subunits exist as a tightly associated unit and dissociate only after denaturation. A negative feedback loop exists due to the intrinsic GTPase activity of the α-subunit which hydrolyzes the GTP to GDP. If GTPγS, a nonhydrolyzable GTP analog, binds to the G_α subunit, G_α becomes continuously active due to the inability of the GTPase activity of the α-subunit to hydrolyze this form of GTP. Although the G_α subunit serves as a major regulator of many cellular proteins, it is also well-documented that $G_{\beta\gamma}$ can also positively regulate effector proteins, such as K^+ channels, adenylate cyclase, PLCβ, phospholipase A_2, phosphoinositide 3-kinase, and β-adrenergic receptor kinases.[18] To date, more than 20 different G_α subunits, five G_β subunits, and six G_γ subunits have been cloned.[18,19]

As stated earlier, the ET receptors belong to the rhodopsin superfamily of G-protein-coupled receptors.[20,21] In membranes prepared from A10 cells, an aortic smooth muscle cell line, ET-1-stimulated phosphoinositide hydrolysis was dependent on the presence of GTPγS and the binding of ET-1 to these membranes was inhibited by GTPγS in a dose-dependent manner.[22] Additionally, two different bacterial exotoxins have proven to be useful tools to characterize the G-proteins involved in ET signal transduction. The first is *Bordetella pertussis* toxin (PTX) which catalyzes the ADP-ribosylation of a cysteine located four amino acids from the carboxyl terminus of α-subunits belonging to the G_i family, G-proteins originally described to be involved in mediating adenylate cyclase inhibition.[19] This covalent modification disrupts G_α function, thus blocking the signaling pathway. PTX-sensitive G-proteins have been implicated in ET-1 signaling in some, but not all, VSMC types. For example, PTX significantly inhibits ET-1-induced phosphoinositide hydrolysis in cultured VSMCs[23] and mesangial cells,[24,25] as well as ET-1-induced contraction and $^{45}Ca^{2+}$ uptake in pig coronary artery.[26] By contrast, PTX-insensitive phosphoinositide turnover has been reported in cultured VSMCs[22] or A10 cells[27] and in pig coronary artery.[26] Pertussis toxin also failed to suppress the contraction of rat mesenteric arteries.[28]

The second toxin is *Vibrio cholera* toxin (CTX) which catalyzes the ADP ribosylation of specific arginine residues 201 and 174 in the α subunits G_s and G_t, G-proteins originally described to be involved in mediating adenylate cyclase and transducin stimulation, respectively.[19,29] This event inhibits the intrinsic GTPase activity of the α-subunit, thus locking it into the activated mode. ET-1 has been reported in rat VSMCs containing ET_A receptors to cause a dose-dependent increase in cAMP production which is PTX-insensitive and CTX-sensitive.[30] In contrast, ET_B receptor stimulation in endothelial cells leads to PTX-sensitive increase in cyclic AMP levels.[30] These studies confirm results using ET_A and ET_B receptor-transfected Chinese hamster ovary cells.[31] Therefore, the activation of adenylate cyclase in these preparations most likely involves G_s. Collectively, these studies suggest that in different types of VSMCs, ET receptors appear to be connected through multiple G-proteins. Experiments with ET receptors stably transfected in Chinese hamster ovary cells have shown that the second and third intracellular loops are important in the specificity of the interaction of the receptors with various G-proteins.[32] Further experiments will be needed to elucidate the exact nature of the G-proteins involved in ET receptor activation in human vascular and other tissues.

Cytosolic Calcium Signaling Mechanisms

Phospholipase C-Mediated Phosphoinositide Hydrolysis

Eight mammalian inositol phospholipid-specific phospholipase C isozymes have been identified and can be divided into three structural classes (PLCβ1-4, γ1,2, δ1,2) at the cDNA level.[33] When stimulated, PLC primarily hydrolyzes phosphatidylinositol 4,5-bisphosphate (PIP_2) resulting in the activation of two ubiquitous second messengers, inositol 1,4,5-triphosphate (IP_3) and sn1,2-diacylglycerol (DAG). IP_3 activates specific receptors on intracellular Ca^{2+} stores resulting in the release of Ca^{2+}, while DAG plays an integral role in protein kinase C (PKC) activation.[34] Although all three classes of PI-PLCs show dependence on Ca^{2+} in vitro, the Ca^{2+} requirement may not be physiologically important since agonists can increase IP_3 production in the presence of low levels of intracellular calcium in vivo. The PLC isoforms also differ in their regulation by heterotrimeric G-proteins and protein tyrosine kinases. Receptor-activated G-proteins have been shown to activate PLC-β1, PLC-β2, and PLC-β3 using the G_q family of PTX-insensitive G-proteins.[33] Interestingly, PLC-β can be activated by both α_q and $G_{\beta\gamma}$ subunits. The rank order of PLC activation by α_q is PLC-β1 $\geq$ PLC-β3 > PLC-β4 > PLC-β2 while the rank order of activation by $G_{\beta\gamma}$ is PLC-β3 $\geq$ PLC-β2 > PLC-β1.[33] $G_{\beta\gamma}$ does not activate PLC-β4. It has been suggested recently that PTX-sensitive PLC-β activation may result from $G_{\beta\gamma}$. PLC-γ is involved in the signaling cascade initiated by growth factor receptor tyrosine kinases (see ref. 33). In general, PLC-β-induced $[Ca^{2+}]_i$ increases generated by G-protein activation occur more rapidly and transiently as compared to PLC-γ-induced $[Ca^{2+}]_i$ increases generated by growth factor receptor-induced tyrosine phosphorylation.

Early experiments demonstrated that ET-1 caused a rapid rise in IP_3 levels in isolated canine[35] and porcine[36,37] coronary arteries which was paralleled by an increase in $^{45}Ca^{2+}$ uptake followed by vascular contraction. Similar studies of VSMCs in culture[27,38-41] and isolated vessels[42,43] have shown a similar rapid activation of PI turnover following ET-1 stimulation. This increase in PI hydrolysis has been shown to lead to increased IP_3 levels within 10 seconds of stimulation.[42] The rise in IP_3 induces Ca^{2+} release from intracellular storage sites (see below). The experiments with PTX discussed previously indicate that there are two different components that can activate PLC in response to ET. Thus, the PTX-insensitive pathway most likely involves the activation of PLC-β by a member of the G_q family. While PLC-β can be activated by the $G_{\beta\gamma}$ dimer, a much higher concentration is required than for activation by α_q. Receptor-mediated G_i activation, the high abundance PTX-sensitive G-protein family, would release $G_{\beta\gamma}$ subunits which in turn could activate PLC-β. This latter pathway may explain the PTX-sensitive pathways for ET-1-stimulated PIP_2 hydrolysis. It is most likely that the PLC-$\beta3$ form is responsible for both the PTX-sensitive and -insensitive ET-stimulated PIP_2 hydrolysis in rabbit vascular smooth muscle since PLC-$\beta1$ was not detected by Northern or Western blotting.[44]

Calcium Flux Pathways

Endothelin-1-induced contractions are more slowly developing, are maintained for longer periods of time, and are more resistant to agonist removal when compared with contractions in response to classical G-protein coupled agonists, such as angiotensin II or the α_1-receptor coupled agonist, phenylephrine. There is evidence that interactions among G-proteins, Ca^{2+}, and protein kinase C are important stimuli to maintain the tonic contraction phase in vascular smooth muscle.[45,46] It is well-documented that binding of ET to its receptor leads to a biphasic increase in $[Ca^{2+}]_i$, in which an initial rapid, transient phase is followed by a sustained elevated plateau phase. The rapid phase is due to Ca^{2+} release from the sarcoplasmic reticulum which is the major source and sink of Ca^{2+} in vascular smooth muscle, while the sustained phase has been shown to be dependent on Ca^{2+} influx from the extracellular space.[45]

Numerous studies utilizing both intact arteries and cultured VSMCs have shown that, in the absence of extracellular Ca^{2+} or in the presence of L-type Ca^{2+} channel blockers,[8,47,48] the transient rise in $[Ca^{2+}]_i$ and the contractile responses to ET-1 are shorter and return to baseline more rapidly than when Ca^{2+} is present in the external solution. In these cases, the sustained contractile phase is greatly attenuated or abolished. These results demonstrate that the initial ET-1-stimulated $[Ca^{2+}]_i$ rise and contraction is attributable to Ca^{2+} release from intracellular storage sites. In rat aortic smooth muscle cells, ET-1 and ET-3 both cause a transient increase in $[Ca^{2+}]_i$, but only ET-1 causes an increase in IP_3 production.[49]

Significant increases in IP_3 levels occur within 10 seconds following ET-1 stimulation and it is well-documented that IP_3 production is responsible for intracellular Ca^{2+} mobilization in VSMCs.[8,47,48] Intracellular Ca^{2+} release can occur through two distinct ion channels in VSMCs, the inositol 1,4,5-trisphosphate receptor channel and the ryanodine receptor channel, that are found in the sarcoplasmic reticulum.[50,51] The IP_3 receptor is a relatively nonselective cation channel consisting of four identical 310 kDa subunits, exhibits a biphasic sensitivity to Ca^{2+}, is desensitized by IP_3 and blocked by heparin, and is sensitive to protein kinase A phosphorylation.[50] Additionally, the ryanodine receptor, a homotetramer of four 560 kDa subunits, may provide an additional mechanism for Ca^{2+} release in smooth muscle cells. Kai et al[52] have documented that the initial ET-1-induced rise in $[Ca^{2+}]_i$ could also be attenuated or blocked completely by ryanodine,[53] or by caffeine pretreatment, a selective agonist for the ryanodine-sensitive Ca^{2+} release channel. The ryanodine receptor can be gated either by electromechanical coupling,[45,54] by Ca^{2+}-induced Ca^{2+} release,[45,54] or by cyclic ADP-ribose, a molecule which has been shown to be involved in regulating the ryanodine Ca^{2+} release channel in a variety of cells and tissues [see ref. 55 for a review]. Thus, these results imply that ET-induced intracellular Ca^{2+} mobilization in vascular smooth muscle is due to the activation of both IP_3 and ryanodine receptor-sensitive Ca^{2+} pools.

Extracellular Ca^{2+} Influx

Early experiments showed that endothelium-derived contracting factor-induced contractions in isolated blood vessels are dependent on extracellular Ca^{2+} and are attenuated by verapamil.[1] Numerous studies have since verified that the sustained contractile actions of ET-1 are dependent on extracellular Ca^{2+} and that ET-1 stimulates[45] Ca^{2+} entry into VSMCs.[8,47,48] The mechanism by which extracellular Ca^{2+} enters the cell has been studied extensively, but the data remain contradictory. L-type Ca^{2+} channel antagonists with different chemical structures were found to attenuate ET-1 induced vasoconstriction in vivo and in isolated arteries in many different species.[36,37,56-59] However, other studies have found little if any effect of the same Ca^{2+} channel antagonists on ET-1 stimulated contraction in similar preparations.[60-63] Similar contradictory results were seen upon examination of Ca^{2+} influx and the sustained increases in $[Ca^{2+}]_i$ in vascular smooth muscle. Ca^{2+} channel blockers reportedly caused significant inhibition of the sustained $[Ca^{2+}]_i$ rise or $^{45}Ca^{2+}$ uptake in isolated primary cultures of VSMCs.[52,59,64,65,78] However, L-type Ca^{2+} channel blockers had little or no effect on ET-1 induced $^{45}Ca^{2+}$ uptake or sustained $[Ca^{2+}]_i$ increases in other studies.[27,53,61,66,67] Huang et al[68] reported that dihydropyridine Ca^{2+} antagonists inhibited both the transient and sustained Ca^{2+} increase in cultured A7r5 smooth muscle cells. This effect was also observed in human umbilical artery SMCs when verapamil and nicardipine were used.[64] These observations imply that there is a component of the ET response which is dependent on extracellular Ca^{2+} influx. It has been postulated that the secondary influx of extracellular Ca^{2+} can stimulate additional Ca^{2+} release from the ryanodine-sensitive sarcoplasmic reticulum pool through a mechanism involving Ca^{2+}-induced Ca^{2+} release[75] or a diffusible second messenger, such as cADP-ribose.[55] For a more in-depth examination of the role of ion channels in the calcium influx pathway, see the chapter 9.

Phospholipase A_2 Activity

The role of phospholipase A_2 (PLA_2) in the ET signaling cascade remains a largely unexplored area. PLA_2 can liberate free fatty acids such as arachidonic acid (AA) from membrane phospholipids. Arachidonic acid can be metabolized by lipoxygenase, cyclooxygenase, and epoxygenase to form leukotrienes, hydoxy-eicosatetraenoic acids, prostaglandins (PGs), thromboxanes, and epoxides.[69-72] Endothelin-1 stimulation appears to activate the arachidonate cascade through a PKC-sensitive pathway in vascular smooth muscle.[73] Activation could be through the PLA_2 pathway which has been shown to be activated by ET-1 in mesangial cells.[74]

Human ET_A receptors transfected into Chinese hamster ovary cells activated PLA_2 through a PTX sensitive G-protein mechanism which is dependent on a calcium influx.[54] The generation of arachidonic acid metabolites may have multiple outcomes in VSMCs, such as modulation of ion channel activities[75] or activation of gene expression.[74]

Protein Kinase C Signaling Mechanisms

Endothelin stimulation also leads to a rapid, biphasic rise in diacylglycerol (DAG) levels which is sustained for up to 20 minutes.[27,76,77] Diacylglycerol levels can increase directly via PI hydrolysis or via phosphatidylcholine hydrolysis by PLC and/or phosphatidylcholine hydrolysis by phospholipase D (PLD) and phosphatidic acid phosphohydrolase. It has been postulated for other vasoconstrictors that the biphasic nature of the DAG signal results initially from PI hydrolysis, followed by a slower and more long-lasting DAG production from phosphatidylcholine. These sustained levels of DAG correspond temporally with sustained PLD activation[78,79] and play an important role in PKC activation. Although the mechanisms regulating PLD activity are unclear at the present time, there are reports that PLD can be activated via a PKC-dependent[79] and independent pathways,[80] as well as a protein tyrosine kinase pathway.[81] Phospholipase D is also postulated to play an important role in the mitogenic response observed following stimulation by ET,[82] however, this point is controversial (see ref. 83).

Although it is well-documented that ET-1 causes activation of protein kinase C (PKC), the specific role of PKC in ET-induced signaling response is complex. Much of this complexity arises from the diversity of the PKC isoforms. At least 11 different isozymes of PKC have been isolated from mammalian tissues.[84,85] These enzymes can be classified into three different groups: (1) "classical" PKCs (cPKC; isoforms α, β_I, β_{II}, γ) that are activated by Ca^{2+}, phosphatidylserine, and diacylglycerol or phorbol esters; (2) "new" PKCs (nPKC; isoforms δ, ϵ, η, θ) that differ from cPKCs only by not requiring Ca^{2+} for activation; and (3) "atypical" PKCs (aPKC; isoforms ξ, ι) which are dependent on phosphatidylserine for activation, but are not affected by Ca^{2+}, diacylglycerol or phorbol esters. The various members of the PKC family most likely differ in their sensitivity to diverse combinations of Ca^{2+}, phosphatidylserine, diacylglycerol, and other phospholipid breakdown products. Of the 11 PKC isoforms isolated to date, only the PKC isoforms α, β, ϵ, and ξ have been found in vascular smooth muscle. PKC-α was purified from A7r5 cells[86] and detected by immunoblot assay in swine carotid artery.[87] PKC-ϵ and PKC-ξ have been detected in aortic smooth muscle.[88-90] Endothelin-1 stimulates PKC translocation from the cytosol to the membrane in cultured bovine VSMCs.[91] Griendling et al[76] demonstrated using cultured rat aortic VSMCs that ET-1 stimulates the rapid and sustained phosphorylation of a PKC specific protein substrate.

Acute PKC activation using phorbol esters appears to enhance activation of L-type Ca^{2+} channels in A7r5 cells.[92,93] Inhibition of endogenous PKC by staurosporine and phloretin blocks both ET-1 and phorbol ester-induced contractions.[62,63] Furthermore, down regulation of PKC activity by preincubation with phorbol 12,13-dibutyrate (PDBu) for 24 hours prevented ET-1 induced contractions in porcine coronary arteries.[94] By contrast, Shimamoto and co-workers[95] using a more specific inhibitor of PKC action, calphostin C, found that it blocked phorbolesterinduced contraction of rat aortic rings, but had minimal effect (13% inhibition) on ET-1-induced contractions. This may be related to an inhibitory effect of PKC on L-type calcium channels in ET-1-stimulated VSMCs.[78]

Protein kinase C also has a prominent role in the regulation of PLC and PLD activities in many tissues.[34] It has been postulated that PKC may control phosphoinositide hydrolysis in vascular smooth muscle by acting as a switch that inhibits PLC-mediated PIP_2 hydrolysis and promotes PLD-mediated phosphatidylcholine hydrolysis. Protein kinase C activation blunts ET-1-induced PI hydrolysis in cultured cells.[96,97] Additionally, pretreatment with phorbol 12,13 myristate acetate causes a 33% decrease in phosphoinositide turnover, while pretreatment with staurosporine, a PKC inhibitor, caused a 75% increase in PI turnover in intact canine coronary artery.[98] These data suggest that PKC activation may negatively feedback to limit ET-1-induced PI hydrolysis. Ryu et al[99] have shown that PKC phosphorylates PLC-β, but does not affect PLC-β activity. These authors suggested that PLC-β phosphorylation by PKC may affect its ability to interact with its G-protein. As mentioned above, PKC is also involved in regulating PLD activity under certain conditions. In sheep pulmonary artery and the A10 VSMC line, Ro-318220 (a PKC inhibitor) blocked stimulation of PLD activity by ET-1.[80,81] Endothelin stimulation of arachidonic acid metabolism pathway was also blocked by staurosporine[79] indicating that PKC may be involved. These results suggests that PKC plays an important role in switching the production of DAG from a PLC pathway to a PLD pathway.

Several studies have documented that ET-1 activates the Na^+-H^+ exchanger in VSMCs.[8,47,48] Although this stimulation appears to be mediated in part via a PKC-dependent mechanism,[8,47,48] the physiological consequences of this increased Na^+-H^+ exchanger activity in VSMCs remain to be determined. Protein kinase C-induced increases in Na^+-H^+ exchanger activity in VSMCs would be expected to increase intracellular pH (pH_i),[46-48] as well as to increase intracellular Na^+ content,[100] two important events in regulating vascular contractility and $[Ca^{2+}]_i$. However, ET-1-induced increases in pH_i can only be shown using bicarbonate-free buffers in cultured VSMCs.[47,101] An increase in intracellular Na^+ content could lead to increases in $[Ca^{2+}]_i$ via an increase in Na^+-Ca^{2+} exchanger activity.[100] Endothelin-1-induced $[Ca^{2+}]_i$ transients are not affected by inhibiting Na^+-H^+ exchanger activity,[101] suggesting that Na^+-Ca^+ exchange is not involved in mediating the effects of ET-1 on $[Ca^{2+}]_i$. Hubel and Highsmith[101] have also recently shown that ET-1-stimulated Na^+-H^+ exchange is secondary to an increase in $[Ca^{2+}]_i$. Although external Na^+ removal inhibits ET-1-induced contraction by 50% in rat aorta,[102] this phenomenon is most likely explained if Na^+ influx causes membrane depolarization leading to a secondary opening of L-type Ca^{2+} channels and Ca^{2+} influx.

Calcium Sensitization of Contractile Proteins

It is well-documented that ET-1[59,103,104] and other agonist-induced[105-107] contraction of vascular smooth muscle is accompanied by an increase in the Ca^{2+} sensitivity of contractile proteins. When Ca^{2+} is removed or reduced to very low levels in the extracellular solution, ET-1 can stimulate a small, but well-maintained vascular contraction.[43,108-112] Application of low ET-1 doses causes a leftward shift in the KCl dose-response curve in isolated coronary arteries[58] and pulmonary arteries and veins.[108] Both PKC[59,113] and monomeric G-proteins (p21^{rhoA}, p21^{ras})[114-116] have been implicated in this phenomenon. Experiments using L-toxin permeabilized rabbit mesenteric arteries[113] or chemically-skinned SMCs from the resistance arteries of the rabbit mesentery[59] showed a significant increase in myofilament Ca^{2+} sensitivity in the presence of ET-1 and GTP suggesting a G-protein-dependent pathway. This increase in sensitivity was attenuated, but not abolished, by using several PKC inhibitors, including staurosporine and chelreythrine,[113] as well as PKC_{19-36}, a peptide fragment corresponding to the autoinhibitory domain of PKC.[59]

Application of either GTPγS or AlF_4^-, activators of heterotrimeric and monomeric G-proteins, to permeabilized vascular smooth muscle also increases Ca^{2+} sensitivity of contractile proteins.[115-117] Prolonged treatment with GTPγS abolishes Ca^{2+} sensitization induced by ET-1 in permeabilized rabbit portal vein.[115] However, prolonged incubation with phorbol esters does not inhibit GTPγS induced Ca^{2+} sensitization in this preparation, thus suggesting that PKC and G-protein-mediated Ca^{2+} sensitization may involve separate upstream signaling pathways. Thus, it has been postulated that monomeric, cytoplasmic G-proteins, such as $p21^{ras}$ and $p21^{rhoA}$ are the key regulators of myofilament Ca^{2+} sensitivity in vascular smooth muscle. Using β-escin skinned mesenteric microarteries, Satoh et al[116] demonstrated that GTP-activated $p21^{H-ras}$ increased myofilament Ca^{2+} sensitivity and that this process could be partially reversed using tyrphostin, a tyrosine kinase inhibitor. Using an alternative approach, Hirata et al[114] has shown that another member of the *ras* superfamily, $p21^{rhoA}$, can regulate myofilament Ca^{2+} sensitivity. In this study, exoenzymes from *Staphylococcus aureus* and *Clostridium botulinum* named EDIN and C3, respectively, that ADP-ribosylate and inactivate $p21^{rhoA}$, completely abolished GTPγS-enhanced Ca^{2+} sensitivity in skinned smooth muscle preparations. Moreover, Gong et al[115] went on to demonstrate that prolonged GTPγS exposure does not decrease total levels of certain heterotrimeric ($G_{αq/11}$, $G_{αi3}$, and $G_β$) and monomeric ($p21^{rhoA}$, $p21^{ras}$) G-proteins or of PKC subtypes. Interestingly, the amount of immunoprecipitable $p21^{rhoA}$, as well as the amount of $p21^{rhoA}$ that could be ADP ribosylated by botulinum C3 toxin, was decreased following GTPγS exposure in these studies. These authors interpret their data as evidence that prolonged GTPγS exposure induces either a conformational change in $p21^{rhoA}$ or a downregulation of an unidentified effector molecule.

Although the underlying mechanisms remain unclear, it is believed that the increased Ca^{2+} sensitivity of myofilaments may involve the phosphorylation and subsequent inhibition of (1) smooth muscle myosin phosphatase-1 activity, resulting in prolonged 20 kDa myosin light chain phosphorylation;[107] or (2) calponin activity, a thin filament-associated protein that inhibits actin-stimulated myosin ATPase, resulting in increased interaction of phosphorylated cross bridges with actin.[106] Kimuara et al[118] have shown that activated Rho interacts with Rho-associated kinase to phosphorylate and inactivate the myosin binding subunit of smooth muscle myosin phosphatase. Moreover, recent studies indicate that ET-1 can phosphorylate calponin via a PKC-dependent mechanism.[119] Thus, these results indicate that ET-1 may induce Ca^{2+} sensitization of contractile proteins in vascular smooth muscle through at least two different mechanisms, namely PKC and monomeric G-proteins.

Summary

Since the discovery of an endothelium-derived contracting factor more than ten years ago, there has been an explosive growth of knowledge concerning the ET system in the cardiovascular system. Depending on the vascular bed, we now know that smooth muscle cells can contain either ET_A receptors, ET_B receptors or a mixture of both subtypes and their activation results in vasoconstriction. ET_B receptor activation on vascular endothelium is linked to prostacyclin and/or nitric oxide release and vasodilatation. Although Ca^{2+} is an integral component of the ET signaling cascade leading to contraction and secretion, it is clear that ET's cellular actions are complex and still incompletely understood.

References

1. Hickey, KA, Rubanyi GM, Paul RJ et al. Characterization of a coronary vasoconstrictor produced by cultured endothelial cells. Am J Physiol 1985; 248:C550-C556.
2. Yanagisawa M, Kurihara H, Kimura S et al. A novel potent vasoconstrictor peptide produced by vascular endothelial cells. Nature London 1988; 332:411-415.
3. Block KD, Eddy RL, Shows TB et al. Structural organisation and chromosomal assignment of the gene encodong endothelin. J Biol Chem 1989; 264:10851-10857.
4. Hoehe MR, Ehrenreich H, Otterud B et al. The human endothelin-1 gene (EDN1) encoding a peptide with potent vasoactive properties maps distal to HLA on chromosome arm 6p in close linkage to D6S89. Cytogenet Cell Genet 1993; 62:631-635.
5. Block KD,Hong CC, Eddy RL et al. cDNA cloning and chromosomal assignment of the endothelin 2 gene: vasoactive intestinal contractor peptide is rat endothelin 2 Genomics 1991; 10:236-242.
6. Block KD, Eddy RL, Shows TB et al. cDNA cloning and chromosomal assignment of the gene encoding endothelin 3. J Biol Chem 1989; 264:18156-18161.
7. Sokolovsky M. Endothelins and sarafotoxins: receptor heterogenity. Int J Biochem 1994; 26:335-340.
8. Rubanyi GM, Polokoff MA. Endothelins: Molecular biology, biochemistry, pharamacology, physiology, and pathophysiology. Pharmacol Rev 1994; 46:325-414.
9. Arai HS, Nori I., Amori, H et al. Cloning and expression of a cDNA encoding an endothelin receptor. Nature 1990; 348:730-732.
10. Sakurai T, Yanagisawa M, Takuwa Y et al. Cloning of a cDNA encoding a non-isopeptide-selective subtype of the endothelin receptor. Nature 1990; 348, 732-735.
11. Karne S, Ayawickreme CK, Lerner MR. Cloning and characterization of an endothelin-3 specific receptor (ET_C) receptor from *Xenopus Laevis* dermal melanophores. J Biol Chem 1993; 268:19126-19133.
12. Bax WA, Saxena PR. The current endothelin receptor classification: time for reconsideration? TIPS 1994; 15:379-386.
13. Sokolovsky M. Endothelin receptor subtypes and their role in transmembrane signaling mechanisms. Pharmac Ther 1995; 68:435-471.
14. Warner TD, Allcock GH, Corder R et al. Use of the endothelin antagonists BQ-123 and PD 142893 to reveal three endothelin receptors mediating smooth muscle contraction and the release of EDRF. Br J Pharmacol 1993; 110:777-782.
15. Battistini B, O'Donnell LJ, Warner TD et al. Characterization of endothelin (ET) receptors in isolated gall bladder of guinea pig: evidence for an additional ET receptor subtype. Br J Pharmacol 1994; 112:1244-1250.
16. Teerlink JR, Breu V, Sprecher U et al. Potent vasoconstriction mediated by endothelin ETB receptors in canine coronary arteries. Circ Res 1994; 74:105-114.
17. Haleen SJ, Cheng XM. Cardiopulmonary indications for endothelin receptor antagonists: review of recent efficacy trials. Exp Opin Invest Drugs 1997; 6:475-487.
18. Neer EJ. Heteromeric G-proteins: organizers of transmembrane signals. Cell 1995; 80:249-257.
19. Hepler JR, Gilman AG. G-proteins. TIBS 1992; 17:383-387.

20. Arai HS, Nori I., Amori, H et al. Cloning and expression of a cDNA encoding an endothelin receptor. Nature 1990; 348:730-732.
21. Sakurai T, Yanagisawa M, Takuwa Y et al. Cloning of a cDNA encoding a non-isopeptide-selective subtype of the endothelin receptor. Nature 1990; 348, 732-735.
22. Takuwa Y, Kasuya Y, Kudo N et al. Endothelin receptor is coupled to phospholipase C via a pertussis toxin insensitive guanine nucleotide binding regulatory protein in vascular smooth muscle cells. J Clin Invest 1990; 85:653-658.
23. Reynolds EE, Mok LL, Kurokawa S. Phorbol ester dissociates endothelin-stimulated phosphoinositide hydrolysis and arachidonic acid release in vascular smooth muscle cells. Biochem Biophys Res Comm 1989; 160:868-873.
24. Simonson MS, Dunn MJ. Cellular signaling by peptides of the endothelin gene family. FASEB J 1990; 4:2989-3000.
25. Simonson MS, Dunn MJ. Endothelin 1 stimulates contraction of rat glomerular mesangial cells and potentiates beta adrenergic mediated cyclic adenosine monophosphate accumulation. J Clin Invest 1990; 85:790-797.
26. Kasuya Y, Takuwa Y, Yanagisawa M et al. A pertussis toxin sensitive mechanism of endothelin action in porcine coronary artery smooth muscle. Br J Pharmacol 1992; 107:456-462.
27. Muldoon LL, Rodland KD, Forsythe ML et al. Stimulation of phosphatidylinositol hydrolysis, diacylglcerol release and gene expression in response to endothelin, a potent new agonist for fibroblasts and smooth muscle cells. J Biol Chem 1989; 264:8529-8536.
28. Vigne P, Breittmayer JP, Marsault R et al. Endothelin mobilizes Ca^{2+} from a caffeine and ryanodine insensitive intracellular pool in rat atrial cells. J Biol Chem 1990; 265:6782-6787.
29. Nestler EJ, Duman RS. G-proteins and cyclic nucleotides in the nervous system. In: Siegel GJ et al., eds Basic Neurochemistry: Molecular, Cellular, and Medical Aspects, 5th Ed., Raven Press, Ltd, New York, 1994.
30. Eguchi S, Hirata Y, Imai T et al. Endothelin receptor subtypes are coupled to adenylate cyclase via different guanyl nucleotide-binding proteins in vasculature. Endocrinology 1993; 132:524-529.
31. Aramori I, Nakanishi S. Coupling of two endothelin receptor subtypes to differeing signal transduction in transfected Chinese hamster ovary cells. J Biol Chem 1992; 267:12468-12474.
32. Ninomiya H, Takagi Y, Miwa S et al. Distinct roles of second and third intracellular loops of human endothelin receptors in the selective activation of G alpha s/G alpha I. J Cardiovascular Pharmacology 1995; 16(Suppl 3):S254-S257.
33. Lee SB, Rhee SG. Significance of PIP2 hydrolysis and regulation of phospholipase C isozymes. Curr Opin Cell Biol 1995; 7:183-189.
34. Divecha N, Irvine RF. Phospholipid signaling. Cell 1995; 80:269-278.
35. Pang DC, John A, Patterson K et al. Endothelin-1 stimulates phosphatidyinositol hydrolysis and calcium uptake in isolated canine coronary arteries. J Cardiovasc Pharmacol 1989; 13(Suppl. 5):S75-S79.
36. Kasuya Y, Ishikawa T, Yanagisawa M et al. Mechanism of contraction to endothelin in isolated porcine coronary artery. Am J Physiol 1989; 257:H1828-H1835.
37. Kasuya Y, Takuwa Y, Yanagisawa M et al. Endothelin-1 induces vasocontriction through two functionally distinct pathways in porcine coronary artery:contribution of phosphoinositide turnover. Biochem Biophys Res Comm 1989; 161:1049-1055.
38. Van Renterghem CP, Vigne J, Barhanin A et al. Molecular mechanism of action of the vasoconstrictor peptide endothelin. Biochem Biophys Res Comm 1988; 157:977-985.

39. Resink TJ, Scott-Burden T, Buhler FR. Endothelin stimulates phospholipase C in cultured vascular smooth muscle cells. Biochem Biophys Res Comm 1988; 157:1360-1368.

40. Araki S, Kawahara Y, Kariya K et al. Stimulation of phospholipase C mediated hydrolysis of phosphoinositides by endothelin in cultured rabbit aortic smooth muscle cells. Biochem Biophys Res Commun 1989; 159:1072-1079.

41. Sugiura M, Inagami T, Hare GM et al. Endothelin action: Inhibition by a protein kinase C inhibitor and involvement of phosphoinositols. Biochem Biophys Res Comm 1989; 158:170-176.

42. Rapoport RM, Stauderman KA, Highsmith RF. Effects of EDCF and endothelin on phosphatidylinositol hydrolysis and contraction in rat aorta. Am J Physiol 1990; 258:C122-C131.

43. Ohlstein EH, Horohonich S, Hay DW. Cellular mechanisms of endothelin in rabbit aorta. J Pharmacol Exp Ther 1989; 250:548-555.

44. Homma Y, Sakamot H, Tsunda M et al. Evidence for involvement of phospholipase C-δ2 in signal transduction of platelet-derived growth factor in vascular smooth muscle cells. Biochem J 1993; 290:649-653.

45. Somlyo AP, Somlyo AV. Signal transduction and regulation in smooth muscle. Nature London 1994; 372:231-236.

46. Lee MW, Severson DL. Signal transduction in vascular smooth muscle: diacylglycerol second messengers and PKC action. Am J Physiol 1994; 267:C659-C678.

47. Brock TA, Danthuluri NR. Cellular actions of endothelin in vascular smooth muscle. In: Ruybanyi G, ed. Endothelin. Clinical Physiology Series, American Physiological Society, Oxford University Press. 1992:103-124.

48. Pollock DM, Keith TL, Highsmith RF. Endothelin receptors and calcium signaling. FASEB J 1995; 9:1196-1204.

49. Little PJ, Neylon CB, Tkachuk VA et al. Endothelin-1 and endothelin-3 stimulate calcium mobilization by different mechanisms in vascular smooth muscle. Biochem Biophys Res Commun 1992; 183:694-700.

50. Clapham DE Calcium signaling. Cell 1995; 80:259-268.

51. Marks AR. Calcium channels expressed in vascular smooth muscle. Circulation 1992; 86(Suppl 3):III-61-III-67.

52. Kai H, Kanaide H, Nakamura M. Endothelin-sensitive intracellular Ca^{2+} store overlaps with a caffeine-sensitive one in rat aortic smooth muscle cells in primary culture. Biochem Biophys Res Commun 1989; 158:235-243.

53. Wagner-Mann C, Bowman L, Sturek M. Primary action of endothelin on Ca release in bovine coronary artery smooth muscle cells. Am J Physiol 1991; 260:C763-C770.

54. Kruger H, Carr S, Brennand JC et al. Activation of phospholipase A2 by the human endothelin eceptor in chinese hamster ovary cells involves Gi protein-mediated calcium flux. Biochem Biophys Res Comm 1995; 217:52-58

55. Lee, HC. Modulator and messenger function of cyclic ADP-ribose in calcium signaling. Recent Progress in Hormone Signaling 1996; 51:355-389.

56. Goto K, Kasuya Y, Matsuki N et al. Endothelin activates the dihydropyridine-sensitive voltage-dependent Ca^{2+} channel in vascular smooth muscle. Proc Natl Acad Sci USA 1989; 86:3915-3918.

57. Takenaka T, Epstein M, Forster H et al. Attenuation of endothelin effects by a chloride channel inhibitor, indanyloxyacetic acid. Am J Physiol 1992; 262:F799-F806.

58. Inoue Y, Oike M, Nakao K et al. Endothelin augments unitary calcium channel currents on the smooth muscle cell membrane of guinea pig portal vein. J Physiol (London) 1990; 423:171-191.

59. Yoshida M, Suzuki A, Itoh T. Mechanisms of vasoconstriction induced by endothelin-1 in smooth muscle of rabbit mesenteric artery. J Physiol 1994; 477:253-265.

60. Blackburn K, Highsmith RF. Nickel inhibits endothelin induced contractions of vascular smooth muscle. Am J Physiol 1990; 258:C1025-C1030.

61. Charbrier PE, Auget M, Roubert P et al. Vascular mechanisms of action of endothelin-1: effect of Ca^{2+} antagonists. J Cardiovasc Pharmacol 1989; 13 (Suppl. 5):32-35.

62. Steffan M Russell JA. Signal transduction in endothelin-induced contraction of rabbit pulmonary vein. Pulm Pharmacol 1990; 3:1-7.

63. D'Orleans-Juste P, DeNucci G. Vane JR. Endothelin-1 contracts isolated vessels independently of dihydropyridine-sensitive Ca^{2+} channel activation. Eur J Pharmacol 1989; 165:289-295.

64. Gardner JP, Tokudome G, Tomonari H et al. Endothelin induced calcium responses in human vascular smooth muscle cells. Am J Physiol 1992; 262:C148-C155.

65. Xuan YT, Whorton AR, Watkins WD. Inhibition by nicardipine on endothelin-mediated inositol phosphate formation and Ca^{2+} mobilization in smooth muscle cells. Biochem Biophys Res Commun 1989; 160:758-764.

66. Mitsuhashi T, Morris, Jr. RC, Ives HE. Endothelin-induced increases in vascular smooth muscle Ca^{2+} do not depend in dihydropyridine-sensitive Ca^{2+} channels. J Clin Invest 1989; 84:635-639.

67. Simpson AW Ashely CC. Endothelin evoked Ca^{2+} transients and oscillations in A1 vascular smooth muscle cells. Biochem Biophys Res Commun 1989; 163: 1223-1229.

68. Huang S, Simonson MS, Dunn MJ. Mandidipine inhibits endothelin-1 induced $[Ca^{2+}]_i$ signaling but potentiates endothelin's effect on *c-fos* and *c-jun* induction in vascular smooth muscle and glomerular mesangial cells. Am Heart J 1993; 125:589-597.

69. Axelrod J, Burch RM, Jelsema CL. Receptor-mediated acivation of phospholipase A2 via GTP-binding proteins: arachidonic acid and its metabolites as second messengers. Trends Neurosci 1988; 11:117-123.

70. Kim D, Lewis DL, Graziadei L et al. G-protein βγ-subunits activate the cardiac muscarinic K^+-channel via phospholipase A2. Nature London 1989; 337:557-560.

71. Needleman P, Turk J, Jakshik BA et al. Arachidonic acid metabolism. Ann Rev Biochem 1986; 55:66-102.

72. Shimizu Y, Wolfe LS. Arachidonic acid cascade and signal transduction. J Neurochem 1990; 55:1-15.

73. Liu Y, Geisbuhler B, Jones AW. Activation of multiple mechanisms including phospholipase D by endothelin-1 in rat aorta. Am J Physiol 1992; 262:C941-C949.

74. Schramek H, Wang Y, Konieczkowski M et al. Endothelin-1 stimulates cytosolic phospholipase A2 activity and gene expression in rat glomerular mesangial cells. Kidney Int 1994; 46:1644-1652.

75. McDonald TF, Pelzer S, Trautwein W et al. Regulation and modulation of calcium channels in cardiac, skeletal and smooth muscle cells. Physiol Rev 1994; 74:365-507.

76. Griendling, K. K., T. Tsuda, and R. W. Alexander. Endothelin stimulates diacylglycerol accumulation and activates proein kinase C in cultured vascular smooth muscle cells. J Biol Chem 264:8237-8240, 1989.

77. Sunaka M, Kawahara Y, Kariya K, Tsuda T, Yokoyama M, Fukuzaki H, Takai Y. Mass analysis of 1,2 diacylglycerol in cultured rabbit vascular smooth muscle cells.

Comparison of stimulation by angiotensin II and endothelin. Hypertension 1990; 15:84-88.

78. Xuan YT, Wang OL, Wharton AR. Regulation of endothelin-induced Ca^{2+} mobilization in smooth muscle cells by protein kinase C. Am J Physiol 1994; 266:C1560-C1567.

79. Liu Y, Geisbuhler B, Jones AW. Activation of multiple mechanisms including phospholipase D by endothelin-1 in rat aorta. Am J Physiol 1992; 262:C941-C949.

80. Plevin R, Kellock NA, Wakelam MJ et al. Regulation by hypoxia of endothelin-1 stimulated phospholipase D activity in sheep pulmonary artery cultured smooth muscle cells. Br J Pharmacology 1994; 112:311-315.

81. Wilkes LC, Patel V, Purkiss JR et al. Endothelin-1 stimulated phospholipase D in A10 vascular smooth muscle derived cells is dependent on tyrosine kinase. FEBS Lett 1993; 322:147-150.

82. Boarder MR. A role for phospholipase D in control of mitogenesis. TIPS 1994; 15:57-62.

83. Paul A, Plevin R. Evidence against a role for phospholipase D in mitogenesis. TIPS 1994; 15:174-175.

84. Bell RM, Burns DJ. Lipid activation of protein kinase C. J Biol Chem 1991; 266:4661-4664.

85. Niskizuka Y. Protein kinase C and lipid signaling for sustained cellular responses. FASEB J 1995; 9:484-496.

86. Stauble B, Boscoboinik D, Azzi A. Purification and kinetic properties of protein kinase C from cultured smooth muscle cells. Biochem Mol Biol Int 1993; 20:203-211.

87. Singer HA, Oren JW, Benscoter H. Myosin light chain phosphorylation in ^{32}P-labelled rabbit aorta stimulated by phorbol-12,13-dibutyrate and phenylephrine. J Biol Chem 1989; 264:21215-21222.

88. Andrea JE, Walsh MP. Protein kinase C of smooth muscle. Hypertension 1992; 20:585-595.

89. Khalil RA, Lajoie C, Resnick MS et al. Ca^{2+}-independent isoforms of protein kinase C differentially translocated in smooth muscle. Am J Physiol 1992; 263:C714-C719.

90. Walsh MP, Andrea JE, Allen BG et al. Smooth muscle protein kinase C. Can J Physiol Pharmacol 1994; 72:1392-1399.

91. Lee TS, Chao T, Hu KQ et al. Endothelin stimulates a sustained 1,2 diacylglycerol increase and protein kinase C activition in bovine aortic smooth muscle cells. Biochem Biophys Res Commun 1989; 162:381-386.

92. Sperti G, Colucci WS. Phorbol ester-stimulated bidirectional transmembrane calcium flux in A7r5 vascular smooth muscle cells. Mol Pharmacol 1987; 32:37-42.

93. Fisher RD, Speriti G, Colucci WS et al. Phorbol ester increases the dihydropyridine-sensitive calcium conductance in a vascular smooth muscle cell line. Circ Res 1988; 62:1049-1054.

94. Marala RB, Mustafa SJ. Adenosine analogues prevent phorbol ester-induced PKC depletion in porcine coronary atery via A1 receptor. Am J Physiol 1995; 268:H271-H277.

95. Shimamoto H, Shimamoto Y, Kwan C-Y et al. Participation of protein kinase C in endothelin-1 induced contraction in rat aorta: studies with a new tool, calphostin C. Br J Pharmacol 1992; 107:282-287.

96. Resink TJ, Scott-Burden T, Weber E et al. Phorbol ester promotes a sustained down-regulation of endothelin receptors and cellular responses to endothelin in human vascular smooth muscle cells. Biochem Biophys Res Commun 1990; 166:1213-1219.

97. Reynolds EE, Mok LLS, Kurokawa S. Phorbol ester dissociates endothelin-stimulated phosphoinosittde hydrolysis and arachidonic acid release in vascular smooth muscle cells. Biochem Biophys Res Commun 1989; 160:868-873.

98. Calderone A, Rouleau de Champlain JL, Belichard P et al. Regulation of the endothelin-1 transmembrane signaling pathway: the potential role of agonist-induced desensitization in the coronary artery of the rapid ventricular pacing-overdrive dog model of heart failure. J Mol Cell Cardiol 1993; 25:895-903.

99. Ryu SH, Kim U, Wahl MI et al. Feedback regulation of phospholipase C-β by protein kinase C. J Biol Chem 1990; 265:17941-17945.

100. Mulvaney MJ, Aalkjaer C, Jensen PE. Sodium-calcium exchange in vascular smooth muscle. Ann NY Acad Sci 1991; 639:498-504.

101. Hubel CA, Highsmith RF. Endothelin-induced changes in intracellular pH and Ca^{2+} in coronary smooth muscle: role of Na^+-H^+ exchange. Biochem J 1995; 310:1013-1020.

102. Zagulova DV, Pinelis VG, Markov KM et al. The role of extracellular calcium in the vasocontriction evoked by endothelin-1. Biull Eksp Biol Med 1993; 116:258-260 (abstract).

103. Nishimura J, Moreland S, Ahn HY et al. Endothelin increases myofilament Ca^{2+} sensitivity in alpha toxin permeabilized rabbit mesenteric artery. Circ Res 1992; 951-959.

104. Kodama M, Yamamoto H, Kanaide H. Myosin phosphorylation and Ca^{2+} sensitization in porcine coronary arterial smooth muscle stimulated with endothelin-1. Eur J Pharmacol 1994; 288:69-77.

105. Morgan JP, Morgan, KG. Stimulus-specific patterns of intracellular calcium levels in smooth muscle of ferret portal vein. J Physiol (Lond) 1984; 51:155-167.

106. Rokolya A, Ahn HY, Moreland S et al. A hypothesis for the mechanism of receptor and G-protein-dependent enhancement of vascular smooth muscle myofilament Ca^{2+} sensitivity. Can J Physiol Pharmacol 1994; 72:1420-1426.

107. Somolyo AP, Somlyo AV. Signal transduction and regulation in smooth muscle. Nature 1994; 372:23-236.

108. Cardell LO, Uddman R, Edvinsson L. Analysis of endothelin-1 induced contractions of guinea pig trachea, pulmonary veins and different types of pulmonary arteries. Acta Physiol Scand 1990; 139:103-111.

109. Huang XN, Hisayama T, Takayanagi L. Endothelin-1 induced contraction of rat aorta: contributions made by Ca^{2+} influx and activation of contractile apparatus associated with no change in cytoplasmic Ca^{2+} level. Naunyn Schmiedibergs Arch Pharmacol 1990; 341:80-87.

110. Kodoma M, Kanaide H, Abe S et al. Endothelin induced Ca independent contraction of the porcine coronary artery. Biochem Biophys Res Commun 1989; 160:1302-1308.

111. Ozaki H, Sato K, Sakata K et al. Endothelin dissociates muscle tension from cytosolic Ca^{2+} in vascular smooth muscle of rat carotid artery. Jpn J Pharmacol 1989; 50:521-524.

112. Sakata K, Ozaki H, Kwon SC et al. Effects of endothelin on the mechanical activity and cytosolic calcium level of various types of smooth muscle. Br J Pharmacol 1989; 98:483-492.

113. Nishimura J, Moreland S, Ahn HY et al. Endothelin increases myofilament Ca^{2+} sensitivity in alpha toxin permeabilized rabbit mesenteric artery. Circ Res 1992; 71:951-959.

114. Hirata K, Kikuchi A, Saraki T et al. Involvement of rho p21 in the GTP-enhanced calcium ion sensitivity of smooth muscle contraction. J Biol Chem 1992; 267:8719-8722.

115. Gong MC, Fujihara H, Walker LA et al. Down regulation of G-protein-mediated Ca²⁺ sensitizatino in smooth muscle. Mol Biol Cell 1997; 8:279-286.

116. Satoh S, Rensland H Pfitzer G. Ras proteins increase Ca²⁺ responsiveness of smooth muscle contractin. FEBS letters 1993; 324:211-215.

117. Jensen PE, Gong MC, Somlyo AV et al. Separate upstream and convergent downstream pathways of G-protein and phorbol ester mediated Ca²⁺ sensitization of myosin light chain phosphorylation in smooth muscle. Biochem J 1996; 381:469-475.

118. Kimura K, Ito M, Amano M et al. Regulation of myosin phosphatase by rho and rho-associated kinase (rho-kinase). Nature 1996; 273:245-248.

119. Mino T, Yuasa U, Naka Met al. Phosphorylation of calponin mediated by protein kinase C in association with contraction in porcine coronary artery. Biochem Biophys Res Commun 1995; 208:397-404.

Endothelin Regulation of Cardiac Contractility: Signal Transduction Pathways

Meredith Bond

Introduction

When the discovery of endothelin (ET) was first reported in 1988, it was described as a potent vasoconstrictor.[1] Subsequent studies have further characterized the effects of the peptide on cardiovascular contractility. It has been shown both in the vasculature and in myocardial preparations that ET stimulation triggers a maintained contraction which is difficult to reverse.[2-5] Since the first reports of ET action on contractility, genes for three endothelins, ET_1, ET_2 and ET_3, and at least two classes of ET cell surface receptors, ET_A and ET_B, expressed in a tissue-specific manner, have been identified.[2,6] One or more of these related family of peptides is present in virtually all tissues and in all mammalian species.[6] Not only do ETs play important physiological roles in smooth and cardiac muscle contraction, but they are also implicated in kidney function, in neurotransmission and in the regulation of development. New functions of the peptide continue to be reported.

There is accumulating evidence for a role of ET in a large number of pathological conditions. With respect to the cardiovascular system, there are many reports that an increase in local production of ET, an increase in circulating ET levels and/or altered regulation of ET signaling pathways contributes to a variety of cardiovascular disease states, including myocardial ischemia and reperfusion injury, hypertension, congestive heart failure, cardiac hypertrophy, coronary and cerebral vasospasm, restenosis injury and atherosclerosis.[3,5,7] However, in many cases, the evidence still remains largely circumstantial. In some cases, the contribution of ET to cardiac disease states (e.g., hypertension) is highly controversial.[8-10] Nonetheless, in view of the potentially broad significance of ET in the regulation of cardiovascular function, extensive efforts continue to focus upon the elucidation of the diverse signal transduction pathways activated by ETs in the heart and vasculature. In the heart, in particular, the primary mechanism of ET-dependent activation of contraction still remains to be elucidated, although myofilament Ca^{2+} sensitization by one or more pathways, is likely to be necessary for the inotropic response.

Endothelin Receptors and Signaling Mechanisms, edited by David M. Pollock and Robert F. Highsmith. © 1998 Springer-Verlag and R.G. Landes Company.

The theme of this review will be the ET-dependent signaling pathways which mediate the short term actions of ET in the cardiovascular system, i.e., ET regulation of contractile function. In particular, because the primary mechanism of ET-dependent regulation of cardiac function is unresolved, this review will focus primarily on ET-dependent signaling pathways which mediate regulation of contraction in the heart. ET-dependent regulation of contraction in vascular smooth muscle and the associated signaling pathways activated are covered in detail in recent reviews.[2-3,5] As discussed below, ET is a powerful modulator of cardiac contractility, however the importance of ET as a regulator of heart function has not always been adequately recognized, perhaps because it has been overshadowed by the potent vasoconstrictor effects of the peptide and resultant secondary effects of vasoconstriction on heart function as a result of increased *circulating* ET.

Some important issues related to mechanisms of ET dependent inotropic effects in the heart will be addressed: (a) does an increase in cytoplasmic Ca^{2+}, $[Ca^{2+}]_c$, contribute significantly to the ET-dependent contractile response? (b) what is the role of ET in the triggered inotropic response of phospholipase Cβ (PLCβ) dependent pathways, i.e., hydrolysis of phosphatidylinositol bisphosphate (PIP$_2$) to form IP$_3$ and DAG, with subsequent activation of PKC? (c) what is the role of activation of the Na^+ H^+ antiporter in the ET dependent inotropic response in the heart? (d) what is the contribution to the ET-dependent inotropic effect from phosphorylation of myofibrillar proteins? (e) what is the importance of cross-talk with other signaling pathways, notably the β-adrenergic pathway? and finally, (f) what is the contribution of other signaling pathways in the short term actions of ET, e.g., phospholipase D (PLD), protein tyrosine kinase (PTK) and mitogen activated protein kinase (MAPK) signaling cascades?

Inotropic Effects of ET in the Atrium and Ventricle

In isolated ventricular muscle preparations and in isolated paced atria, ET causes a slow and sustained increase in force development.[11-15] In all cardiac preparations studied to date, following ET stimulation, there is an initial lag then a sustained increase in contractility which is difficult to reverse. In fact, in isolated cardiac muscle preparations, a contractile response is still present several minutes after the peptide has been washed out.[16] This effect is reminiscent of the effects of the peptide on vascular smooth muscle.[1] Similarly, in isolated cardiac myocytes, ET increases the intensity of the contraction (measured by increased amplitude of unloaded cell shortening).[17-21] Cardiac inotropic effects of ET are frequently observed at sub-nanomolar concentrations of ET$_1$, however higher peptide concentrations are required to observe an inotropic response to ET$_3$.[22] (A contractile response to ET$_3$ is reportedly not present or is very much reduced as compared with that of ET$_1$ in the ventricle—see section IV for further discussion of receptors for ET$_1$ and ET$_3$ on ventricular myocytes.) These findings indicate that ET$_1$ is one of the most potent modulators of cardiac muscle identified to date.[14,18,22-23] These observations parallel the very potent vasoconstrictor effects reported for ET.[1]

Effect of ET Perfusion on the Intact Heart

Consistent with the potent vasoconstrictor effects of ET, ET infusion in vivo causes an initial decrease, then a sustained increase in blood pressure, which is difficult to reverse.[24] This has been shown to be due primarily to increased peripheral resistance. Some investigators report that the maintained effect observed is a decrease in cardiac output, concomitant with the increase in blood pressure.[24] Decreased cardiac output would not be predicted from the positive inotropic and chronotropic effects of ET observed in isolated cardiac muscle preparations and isolated cardiac myocytes.[11-15,17-21] However, the decrease in cardiac output may be a consequence of the ET-induced increase in vasoconstriction discussed above.[24] It should be noted however that ET is released in close vicinity to the cardiac myocytes which it activates either from endocardial endothelium or vascular endothelium or from the cardiac myocytes themselves.[25-26] It is therefore quite likely that activation of ET receptors by locally released peptide is physiologically more relevant that the effects observed on cardiac function upon infusion of the peptide through the vasculature.

Further observations shed light on the complex effects of infused ET on the heart. ET can increase left ventricular pressure development in isolated perfused hearts.[27-28] When low concentrations of ET are perfused through the coronaries, a slow but sustained increase in cardiac output is observed; but when higher concentrations of ET are used, there is a more rapid increase in cardiac output. Cardiac output then falls as a result of ET-dependent vasoconstriction. This latter effect can be specifically blocked in the presence of Ca^{2+} channel antagonists.[28]

Comparison of Pathways Activated by ET and β-adrenergic Agonists in the Heart

It is useful to compare the signal transduction pathways responsible for the inotropic effect of β-adrenergic stimulation (which have been clearly delineated) with the ET-dependent inotropic response in the heart and the signaling pathways which mediate the short term (contractile) actions of the peptide. The inotropic actions of β-adrenergic agonists are attributed to PKA-phosphorylation of specific myofilament (troponin I, C-protein) and membrane (phospholamban, L-type Ca^{2+} channels, the SR Ca^{2+} release channel) proteins, with a resultant increase in Ca^{2+} induced Ca^{2+} release from the SR and in SR Ca^{2+} reuptake. Beta-adrenergic stimulation also causes an abbreviation of the contraction, primarily as a result of enhanced Ca^{2+} uptake into the SR, due to protein kinase A (PKA) dependent phosphorylation of phospholamban, the SR Ca^{2+} ATPase accessory protein. In addition, phosphorylation of the thin filament regulatory protein, troponin I (Tn-I) by PKA desensitizes the myofilaments to Ca^{2+}. Ca^{2+} desensitization would result in a decrease in developed force or pressure at the same free Ca^{2+} concentration. This would also enhance removal of Ca^{2+} from binding sites on troponin-C and accelerate Ca^{2+} reuptake into the SR. In contrast, a primary observation following ET stimulation is myofilament *desensitization* and a *prolongation* of the twitch contraction.

In distinct contrast to our knowledge of the intracellular pathways activated by β-adrenergic agonists, the primary pathways necessary for induction of an inotropic response to ET have not been unequivocally identified. However, the effects of ET on the amplitude and time course of the cardiac muscle contraction suggest that very different effector pathways are activated, as compared with the downstream signaling pathways activated in response to β-adrenergic stimulation. Thus, no abbreviation of the contraction following ET stimulation of the cardiac myocytes is observed.[11,12,14,29] However, in rabbit papillary muscle, a significant *prolongation* of the rate of relaxation was observed as a function of ET stimulation and in ferret papillary muscle, ET also prolonged the time-course of the contraction.[24,30] Thus it is clear that the signaling pathways activated by ET in cardiac muscle are distinct from the signaling pathways activated by β-adrenergic agonists. As will be subsequently discussed, a requirement for both an increase in $[Ca^{2+}]_c$ as well as a *necessary* role for phosphorylation of proteins in the cardiac muscle cell (by PKC and possibly other kinases) in the short term (contractile) actions of ET is still very much a subject of debate.

$[Ca^{2+}]_c$ and the Positive Inotropic Effect of ET?

In almost all tissues studied, ET stimulation elicits a significant increase in cytoplasmic Ca^{2+}, $[Ca^{2+}]_c$.[4,31] However in cardiac muscle preparations or in isolated myocytes, an increase in cytoplasmic Ca^{2+} has not been consistently observed to accompany the inotropic response to ET.[17-19] A *requirement* for SR Ca^{2+} release could also not be demonstrated.[30] Wang and colleagues observed that although an increase in aequorin luminescence (indicative of an increase in $[Ca^{2+}]_c$) accompanies an increase in developed tension in isolated papillary muscles, pretreatment with ryanodine (to eliminate SR Ca^{2+} release) reduced, but did not eliminate, the inotropic response to ET.[30] These results suggested a dissocation between the rise in $[Ca^{2+}]_c$ and the inotropic response.

Whereas some labs found no significant elevation in $[Ca^{2+}]_c$ with ET stimulation, others *have* reported a concomitant increase in cytosolic Ca^{2+} and increase in contraction both in the ventricle and in the atria.[17-18,25,32-33] The contractile response was also dependent upon a functional intracellular Ca^{2+} store.[33] In many systems, a maximal contractile response to ET is observed at concentrations at or below 1 nM, with a half-maximal inotropic response observed at around 50 pM.[17-18] This is comparable to the binding constant for ET to its receptors.[4] Some of the controversy associated with the question of a role for increased $[Ca^{2+}]_c$ as a mediator of the contractile response to ET could be due to the fact that different concentrations of the peptide were used in different studies. Thus, increases in cytoplasmic Ca^{2+} concentration have been observed where concentrations of ET of 10 nM or higher were used, although an elevation on $[Ca^{2+}]_c$ has also been reported when lower (sub-nanomolar concentrations) were used.[21,25,32] An explanation for some of the divergence in results between different studies—in terms of a role for an increase in $[Ca^{2+}]_c$—may be related to different preparations or the different animal species used.

In this context, an important issue to be resolved is the *local* (as opposed to systemic) concentration of ET in the vicinity of the ET receptor on the cardiac myocyte cell membrane, i.e., the transient local fluctuations in ET concentration, which result from regulated release of the peptide from endocardial endothelial cells or from the cardiac myocytes themselves.[1,26] To date, most measurements of ET concentration have been limited to the concentration of the peptide in the plasma and, with the possible exception of certain pathological conditions, such as heart failure, cardiogenic shock, and myocardial ischemia, circulating ET levels will almost certainly undergo much smaller fluctuations than local concentrations of the peptide near its site of secretion.[7,34-36]

PLCβ Dependent PI Turnover in the Positive Inotropic Response to ET

Messenger RNA for both ET_A (with affinities $ET_1 \cong ET_2 >> ET_3$) and ET_B receptors ($ET_3 = ET_1 = ET_2$) has been reported in the heart.[37-41] Recent ligand binding studies show the presence of only ET_A receptors in rat ventricular myocytes although both ET_A and ET_B receptors are reported to be present in guinea pig atria.[42-44] Consistent with the observation of Hilal-Dandan et al in rat ventricular myocytes, the inotropic effect of ET_3 is very much reduced, as compared with the action of ET_1.[13,22]

The structural similarity between ET receptors and other serpentine receptors, (receptors with 7-membrane spanning domains) which activate downstream signaling pathways via G-proteins suggests a requirement for G-proteins in ET-dependent signaling pathways.[45] This idea is further supported by observations that in cell membrane preparations, ET-dependent PI hydrolysis is activated by addition of the nonhydrolyzable GTP analog, GTPγS.[46-47]

In the different cell types where ET stimulation results in activation of phosphatidylinositol turnover, both PTX-sensitive and PTX-insensitive pathways are reportedly activated.[4] Pertussis toxin insensitive activation of PLCβ occurs primarily via $G\alpha_q$ or $G\alpha_{11}$, whereas PLCβ can also be activated by the βγ subunits of G_i via a PTX-sensitive pathway.[48] Thus, multiple signaling pathways may be triggered under conditions where activation of $G\alpha_i$ stimulates one effector pathway (inhibition of adenylyl cyclase) and the βγ subunits of G_i stimulate a different pathway (activation of PLCβ).[49-50] With respect to the ET_A and ET_B receptors, activation of multiple signaling pathways has been confirmed. In COS cells, transfection of ET_A and ET_B receptors preferentially activated PLC when cells were cotransfected with the PTX-insensitive G-proteins, $G\alpha_q$ or $G\alpha_{11}$, whereas cotransfection of ET receptors and $G\alpha_i$ resulted in PTX-inhibitable cAMP formation.[51] As discussed in more detail in below, the $G\alpha_i$ activation by ET receptor stimulation may contribute to cross-talk between the ET and adenylyl cyclase/cAMP signaling pathways in the heart.

In a large number of different cell types, including vascular smooth muscle cells, glomerular mesangial cells, fibroblasts, atrial and ventricular cardiac myocytes papillary muscle preparations and isolated hearts, ET stimulation results in activation of phosphatidylinositol-(PI) specific PLC, (PLCβ), production of the second messengers IP_3 and DAG, and subsequent activation of PKC.[33,52-59] In some, but not all cells and tissues which respond to ET stimulation, including cardiac myocytes, ET-dependent activation of PLCβ and resultant PI turnover occurs by means of a PTX-resistant pathway, thus implicating a G protein of the G_q family in the signal transduction process.[60-61]

In the heart, activation of PLC-dependent PI hydrolysis by ET shares a number of similarities with activation of PI turnover by agonist stimulation of other 7-transmembrane G-protein coupled receptors, e.g., cardiac α_1 receptors, angiotensin II receptors and muscarinic receptors.[62] Activation of PLC-dependent PI turnover by ET stimulation in cardiac muscle preparations and in isolated myocytes has been reproducibly demonstrated by a number of different labs (see refs. 4 and 5 for review); however, a critical question is whether there is a *requirement* for this PLC-dependent pathway in the inotropic effect of the peptide. On the one hand, studies in rabbit ventricle show that ET-dependent activation of PI hydrolysis (as indicated by [³H]IP_1 accumulation) has a very similar concentration and time-dependence as the positive inotropic effect of the peptide.[57] Accumulation of IP_3 also precedes tension developments as required if activation of the PI pathway is implicated in the ET-dependent inotropic response. However, during prolonged stimulation, IP_3 levels return to baseline while the contractile response is still maintained. This suggests either dissociation between IP_3 production and force development and/or a more prominent role for other components of the PI signaling pathway, such as DAG production with subsequent activation of PKC. The latter is more likely because it still remains to be established whether IP$_3$ contributes to the triggering of SR Ca^{2+} release in cardiac muscle.[63-65]

A further complication in assessing a role for DAG and PKC as mediators of the inotropic response to ET is that phorbol esters, such as PMA, and cell permeant DAG analogs, (powerful activators of endogenous PKC), reportedly exert negative inotropic effects on cardiac muscle preparations and cardiac myocytes.[66-67] Positive inotropic responses to low (sub-nanomolar) concentrations of phorbol esters have been reported.[68] This apparent anomaly between the contractile response to PKC activators and the contractile response elicited by stimulation of ET receptors is addressed in a recent study by Walker and colleagues.[69] In this study, isolated myocytes were stimulated to contract by incubating the cells with caged (photolyzable) DAG analogs then activating the intracellular caged compounds by flash photolysis. Under these conditions, a positive inotropic response was observed.[69] These investigators proposed that the negative inotropic effects previously observed upon addition of phorbol esters or DAG analogs to the extracellular solution could be due to membrane effects or extracellular actions of these highly lipophilic agents. On the other hand, inhibitory effects on contraction resulting from PKC-dependent phosphorylation of myofibrillar proteins (see below) and negative inotropic effects resulting from phosphorylation of ion channels have been reported.[70-73]

In response to ET stimulation, PKC-dependent phosphorylation of a range of substrates in cardiac myocytes has been reported, e.g.,by some but not all investigators, the Na$^+$ H$^+$ antiporter (see below), contractile proteins (myosin light chain 2, troponin I and troponin T) and ion channels, including the transient outward K$^+$ channel, L-type Ca^{2+} channel and the PKA-regulated Cl$^-$ channel.[20,70-77] The action of ET on ion channels is covered in detail in the chapter by Highsmith. The potential role of PKC-dependent phosphorylation in the activation of the Na$^+$ H$^+$ antiporter and its role in the ET-dependent inotropic response is discussed in the following section (see also chapter 9) and the actions of PKC in the phosphorylation of myofibrillar proteins is considered later in this chapter.

ET-dependent Intracellular Alkalinization via Activation of the Na$^+$ H$^+$ Antiporter

In the absence of a significant rise in [Ca^{2+}]$_c$, one proposed explanation for the ET-dependent Ca^{2+} sensitization in cardiac muscle cells is ET-dependent activation of the Na$^+$ H$^+$ antiporter. Consistent with observations in other cell types, ET stimulation of cardiac myocytes is accompanied by intracellular alkalinization.[54,78] Since increased intracellular pH desensitizes the myofilaments to Ca^{2+} following ET stimulation, in the absence of a significant rise in [Ca^{2+}]$_c$, the resultant intracellular alkalinization could contribute to the inotropic response. Because the Na$^+$ H$^+$ antiporter inhibitor, amiloride, prevents the intracellular alkalinization, increased intracellular pH was attributed to Na$^+$ H$^+$ antiporter activation.[18] Interestingly though, it was reported that the increase in pH occurred more rapidly than the increase in the contractile response and that the EC$_{50}$ for the pH change was higher than the EC$_{50}$ of the inotropic response.[18] These findings suggests that activation of the Na$^+$ H$^+$ antiporter contributes to, but is not essential for, the ET-dependent inotropic response. This idea is supported by the fact that amiloride reduced, but did not eliminate, the inotropic response to ET.[18]

As discussed previously, the inotropic response to ET is inhibited by pretreatment of cells with pertussis toxin (PTX), thus implicating a G protein-dependent signal transduction pathway that involves either (a) coupling of the ET receptor to G_o or G_i, (most likely G_i since ET activation inhibits adenylyl cyclase activity; see section VII, below) and then subsequent activation of downstream effectors and/or (b) PLCβ activation by the βγ subunits of G_i.[60] One group of investigators reported that the alkalinization resulting from ET-dependent inhibition of the Na^+ H^+ antiporter is mediated by a PKC dependent pathway, i.e., that the activation of the antiporter could be inhibited by PKC antagonists and that this ET dependent, PKC-mediated, activation of the Na^+ H^+ antiporter was not significantly affected by PTX inhibition.[18] This suggested that the pathway most likely involves coupling to downstream effector systems via G_q or G_{11}. It should be noted that the ability of the Na^+ H^+ antiporter to be phosphorylated by PKC is debated, and that some investigators have not demonstrated a role for PKC-dependent activation of the antiporter by ET or by other agonists, such as by α-adrenergic agonists or in response to activation of purinergic receptors.[78-79] Some of the confusion in this field may be explained by the results of a recent study which showed that phosphorylation of the Na^+ H^+ antiporter could be demonstrated in platelets following activation of the cells by PMA, but that a similar level of activation of the Na^+ H^+ antiporter by growth factors known to directly activate the MAPK pathway occurred in the absence of phosphorylation of the antiporter.[80]

Taken together, these findings suggest that sensitization of the myofibrils to Ca^{2+} as a result of ET-dependent, PKC-mediated activation of the Na^+ H^+ antiporter could contribute to, but not fully explain, the inotropic response to ET.

ET-dependent PKC Phosphorylation of Myofibrillar Proteins

The principal PKC isoform activated in response to ET stimulation in the heart, at least in the context of PKC-dependent phosphorylation of the myofibrillar proteins, MLC_2 and TnI, appears to be a Ca^{2+} insensitive PKC isoform.[20] This was demonstrated by the fact that the extent of phosphorylation of MLC_2 and of TnI by ET was unaffected by Ca^{2+} depletion of the intracellular Ca^{2+} pools and by prolonged incubation of the cells in a Ca^{2+} free medium.[20] A primary role for the Ca^{2+} insensitive isoform, PKCε is further indicated by the fact that, in response to activation by phorbol esters or by α-adrenergic agonists, PKCε is selectively translocated from the soluble to the particulate fraction upon ET stimulation.[23,81-82] In studies from the laboratory of Mochly-Rosen, immunofluorescence staining showed increased $PKC_ε$ localization to the myofilaments, giving a periodic cross-striated pattern.[83] In studies from the same group, translocation of PKCε, was shown to be necessary for the regulation of contraction by $α_1$-adrenergic receptor stimulation in neonatal ventricular myocytes.[84] This was demonstrated by incubation of permeabilized cells with a short peptide derived from the V1-variable region of PKCε which had been shown by these investigators to inhibits translocation.

Phosphorylation of TnI and MLC_2 as a result of ET stimulation of ventricular myocytes can be prevented by pretreatment of the cells with the PKC-specific inhibitor, calphostin C, and is mimicked by incubation of the cells with PMA. Earlier studies by Kuo and collaborators using isolated myofibrillar preparations or permeabilized myocytes with PKC or PMA demonstrated an increase in phosphorylation of TnI, TnT and MLC_2.[75-76,85] Phosphorylation of the thin filament proteins, TnT and TnI, has been associated with inhibition of maximal actomyosin

ATPase activity but with no significant change in myofilament sensitivity to Ca^{2+}.[76] In contrast, phosphorylation of MLC_2 increases maximal actomyosin ATPase activity and also reportedly increases myofilament Ca^{2+} sensitivity.[75,86-87] Myosin light chain 2 can be phosphorylated by myosin light chain kinase (MLCK) and by PKC at the same sites; however, rapid turnover of $^{32}P_i$ on MLC_2 in cardiac myocytes appears to be attributable primarily to phosphorylation by PKC.[75,85] It is possible that MLC_2 phosphorylation plays a role in the Ca^{2+} sensitization of the myofilaments in response to ET stimulation.[18,79] There may also be a counteracting inhibitory effect on actomyosin ATPase as a result of PKC-dependent phosphorylation of TnI and possibly of TnT. This may contribute to the negative inotropic effect observed by a number of investigators by stimulation of cardiac myocytes with PKC activators such as phorbol esters. A more recent report shows that the pattern of TnI phosphorylation differs for different PKC isozymes and also that most PKC isozymes, including PKCe (but not PKCz) preferentially phosphorylate TnI rather than TnT.[88] This is consistent with observations in intact myocytes where increases only in TnI and MLC_2 phosphorylation occurred upon ET stimulation.[20]

Inhibitory Effects of ET Stimulation on the β-adrenergic Pathway in Cardiac Myocytes: Role of G_i Dependent Pathways

In the heart in vivo, regulation of cardiac contractility by ET occurs against a background of chronic basal stimulation by the sympathetic nervous system, resulting primarily in activation of the β-adrenergic pathway by norepinephrine and other catecholamines. Moreover, under various pathophysiological conditions, such as heart failure and in hypertension, there is a chronic elevation of sympathetic tone. It is, therefore, important to not only consider the ET dependent cardiac actions and signaling pathways in isolation, but to also assess the physiologically more relevant interaction between the ET-dependent signaling pathways and the β-adrenergic pathway. Pretreatment of cardiac muscle preparations with ET exerts a negative modulatory action on the subsequent positive inotropic effect of β-adrenergic agonists on cardiac muscle preparations.[89] Other evidence of an ET-dependent inhibitory effect on events triggered by activation of the β-adrenergic pathway includes a reduction of the increased heart rate resulting from β-adrenergic stimulation, an abbreviation of the action potential in atrial cells as well as inhibition of the PKA-activated ion channels in cardiac myocytes.[70-71,73]

There is now considerable evidence from biochemical studies that ET exerts this inhibitory effect on the β-adrenergic pathway via a decrease in cAMP production by adenylyl cyclase inhibition.[59,71,73,90] This effect on cAMP levels as well as the observed actions of ET on PKA-activated ion channels was shown to result from a pertussis-toxin (PTX) sensitive inhibition of adenylyl cyclase activity.[59] This effect is similar to the inhibitory actions of muscarinic receptor activation on ion channels, or activation of $α_2$-adrenergic receptors which are also mediated by the PTX-sensitive G-protein, G_i.[91]

ET-dependent Activation of PKC and Cross-talk with Other Signaling Pathways

In response to ET stimulation of cardiac myocytes, either direct activation, or activation by PKC, of other signaling pathways, is now well-documented, e.g., PKC phosphorylation and subsequent activation of PLD and PKC-dependent activation of PTK and MAPK signaling pathways.[23,92-94,97,99] The activity of other protein kinases, notably c-Raf and A-Raf, which are located upstream of MEK and MAPK in the MAPK signaling cascades, are also reported to be activated by ET stimulation of ventricular myocytes.[100] Activation of these pathways, through stimulation of growth factor/PTK receptors and by stimulation of G-protein coupled receptors (such as ET receptors), activation of PLCβ and PLCβ hydrolysis of PI has been clearly shown to mediate the effects of ET on transcription of immediate early genes and stimulation of pathways which contribute to development of cardiac hypertrophy.[23,94-95,98,101-102] However, it remains to be determined whether activation of PTK and MAPK signaling pathways, as a result of ET activation of PKC, also contribute to the short-term actions of the peptide, i.e., the inotropic effects. Some insight into this question could be gained once those membrane-associated, myofibrillar and/or cytoplasmic proteins phosphorylated in response to activation of PTK/ MAPK pathways are identified. This would then shed light on the possible role, if any, of these effector pathways in the short term actions of ET stimulation of cardiac myocytes. The role of ET as a mitogen, the ET dependent signal transduction pathways which participate in the activation of nuclear transcription factors and the regulation of gene transcription are covered in detail in chapter 12.

There is now widespread evidence in a number of different cells and tissues that activation of PLD, and PLD-dependent hydrolysis of phosphatidyl choline (PC) occurs in response to stimulation by PKC.[103-105] Stimulation of PLD-dependent PC hydrolysis produces choline and phosphatidic acid (PA). Phosphatidic acid can then be dephosphorylated to form DAG, by PA phosphohydrolase, although PA itself may exert specific downstream effects in the cell There is now strong evidence that, in addition to PLCβ dependent PI hydrolysis, in many cell types DAG production, resulting from PLD activity, is an important intracellular effector pathway and is activated in response to ET stimulation.[106-107] In other words, DAG production does not result from PLC-dependent PI breakdown alone. (In addition to PA formation by PLD-dependent PC hydrolysis, it should be noted that PA can also be produced by DAG kinase dependent-phosphorylation of DAG, the latter formed by PLC hydrolysis of PI. Thus the formation of PA, *per se*, is not necessarily an indicator of PLD activity. However, evidence for a role for PLD can also be demonstrated by agonist-dependent hydrolysis of a phosphatidyl alcohol).

PLD activity has been reported in the heart and activation of PLD activity was observed to occur in response to activation by phorbol esters.[93] In addition, in cardiac myocytes, ET stimulation has been reported to stimulate PA formation via a PLD-dependent pathway; in this same study, ET-dependent P-butanol hydrolysis was also reported.[92] Whether or not ET-dependent PLD activation in cardiac myocytes occurs as a result of PKC-activation of PLD in the heart is questionable, since the only PKC isozymes reported to activate PLD are PKCα and PKCβ.[103-104] However, PKCβ is reportedly absent from adult cardiac myocytes and only faint PKCα immunoreactivity has been detected in adult ventricular myocytes, though present in neonatal myocytes.[108] It is possible that other mechanisms for activation of PLD in cardiac myocytes may need to be invoked, such as PLD activation by PIP_2.[104,109] Finally, as is the case for ET-dependent activation of MAPK signaling pathways, the role of PA and DAG production in the short term actions of ET remains to be explored.

References

1. Yanagisawa M, Kurihara H, Kimura S et al. A novel potent vasoconstrictor peptide produced by vascular endothelial cells. Nature 1988; 332(31):411-415.
2. Masaki T, Yanagisawa M. Physiology and pharmacology of endothelins. Medicinal Res Rev 1992; 12(4):391-421.
3. Rubanyi GM, Polokoff MA. Endothelins: molecular biology, biochemistry, pharmacology, physiology, and pathophysiology. Pharmacol Rev 1994; 46(3):325-414.
4. Simonson MS. Endothelins: multifunctional renal peptides. Physiol Rev 1993; 73:375-411.
5. Grossman JD, Morgan JP. Cardiovascular effects of endothelin. News Physiol Sci 1997; 12:113-117.
6. Sokolovsky M. Endothelin receptor heterogeneity, G-proteins, and signaling via cAMP and cGMP cascades. Cell Mol Neurobiol 1995; 15(5):561-571.
7. Hasdai D, Kornowski R, Battler A. Endothelin and myocardial ischemia. Cardiovasc Drugs Ther 1994; 8:589-599.
8. Schiffrin EL. Endothelin in hypertension. Curr Opin Cardiol 1995; 10:485-494.
9. Lüscher TF, Seo B, Bühler FR. Potential role of endothelin in hypertension. Hypertension 1993; 21(6):752-757.
10. Vanhoutte PM. Is endothelin involved in the pathogenesis of hypertension? Hypertension 1993; 21(6):747-751.
11. Moravec CS, Reynolds EE, Stewart RW et al. Endothelin is a positive inotropic agent in human and rat heart in vitro. Biochem Biophys Res Commun 1989; 159(1):14-18.
12. Watanabe T, Kusumoto K, Kitayoshi et al. Positive inotropic and vasoconstrictive effects of endothelin-1 in in vivo and in vitro experiments: characteristics and the role of L-type calcium channels. J Cardiovasc Pharmacol 1989; 13(Suppl 5):S108-S111.
13. Endoh M, Norota I, Yang H et al. The positive inotropic effect and the hydrolysis of phosphoinositide induced by endothelin-3 in rabbit ventricular myocardium: inhibition by a selective antagonist of ET_A receptors, FR139317. J Pharmacol Exp Ther 1996; 277(1):61-70.
14. Ishikawa T, Yanagisawa M, Kimura S et al. Positive inotropic action of novel vasoconstrictor peptide endothelin on guinea pig atria. Am J Physiol 1988; 255:H970-H973.
15. Reid JJ, Lieu At, Rand MJ. Interactions between endothelin-1 and other chronotropic agents in rat isolated atria. Eur J Pharmacol 1991; 194:173-181.
16. Krämer BK, Nishida M, Kelly RA et al. Myocardial actions of a new class of cytokines. Circulation 1992; 85(1):350-356.
17. Kelly RA, Eid H, Krämer BK et al. Endothelin enhances the contractile responsiveness of adult rat ventricular myocytes to calcium by a pertussis toxin-sensitive pathway. J Clin Invest 1990; 86:1164-1171.
18. Krämer BK, Smith TW, Kelly RA. Endothelin and increased contractility in adult rat ventricular myocytes. Circ Res 1991; 68(1):269-279.
19. Kohmoto O, Ikenouchi H, Hirata Y et al. Variable effects of endothelin-1 on $[Ca^{2+}]i$ transients, pH_i, and contraction in ventricular myocytes. Am J Physiol 1993; 265:H793-H800.
20. Damron DS, Darvish A, Murphy L et al. Arachidonic acid-dependent phosphorylation of troponin I and myosin light chain 2 in cardiac myocytes. Circ Res 1995; 76(6):1011-1019.
21. Fujita S, Endoh M. Effects of endothelin-1 on $[Ca^{2+}]_i$-shortening trajectory and Ca^{2+} sensitivity in rabbit single ventricular cardiomyocytes loaded with indo-1/

AM: comparison with the effects of phenylephrine and angiotensin II. J Card Fail 1996; 2(4S)S45-S57.

22. Ishikawa T, Liming L, Shinmi O et al. Characteristics of binding of endothelin-1 and endothelin-3 to rat hearts. Developmental changes in mechanical responses and receptor subtypes. 1991; Circ Res 69(4):918-926.

23. Jiang T, Pak E, Zhang H. Endothelin-dependent actions in cultured AT-1 cardiac myocytes. The role of the ε isoform of protein kinase C. Circ Res 1996; 78(4)724-736.

24. Li K, Stewart DJ, Rouleau J. Myocardial contractile actions of endothelin-1 in rat and rabbit papillary muscles. Role of endocardial endothelium. Circ Res 1991; 69(2)301-312.

25. Mebazaa A, Mayoux E, Maeda K et al. Paracrine effects of endocardial endothelial cells on myocyte contraction mediated via endothelin. Am J Physiol 1993; 265:H1841-H1846.

26. Suzuki T, Kumazaki T, Mitsui Y. Endothelin-1 is produced and secreted by neonatal rat cardiac myocytes in vitro. Biochem Biophys Res Comm 1993; 191(3): 823-830.

27. Baydoun AR, Peers SH, Cirino G et al. Effects of endothelin-1 on the rat isolated heart. J Cardiovasc Pharmacol 1989; 13(Suppl 5):S193-S196.

28. Firth JD, Roberts AFC, Raine AEG. Effect of endothelin on the function of the isolated perfused working rat heart. Clin Sci 1990; 79:221-226.

29. Shah AM, Lewis MJ, Henderson AH. Inotropic effects of endothelin in ferret ventricular myocardium. Eur J Pharmacol 1989; 163:365-367.

30. Wang J, Paik G, Morgan JP. Endothelin 1 enhances myofilament Ca^{2+} responsiveness in aequorin-loaded ferret myocardium. Circ Res 1991; 69(3):582-589.

31. Highsmith RF, Blackburn K, Schmidt DJ. Endothelin and calcium dynamics in vascular smooth muscle. Ann Rev Physiol 1992; 54:257-277.

32. Lauer MR, Gunn MD, Clusin WT. Endothelin activates voltage-dependent Ca^{2+} current by a G protein-dependent mechanism in rabbit cardiac myocytes. J Physiol 1992; 448:729-747.

33. Vigne P, Lazdunski M, Frelin C. The inotropic effect of endothelin-1 on rat atria involves hydrolysis of phosphatidylinositol. FEBS Lett 1989; 249(2):143-146.

34. Stewart DJ, Cernacek P, Costello KB et al. Elevated endothelin 1 in heart failure and loss of normal response to postural change. Circulation 1992; 85:510-517.

35. Cavero PG, Miller WL, Heublein DM et al. Endothelin in experimental congestive heart failure in the anesthetized dog. Am J Physiol 1990; 259:F312-F317.

36. Lerman A, Hildebrand FL, Aarhus LL et al. Endothelin has biological actions at pathophysiological concentrations. Circulation 1991; 83:1808-1814.

37. Arai H, Hori S, Aramori I et al. Cloning and expression of a cDNA encoding an endothelin receptor. Nature 1990; 348:730-732.

38. Cyr C, Heubner K, Druck T et al. Cloning and chromosomal localization of a human endothelin E_A receptor. Biochem Biophys Res Comm 1991; 181:184-190.

39. Sakurai T, Yanagisawa M, Takuwa Y et al. Cloning of a cDNA encoding a non-isopeptide-selective subtype of the endothelin receptor. Nature 1990; 348:732-735.

40. Aramori I, Nakanishi S. Coupling of two endothelin receptor subtypes to differing signal transduction in transfected Chinese hamster ovary cells. J Biol Chem 1992; 267(18):12468-12474.

41. Molenaar P, O'Reilly G, Sharkey A et al. Characterization and localization of endothelin receptor subtypes in the human atrioventricular conducting system and myocardium. Circ Res 1993; 72(3):526-538.

42. Hilal-Dandan R, Merck DT, Lujan JP et al. Coupling of the type A endothelin receptor to multiple responses in adult rat cardiac myocytes. Mol Pharmacol 1994; 45(6):1183-1190.

43. Hilal-Dandan R, Ramirez MT, Villegas S et al. Endothelin ET_A receptor regulates signaling and ANF gene expression via multiple G protein-linked pathways. Am J Physiol 1997; 272(1):H130-H137.

44. Ono K, Eto K, Sakamoto A et al. Negative chronotropic effect of endothelin 1 mediated through ET_A receptor in guinea pig atria. Circ Res 1995; 76(2):284-292.

45. Dohlman HG, Caron MG, Lefkowitz RJ. A family of receptors coupled to guanine nucleotide regulatory proteins. Biochemistry 1987; 26:2657-2664.

46. Takuwa Y, Kasuya Y, Takuwa N et al. Endothelin receptor is coupled to phospholipase C via a pertussis toxin-insensitive guanine nucleotide regulatory binding regulatory protein in vascular smooth muscle cells. J Clin Invest 1990; 85:653-658.

47. Thomas CP, Kester M, Dunn MJ. A pertussis toxin-sensitive GTP-binding protein couples endothelin to phospholipase C in rat mesangial cells. Am J Physiol 1991; 260:F347-F352.

48. Berstein G, Blank JL, Smrcka AV et al. Reconstitution of agonist-stimulated phosphatidylinositol 4,5-bisphosphate hydrolysis using purified m1 muscarinic receptor, Gq/11, and phospholipase C-beta 1. J Biol Chem 1992; 267:8081-8088.

49. Camps M, Carozzi A, Schnabel P. Isozyme-selective stimulation of phospholipase C-β2 by G protein $\beta\gamma$-subunits. Nature 1992; 360:684-686.

50. Katz A, Wu D. Simon MI. Subunits $\beta\gamma$ of heterotrimeric G protein activate β2 isoform of phospholipase C. Nature 1992; 360:686-689.

51. Takigawa M, Sakurai T, Kasuya Y et al. Molecular identification of guanine-nucleotide-binding regulatory proteins which couple to endothelin receptors. Eur J Biochem 1995; 228:102-108.

52. Marsden PA, Danthuluri NR, Brenner BM et al. Endothelin action on vascular smooth muscle involves inositol trisphosphate and calcium mobilization. Biochem Biophys Res Commun 1989; 158:86-93.

53. Griendling KK, Tsuda T, Alexander RW. Endothelin stimulates diacylglycerol accumulation and activates protein kinase C in cultured vascular smooth muscle cells. J Biol Chem 1989; 264:8237-8240.

54. Simonson MS, Wann S, Mene P et al. Endothelin stimulates phospholipase C, Na^+-H^+ exchange, c-fos expression, and mitogenesis in rat mesangial cells. J Clin Invest 1989; 83:708-712.

55. Muldoon L, Rodland KD, Forsythe ML et al. Stimulation of phosphatidylinositol hydrolysis, diacylglycerol release, and gene expression in response to endothelin, a potent new agonist for fibroblasts and smooth muscle cells. J Biol Chem 1989; 264:8529-8536.

56. Galron R, Kloog Y, Bdolah A et al. Functional endothelin/sarafotoxin receptors in rat heart myocytes: structure-activity relationships and receptor subtypes. Biochem Biophys Res Comm 1989; 163:936-943.

57. Takanashi M, Endoh M. Concentration- and time-dependence of phosphoinositide hydrolysis induced by endothelin-1 in relation to the positive inotropic effect in the rabbit ventricular myocardium. J Pharmacol Exp Ther 1992; 262(3):1189-1194.

58. Prasad MR. Endothelin stimulates degradation of phospholipids in isolated rat hearts. Biochem Biophys Res Comm 1991; 174(2):952-957.

59. Hilal-Dandan R, Urasawa K, Brunton LL. Endothelin inhibits adenylate cyclase and stimulates phosphoinositide hydrolysis in adult cardiac myocytes. J Biol Chem 1992; 267(15):10620-10624.

60. Lee CH, Parks D, Wu D et al. Members of the G_q α subunit gene family activate phospholipase C β isozymes. J Biol Chem 1992; 267(23):16044-16047.

61. Neer EJ. Heterotrimeric G proteins: organizers of transmembrane signals. Cell 1995; 80:249-257.

62. Lamers JMJ, De Jonge HW, Panagia V et al. Receptor-mediated signalling pathways acting through hydrolysis of membrane phospholipids in cardiomyocytes. Cardioscience 1993; 4(3):121-131.

63. Kentish J, Barsotti R, Lea T et al. Calcium release from cardiac sarcoplasmic reticulum induced by photorelease of calcium or Ins(1,4,5)P$_3$. Am J Physiol 1990; 258:H610-H615.

64. Movsesian M, Thomas A, Selak M et al. Inositol triphosphate does not release Ca^{2+} from permeabilized cardiac myocytes and sarcoplasmic reticulum. FEBS Lett 1985; 185:328-332.

65. Nosek TM, Williams MF, Zeigler ST et al. Inositol triposphate enhances calcium release in skinned cardiac and skeletal muscle. Am J Physiol 1986; 250:C807-C811.

66. Yuan S, Sunahara FA, Sen AK. Tumor-promoting phorbol esters inhibit cardiac functions and induce redistribution of protein kinase C in perfused beating rat hearts. Circ Res 1987; 61:372-378.

67. Capogrossi MC, Kaku T, Filburn CR et al. Phorbol ester and dioctanoylglycerol stimulate membrane association of protein kinase C and have a negative inotropic effect mediated by changes in cytosolic Ca^{2+} in adult rat cardiac myocytes. Circ Res 1990; 666:1143-1155.

68. MacLeod KT, Harding SE. Effects of phorbol ester on contraction, intracellular pH and intracellular Ca^{2+} in isolated mammalian ventricular myocytes. J Physiol (Lond) 1991; 444:481-498.

69. Pi Y, Sreekumar R, Huang X et al. Positive inotropy mediated by diacylglycerol in rat ventricular myocytes. Circ Res 1997; 81(1):92-100.

70. Washizuka T, Horie M, Watanuki M et al. Endothelin-1 inhibits the slow component of cardiac delayed rectifier K^+ currents via a pertussis toxin-sensitive mechanism. Circ Res 1997; 81(2):211-218.

71. Ono K, Tsujimoto G, Sakamoto A et al. Endothelin-A receptor mediates cardiac inhibition by regulating calcium and potassium currents. Nature 1994; 370:301-304.

72. Kim D. Endothelin activation of an inwardly rectifying K^+ current in atrial cells. Circ Res 1991; 69(1):250-255.

73. James AF, Xie L, Fujitani Y et al. Inhibition of the cardiac protein kinase A-dependent chloride conductance by endothelin-1. Nature 1994; 370:297-300.

74. Habuchi Y, Tanaka H, Furukawa T et al. Endothelin enhances delayed potassium current via phospholipase C in guinea pig ventricular myocytes. Am J Physiol 1992; 262:H345-H354.

75. Venema RC, Raynor RL, Noland Jr TA et al. Role of protein kinase C in the phosphorylation of cardiac myosin light chain 2. Biochem J 1993; 294:401-406.

76. Noland Jr TA, Kuo JF. Protein kinase C phosphorylation of cardiac troponin I and troponin T inhibits Ca^{2+}-stimulated MgATPase activity in reconstituted actomyosin and isolated myofibrils, and decreases actin-myosin interactions. J Mol Cell Cardiol 1993; 25:53-65.

77. Damron DS, Van Wagoner DR, Moravec CS et al. Arachidonic acid and endothelin potentiate Ca^{2+} transients in rat cardiac myocytes via inhibition of distinct K^+ channels. J Biol Chem 1993; 268(36):27335-27344.

78. Meyer-Lehnert H, Wanning C, Predel H et al. Effects of endothelin on sodium transport mechanisms: potential role in cellular Ca^{2+} mobilization. Biochem Biophys Res Commun 2989; 163:458-465.

79. Pucéat M, Clément-Chomienne O, Terzic A et al. α_1-Adrenoceptor and purinoceptor agonists modulate Na-H antiport in single cardiac cells. Am J Physiol 1993; 264:H310-H319.

80. Aharonovitz O, Granot Y. Stimulation of mitogen-activated protein kinase and Na^+/H^+ exchanger in human platelets. Differential effect of phorbol ester and vasopressin. J Biol Chem 1996; 271(28):16494-16499.

81. Pucéat M, Hilal-Dandan R, Strulovici B et al. Differential regulation of protein kinase C isoforms in isolated neonatal and adult rat cardiomyocytes. J Biol Chem 1994; 269(24):16938-16944.

82. Bogoyevitch MA, Glennon PE, Anderson MB et al. Endothelin-1 and fibroblast growth factors stimulate the mitogen-activated protein kinase signaling cascade in cardiac myocytes. The potential role of the cascade in the integration of two signaling pathways leading to myocyte hypertrophy. J Biol Chem 1993; 1110-1119.

83. Disatnik M, Buraggi G, Mochly-Rosen D. Localization of protein kinase C isozymes in cardiac myocytes. Exp Cell Res 1994; 210:287-297.

84. Johnson JA, Gray MO, Chen C et al. A protein kinase C translocation inhibitor as an isozyme-selective antagonist of cardiac function. J Biol Chem 1996; 271(40):24962-24966.

85. Venema RC, Kuo JF. Protein kinase C-mediated phosphorylation of troponin I and C-protein in isolated myocardial cells is associated with inhibition of myofibrillar actomyosin MgATPase. J Biol Chem 1993; 268(4):2705-2711.

86. Morano I, Hofmann F, Zimmer M et al. The influence of P-light chain phosphorylation by myosin light chain kinase on the calcium sensitivity of chemically skinned heart fibres. FEBS Lett 1985; 189(2):221-224.

87. Sweeney HL, Stull JT. Phosphorylation of myosin in permeabilized mammalian cardiac and skeletal muscle cells. Am J Physiol 1986; 250:C657-C660.

88. Jideama NM, Noland Jr TA, Raynor RL et al. Phosphorylation specificities of protein kinase C isozymes for bovine cardiac troponin I and troponin T and sites within these proteins and regulation of myofilament properties. J Biol Chem 1996; 271(38):23277-23283.

89. Reid JJ, Wong-Dusting HK, Rand MJ. The effect of endothelin on noradrenergic transmission in rat and guinea-pig atria. Eur J Pharmol 1989; 168:93-96.

90. Vogelsang M, Broede-Sitz A, Schäfer E et al. Endothelin ET_A-receptors couple to inositol phosphate formation and inhibition of adenylate cyclase in human right atrium. J Cardiovasc Pharmacol 1994; 23:344-347.

91. Van Biesen T, Muttrell LM, Hawes BE et al. Mitogenic signaling via G protein-coupled receptors. Endocr Rev 1996; 17(6):698-714.

92. Ye H, Wolf RA, Kurz T et al. Phosphatidic acid increases in response to noradrenaline and endothelin-1 in adult rabbit ventricular myocytes. Cardiovasc Res 1994; 28:1828-1834.

93. Lindmar R, Löffelhotz K. Phospholipase D in heart: basal activity and stimulation by phorbol esters and aluminum fluoride. Arch Pharmacol 1992; 346607-613.

94. Sadoshima J, Qiu Z, Morgan JP et al. Angiotensin II and other hypertrophic stimuli mediated by G protein-coupled receptors activate tyrosine kinase, mitogen-activated protein kinase, and 90-kD S6 kinase in cardiac myocytes. The critical role of Ca^{2+}-dependent signaling. Circ Res 1995; 76(1):1-15.

95. Sugden PH, Bogoyevitch MA. Intracellular signalling through protein kinases in the heart. Cardiovasc Res 1995; 30:478-492.

96. Bogoyevitch MA, Glennon PE, Sugden PH. Endothelin-1, phorbol esters and phenylephrine stimulate MAP kinase activities in ventricular cardiomyocytes. FEBS Lett 1993; 317(3):271-275.

97. Lazou A, Bogoyevitch MA, Clerk A et al. Regulation of mitogen-activated protein kinase cascade in adult rat heart preparations in vitro. Circ Res 1994; 75(5):932-941.

98. Lembo G, Hunter JJ and Chien KR. Signaling pathways for cardiac growth and hypertrophy. Recent advances and prospects for growth factor therapy. Ann N Y Acad Sci 1995; 752:115-127.

99. Shubeita HE, McDonough PM, Harris AN et al. Endothelin induction of inositol phospholipid hydrolysis, sarcomere assembly, and cardiac gene expression in ventricular myocytes. J Biol Chem 1990; 265(33):20555-20562.

100. Bogoyevitch MA, Marshall CJ, Sugden PH. Hypertrophic angonists stimulate the activities of the protein kinase c-Raf and A-Raf in cultured ventricular myocytes. J Biol Chem 1995; 270(44):26303-26310.

101. Gardner AM, Vaillancourt RR, Johnson GL. Activation of mitogen-activated protein kinase/extracellular signal-regulated kinase kinase by G protein and tyrosine kinase oncoproteins. J Biol Chem 1993; 268:17896-17901.

102. Winitz S, Russell M, Qian N et al. Involvement of Ras and Raf in the Gi-coupled acetylcholine muscarinic m2 receptor activation of mitogen-activated protein (MAP) kinase kinase and MAP kinase. J Biol Chem 1993; 268:19196-19199.

103. Kiss Z. Regulation of phospholipase D by protein kinase C. Chem Phys Lipids 1996; 80:81-102.

104. Exton JH. New developments in phospholipase D. J Biol Chem 1997; 272(25):15579-15582.

105. Friedlaender MM, Jain D, Ahmed Z et al. Endothelin activation of phospholipase D: dual modulation by protein kinase C and Ca^{2+}. Am J Physiol 1993; 264:F845-F853.

106. Billah MM, Eckel S, Mullmann TJ et al. Phosphatidylcholine hydrolysis by phospholipase D determines phosphatidate and diglyceride levels in chemotactic peptide-stimulated human neutrophils. J Biol Chem 1989; 264:17069-17077.

107. Baldi E, Musial A, Kester M. Endothelin stimulates phosphatidylcholine hydrolysis through both PLC and PLD pathways in mesangial cells. Am J Physiol 1994; 266:F957-F965.

108. Rybin VO, Steinberg SF. Protein kinase C isoform expression and regulation in the developing rat heart. 1994; 74(2):299-309.

109. Brown HA, Gutowski S, Moomaw CR et al. ADP-ribosylation factor, a small GTP-dependent regulatory protein, stimulates phospholipase D activity. Cell 1993; 75:1137-1144.

Endothelin Signaling to the Nucleus: Regulation of Gene Expression and Phenotype

Michael S. Simonson

Introduction

Endothelins (ET) were discovered as endothelium-derived vasoconstrictor peptides, and much research has focused on elucidating transmembrane signals by which ET receptors control vasoconstriction. Shortly after the discovery of ET, it became clear that these peptides also stimulate mitogenesis. These results demonstrated that ET peptides regulate not only short-term events (i.e., vasoconstriction) but also long-term actions requiring coordinated and differential regulation of gene expression. The idea that ETs could regulate a cell's genetic program provided a new perspective with which to consider the physiological and pathophysiological actions of ETs.

ET peptides regulate the genetic program of target cells in surprisingly diverse biological systems. As discussed in detail below, ET peptides are potent mitogens for vascular and nonvascular cell types. Particular interest is now being paid to the role of ET-1 in fibroproliferative cardiovascular and renal diseases and in oncogenesis. ET also controls differentiation and development of neural crest-derived cells. Mutations in the ET_B receptor gene have been linked to Hirschsprung disease and related neurocristopathies in humans. But despite recent advances, relatively little is known about transmembrane signals linking ET receptors in the plasma membrane to control of gene expression in the nucleus, i.e., nuclear signaling.

The objective of this chapter is to analyze recent advances in our understanding of nuclear signaling by ET peptides. I first discuss biological systems where ETs regulate gene expression and phenotype in target cells. Second, I review recent experiments implicating nonreceptor protein tyrosine kinases (PTK) as proximal effectors in ET nuclear signaling cascades. Last, I discuss how these proximal PTK signals are propagated to the nucleus to regulate specific *cis*-elements in the *c-fos* immediate early gene, an important genomic target of activated ET receptors. This chapter is by necessity selective and is meant only to introduce the reader to several themes in ET nuclear signaling. For additional depth the reader is referred to several recent reviews.[1-5]

Endothelin Receptors and Signaling Mechanisms, edited by David M. Pollock and Robert F. Highsmith. © 1998 Springer-Verlag and R.G. Landes Company.

Phenotypic Control by ET

Cell Growth

The first evidence that ET regulates gene expression derived from the unexpected discovery that ET-1 is a mitogen for vascular smooth muscle and mesangial cells in culture.[6-8] When added to serum-starved, quiescent (Go) cells in culture, ET-1 increases [³H]thymidine uptake into genomic DNA by 3- to 6-fold. In most cells the increase in DNA synthesis precedes an increase in cell number (i.e., hyperplasia), but in some cells a hypertrophic response ensues (see reference 9 for review). ET-1 is typically a comitogen that requires low concentrations of insulin or serum for full mitogenic action; however, in some cells ET-1 acts independently to stimulate cell growth. ET-1 can also induce other mitogens (i.e., platelet-derived growth factor) that cause delayed entry into the cell cycle. Although first demonstrated in mesangial and vascular smooth muscle cells, it is now clear that the mitogenic effects of ET-1 are widespread and occur in fibroblasts, epithelial cells, endothelial cells, osteoclasts, and sub-populations of glial cells.[9] Depending on the cell type in question, ET_A or ET_B receptors evoke mitogenic signaling.

Initially identified as a mitogen for cultured cells, accumulating evidence suggests that ET-1 is an important mitogen in vivo. Elevated secretion of ET-1 is a hallmark of fibroproliferative cardiovascular, pulmonary and renal diseases such as atherosclerosis, glomerulosclerosis, coronary artery restenosis, transplantation-associated vascular sclerosis and heart failure.[1,4,10] A functional role in these disorders is demonstrated by the ability of ET receptor antagonists to block smooth muscle cell proliferation and neointima formation in restenosis and atherosclerosis; mesangial cell proliferation in glomerulonephritis; and myocardial hypertrophy and fibrosis in cardiac failure.[10-16] A unique transgenic rat model has been established in which renal ET-2 expression is greatly elevated in glomeruli.[17] In these rats glomeruli develop extensive glomerulosclerosis with mesangial matrix expansion and apparent increases in mesangial cell proliferation. Transgenic mice that overexpress ET-1 develop pronounced glomerulosclerosis and interstitial fibrosis.[18] The experiments summarized above have been performed so far only in animal models, but every indication suggests that ET-1 plays a similar role in humans.

In contrast to attention paid to growth stimulation by ET-1, relatively little research has focused on feedback mechanisms that counterbalance ET-1-induced mitogenesis in vitro or in vivo. ET-1 activates atrial natriuretic peptide (ANP) release from cardiac myocytes, and these peptides potently antagonize ET-1-induced contraction.[19-21] ANP also inhibits ET-1-stimulated growth of vascular smooth muscle cells and glomerular mesangial cells in culture, but it is not yet clear how or whether release of ANP by cardiac myocytes would antagonize ET-1-stimulated vascular cell growth in peripheral vessels.[1,22] Nitric oxide (NO) is another candidate for a physiologically relevant counterbalance to ET-1-stimulated vascular cell growth. ET-1 potently activates NO release by endothelial cells, which in turn antagonizes ET-1-induced vasoconstriction in vivo and cell growth in culture.[1] cGMP is most likely the common mediator of the antimitogenic actions of both ANP and NO. However, given the short half-life of NO, it is uncertain whether NO would effectively regulate ET-1-induced mitogenesis in vivo.

Retinoic acid is another candidate inhibitor of ET-1-induced cell growth in vivo. In cultured glomerular mesangial cells, retinoic acid completely inhibits [³H]thymidine uptake in response to ET-1, and in cardiac myocytes retinoic acid inhibits hypertrophy and changes in gene expression associated with ET-1 expression and cardiac failure.[23-25] It is not yet clear how retinoic acid blocks ET-1-stimulated cell growth, but retinoic acid does block activation of the AP-1 transcription factor by ET-1.[23] AP-1 is activated by most growth factors, and AP-1 *cis*-elements are found in several genes associated with cell cycle progression.

Collectively, the experiments summarized above suggest that ET-1 plays an important role in adaptive proliferative responses in vascular injury. Investigation of the signals by which ET-1 evokes the G_0 to G_1 cell cycle transition has become the major model to study how ET-1 regulates gene expression. Several signaling cascades — including phospholipases C, D, and A_2, protein kinase C, mitogen-activated protein kinases (MAPK), nonreceptor protein tyrosine kinases (PTK), and S6 kinases — have been implicated in mitogenic signaling by ET-1 and are discussed in detail later in this chapter. ET-1 serves as an important model for mitogenic signaling by other growth factors that bind to the general class of G protein-coupled receptors.[26]

Development and Differentiation

Gene targeting experiments have firmly established that ET peptides regulate gene expression in development and differentiation of several tissues, particularly in cells derived from the neural crest. Targeted disruption of the ET-1 gene in mice causes massive malformation of pharyngeal arch-derived craniofacial tissues and organs.[27] ET-1$^{-/-}$ mice suffer anoxia and die at birth from respiratory failure. The phenotypic hallmarks are craniofacial abnormalities in the mandible, zygomatic and temporal bones, tympanic rings, hyoid, thyroid cartilage, tongue, and in soft tissues of the neck, palate, and ears.[27] All of these cells derive from the pharyngeal arches, which in turn derive primarily from neural crest ectomesenchymal cells. Thus the ET-1 and ET_A genes direct differentiation of neural crest-derived cells that form craniofacial developmental fields. ET-1 knockout mice also display abnormal development of the thyroid, thymus, heart and great vessels.[28] The molecular mechanisms and genes responsible for the developmental abnormalities in ET-1 knockout mice are unclear.

Gene targeting studies of the ET-3 and ET_B genes in mice reveal a requirement for normal development and migration of neural crest cells to the enteric nervous system (myenteric ganglion neurons) and skin (i.e., epidermal melanocytes).[29,30] The lack of myenteric ganglion neurons and epidermal melanocytes results in aganglionic megacolon and coat color spotting, respectively. These mutations do not complement in crossbreeding studies, suggesting that both ET-3 and ET_B are required.[29] These findings in mice apparently have important implications in humans. Mutations in human ET-3 or ET_B genes predispose afflicted individuals to the genetic disorder of Hirschsprung disease and similar neurocristopathies.[31,32] The major phenotypes in Hirschsprung disease is a lack of enteric neurons in the colon leading to megacolon and a lack of melanocytes leading to abnormal pigmentation. ET-3 is a potent mitogen for early neural crest cell precursors, the majority of which give rise to melanocytes.[33] In patients with Hirschsprung disease, a

G to T missense mutation in *EDNRB* exon 4 substitutes the highly conserved Trp-276 residue in the fifth transmembrane helix of the receptor with a Cys residue that renders the receptor less sensitive to ligand activation.[31,32]

Preliminary data also suggest that ET-1 contributes to the maturation and stabilization of blood vessels. Development of the vasculature depends critically on paracrine interactions between endothelial cells and underlying cells such as vascular smooth muscle, pericytes, and fibroblasts.[34] In immature vessels, endothelial cells secrete signaling molecules that recruit neighboring cells to proliferate and form surrounding cellular and matrix structures within the vessel. In cell culture systems ET-1 is at least one of the peptides responsible for endothelial cell-dependent proliferation of vascular pericytes.[35,36] Endothelial cells enhance pericyte survival and growth in a coculture system, which is blocked by neutralizing antibodies against ET-1. Exogenous ET-1 can in part replace the endothelial cell layer by activating ET$_A$ receptors in underlying pericytes.[36] ET-1 has also been implicated in endothelium-directed formation of myofibroblasts in wound healing.[37,38] In fibroblasts, ET-1 induces genes for smooth muscle α-actin and collagen type I, two essential steps in the transformation of vascular fibroblasts to myofibroblasts that contract a newly synthesized matrix during wound repair.[38] ET-1 secreted by endothelial cells promotes collagen gel contraction by fibroblasts, an important step in wound healing.[37]

One pathophysiological correlate of the ability of ET-1 to regulate blood vessel formation appears to be development of cerebral arteriovenous malformations (AVM) in humans.[39] In this disorder, which occurs in 0.5 % of humans, a tangle of blood vessels form with abnormal vascular cell phenotypes.[40,41] In particular, these vessels typically lack vascular smooth muscle cells and pericytes. We recently demonstrated that the preproET-1 gene is specifically repressed in 17/17 patients with cerebrovascular AVMs.[39] In contrast, ET-1 expression was normal in vessels feeding and draining the AVMs as well as elsewhere in the body. It seems possible that local repression of the preproET-1 gene in AVMs contributes to the lack of vascular smooth muscle cells and pericytes in AVMs.[39] The lack of ET-1 secretion and concomitant vasoconstriction might also account for the inability of AVMs to autoregulate blood flow.[40]

Oncogenesis and Cancer

Given the mitogenic actions of ET-1, it is perhaps not surprising that ET-1 has also been implicated in oncogenesis and cancer. Anchorage-independent growth of cells in culture is a hallmark of the transformed phenotype. Early studies revealed that ET-1 potentiates anchorage-independent growth of Rat-1 fibroblasts and NRK 49F cells in the presence of epidermal growth factor.[42,43] The ability of ET-1 to increase anchorage-independent growth and mitogenesis requires active protein kinase C.

Several studies implicate ET-1 in the development of human tumors. Cell lines derived from human tumors—particularly tumors of the breast, pancreas, colon, and prostate—secrete abundant amounts of ET-1.[44-46] Circulating levels of ET-1 are elevated in patients with hepatocellular carcinoma and metastatic prostate cancer.[45,46] ET-1 apparently derives not only from the extensive tumor vasculature but also from transformed cells themselves.[45,46] At least two roles have been envisioned for ET-1 in the growth and survival of neoplasms. First, ET-1 is a mitogen for several cancer cell lines and might increase tumor growth in an autocrine fashion or stromal cell growth by a paracrine mechanism. Second, ET-1 might also participate in the osteoblastic response of bone to metastatic prostate and other cancers.[46]

Regulation of Other Genes by ETs

ET-1 also induces expression of numerous other genes by mechanisms that have not been characterized. In particular, ET-1 regulates (i.e., induces or represses) genes encoding ANF, renin, aldosterone, catecholamines, and pituitary peptides (see reference 1 for review). Two classes of genes induced by ET-1 are particularly important in fibroproliferative diseases: chemoattractants and extracellular matrix constituents. In monocytes, ET-1 induces genes for the potent chemoattractant interleukin-8 and monocyte chemotactic protein-1.[47] ET-1 also elevates secretion of soluble and insoluble fibronectin by vascular smooth muscle cells.[48] It is not yet clear how ET-1 induces these genes or whether induction of these genes occurs in vivo.

c-Src and Other PTKs in ET-1 Nuclear Signaling

We now shift to a discussion of nuclear signaling by ET-1. When our laboratory and others began investigating transduction of signals from ET-1 receptors to the nucleus, one question stood out: do ET-1 receptors activate PTKs, and if so how do these kinases propagate signals to the nucleus? This question was intriguing because at the time it was well-established that most receptors associated with phenotypic control (i.e., growth and/or differentiation) required PTK activity.[49] Receptors for platelet-derived growth factor and related growth factors have intrinsic PTK activity that phosphorylates specific tyrosine residues on downstream effectors required for mitogenic signaling (i.e., phosphatidyl inositol-3-kinase, ras GAP, phospholipase Cγ, and Raf-1). Some of these effectors are themselves PTKs that set in motion a complicated cascade of PTK signaling. Even cytokine receptors, which lack intrinsic PTK activity, were known to recruit nonreceptor PTKs such as JAK (Janus kinases) or c-Src to propagate signals to the nucleus. Although two early studies suggested that G protein-coupled receptors might somehow activate PTKs, is was generally unclear whether G protein-coupled receptors communicated with PTKs.[50,51]

To determine whether PTKs contribute to mitogenic signaling by ET-1 in vascular cells, we set out to answer three related questions. First, does ET-1 increase tyrosine phosphorylation of cellular proteins? Second, what PTKs are activated by ET-1 and by what mechanisms? Third, do specific PTKs transduce signals to the promoters of genes induced by ET-1? Although most of these experiments have been performed in cultured glomerular mesangial cells and vascular smooth muscle cells (important vascular targets for endothelium-derived ET-1), preliminary results suggest that activation of nonreceptor PTKs is a widespread mechanism for nuclear signaling by ET receptors in other cell types and by a variety of G protein-coupled receptors.

ET-1 elevates tyrosine phosphorylation of cellular proteins The first suggestion that ET-1 activates PTKs was the finding that ET-1 elevates tyrosine phosphorylation (P-Tyr) of cellular proteins. When added to quiescent cells, ET-1 increases P-Tyr of several proteins in whole cell lysates of mesangial cells, fibroblasts, and vascular smooth muscle cells.[52-57] The increase in P-Tyr is apparently biphasic: between 5-10 min after adding ET-1 two to three proteins demonstrate increased P-Tyr, whereas by 20 min six to eight additional proteins show elevated P-Tyr.[57] The P-Tyr proteins increased by ET-1 show a surprising degree of cell specificity, although proteins around pp60 kDa, pp90, pp125, and pp225 are commonly observed. In mesangial cells, the biphasic time course of ET-1-stimulated P-Tyr accumulation contrasts with the rapid and transient induction of P-Tyr by platelet-derived growth factor, suggesting that different mechanisms are involved.[57] It is important to note that ET-1 stimulates accumulation of P-Tyr proteins with molecular mass similar to P-Tyr proteins that accumulate after treatment with platelet-derived growth factor and epidermal growth factor, whose receptors possess intrinsic PTK activity.[52,57] These results imply that ET-1 receptors target many of the same P-Tyr downstream effectors utilized by receptor PTKs.

The next step was to determine the cellular localization of P-Tyr proteins in response to ET-1. Focal adhesions are intracellular, oligomeric complexes of proteins that form when integrin receptors link components of the extracellular matrix to cytoskeletal elements. Focal adhesions are a particularly rich source of P-Tyr proteins that regulate cell cycle, and P-Tyr of focal adhesion proteins is commonly observed following addition of growth factors that bind to receptor PTKs.[58] It was thus important to determine whether ET-1 elevates P-Tyr of proteins in focal adhesions. Immunocytochemical analysis of P-Tyr proteins in quiescent mesangial cells reveals a marked increase in focal adhesion P-Tyr proteins in cells treated with mitogenic concentrations of ET-1.[57] Consistent with activation of focal adhesion proteins by growth factors, ET-1 also increases actin stress fiber formation.[59] ET-1 stimulates P-Tyr of the focal adhesion-associated protein paxillin.[60] Paxillin is a 68 kDa protein that binds to the actin-capping protein vinculin in focal adhesions, and it is a major P-Tyr protein in cells transformed with *v-src* or subjected to integrin activation.[58] These results suggest that P-Tyr of paxillin might play an important role in cell cycle control by ET-1. Other focal adhesion-associated proteins that are P-Tyr in ET-1-treated cells have yet to be identified.

ET-1 also elevates P-Tyr in cytoplasmic proteins.[57] With the exception of c-Src and focal adhesion kinase (FAK) (see below), the specific cytosolic proteins that display elevated P-Tyr following ET-1 are unknown. This is an important gap in our current knowledge of mitogenic signaling by ET-1.

At this step it is important to note that elevation of P-Tyr can occur by increasing PTK activity or by reducing PTPase activity. It is also possible that growth factors increase both PTK and PTPase activity resulting in differential P-Tyr of specific substrates. To determine if ET-1-stimulated P-Tyr results from PTK activity, PTKs have been immunoprecipitated in ET-1-treated cells using anti-P-Tyr antibodies (i.e., almost all PTKs are themselves P-Tyr proteins). Using this technique, it was found that ET-1 does indeed increase PTK activity.[54,57,61] The time course of activation was biphasic and paralleled accumulation of P-Tyr proteins observed in concurrent experiments.[57] The important question of whether ET-1 also increases PTPase activity has not yet been addressed. However, as discussed in the next section, we have made some progress identifying specific PTKs activated by ET-1.

ET-1 Activates Nonreceptor PTKs Src and Focal Adhesion Kinase

To identify specific PTKs activated by ET-1 receptors, immunoprecipitation/in vitro kinase assays were used with antisera for different families of PTKs. Most PTKs can be grouped into one of three families: receptor PTKs in the plasma membrane (e.g., platelet-derived growth factor receptor), nonreceptor PTKs attached to the plasma membrane (e.g., c-Src, c-Yes), and diffusable PTKs (e.g., Janus kinases; focal adhesion kinase, FAK). Attention initially focused on nonreceptor PTKs attached to the plasma membrane (i.e., c-Src) because of their close proximity to ET-1 receptors.

The first evidence that ET-1 activates a specific nonreceptor PTK was the finding that ET-1 increases pp60 c-Src activity in quiescent mesangial cells.[53,62] Immunoprecipitation/in vitro kinase assays with anti-v-Src antibodies revealed rapid activation of c-Src autophosphorylation and PTK activity by ET-1. Mesangial cells also express the Src family member c-Yes, but c-Yes activity is unaffected by ET-1, suggesting that activation of Src-family kinases is tightly regulated (Simonson et al, unpublished and reference 53). Autophosphorylation of c-Src, which typically accompanies c-Src activation, was rapid and transient whereas Src PTK activity was sustained.[53] The dose-response curve for ET-1 stimulated c-Src activity was identical to that for ET-1-stimulated mitogenesis. Depending on the cell type, both ET_A and ET_B (Simonson and Herman, unpublished results) receptors stimulate c-Src activity.[53] These results were among the first to demonstrate cross-talk between a G protein-coupled receptor and a nonreceptor PTK.

Relatively little detailed information is known about signaling mechanisms that link ET-1 receptors to c-Src. Increased c-Src activity in ET-1-stimulated cells requires Ca^{2+} influx but is apparently independent of protein kinase C. Activation of protein kinase C by phorbol ester increases P-Tyr accumulation, suggesting that protein kinase C-dependent pathways can activate PTKs. However, inhibition or depletion of protein kinase C has no effect on P-Tyr accumulation in ET-1-treated mesangial cells.[52,53] In contrast, the Ca^{2+} ionophore, ionomycin, mimics ET-1-stimulated PTK activity, which is also inhibited by chelation of extracellular Ca^{2+} influx.[57] Similar dependence on Ca^{2+} influx is also observed for ET-1-stimulated c-Src activation in mesangial cells.[57] Ca^{2+}-dependent activation of c-Src in mesangial cells is reminiscent of Ca^{2+}-activated c-Src in a genetic program of differentiation in keratinocytes.[63] Although the available evidence points to Ca^{2+}-dependent activation of c-Src by ET-1 receptors, the mechanisms and Ca^{2+}-dependent effectors involved have been difficult to identify.

Another PTK activated by ET receptors is FAK, a p125 kDa cytosolic PTK that forms stable complexes with c-Src and is a major P-Tyr protein in *v-src* transformed fibroblasts.[55,64] It is unknown whether ET-1-activated c-Src is responsible for activation of FAK; in addition, unlike activation of c-Src by ET-1, activation of FAK requires protein kinase C.[64] The role of FAK in mitogenic signaling has not been characterized in detail, but recent experiments with a dominant negative FAK mutant suggest that FAK is required for anchorage-dependent growth of fibroblasts.[65] Given the potential importance of FAK in cell cycle control, FAK is an attractive target PTK activated by ET-1.

Src in Nuclear Signaling by ET-1

Simply demonstrating that ET-1 receptors activate PTKs is not evidence that the PTKs function in nuclear signaling by ET-1. Relatively selective PTK inhibitors (i.e., genistein, herbimycin A) block ET-1-induced *c-fos* mRNA induction and mitogenesis in mesangial cells.[53] These results are consistent with a role for PTKs in ET-1 nuclear signaling, but these inhibitors block a variety of PTKs and might also have nonspecific effects that are difficult to detect. Thus the results with PTK inhibitors need to corroborated by techniques that are more specific for particular families of PTKs.

To specifically inhibit c-Src activity, we used a dominant negative c-Src mutant (Src K-) in which the conserved lysine residue (Lys 295) in the ATP-binding kinase domain is mutated to methionine.[66,67] In Src K-, the domains that link c-Src to upstream and downstream effectors, the SH2 and SH3 domains, are wild type whereas the domain responsible for phosphorylating effectors, the kinase domain, is inactive. Thus SrcK- can interact with proteins that contribute to Src signaling but can not transduce signals by phosphorylation. Overexpression of SrcK- can in theory titrate the action of endogenous c-Src and function as a dominant negative mutant.[68] Indeed Courtneidge and co-workers have shown that SrcK- blocks gene expression and cell growth induced by platelet-derived growth factor, which convincingly establishes a role for c-Src in some pathways of nuclear and mitogenic signaling.[66,67]

We used SrcK- mutants in our experiments to ask whether c-Src contributes to induction of the *c-fos* immediate-early gene promoter by ET-1. As discussed above, we previously showed that PTK antagonists blocked induction of *c-fos* mRNA by ET-1, and thus the *c-fos* gene was chosen as a Src target gene in these experiments.[53] Mesangial cells were transfected with a *c-fos*-luciferase reporter gene and treated with mitogenic concentrations of ET-1. ET-1 stimulated a 3.4-fold increase in *c-fos* promoter activity.[69] Cotransfection with a plasmid expressing SrcK- blocked activation of the *c-fos* promoter by ET-1 whereas a plasmid expressing wild type SrcK+ had no effect. By using point mutants of the *c-fos* promoter we determined that ET-1 requires at least two *cis*-elements to activate the promoter: the serum response element and the Ca^{2+}/cAMP response element.[69] The ET-1-c-Src signaling pathway apparently targets the serum response element but not the Ca^{2+}/cAMP response element, suggesting a divergence of the pathway upstream of Src.[69]

To obtain independent evidence that c-Src propagates an ET-1 signal to the nucleus, we transfected mesangial cells with a plasmid expressing COOH-terminal Src kinase (Csk), which phosphorylates the COOH-terminal tyrosine in Src (i.e., Tyr 527) and repressed Src activity.[70] Gene targeting studies with $Csk^{-/-}$ mice recently confirmed that Csk negatively regulates c-Src in vivo.[71] We found that overexpression of Csk also blocked activation of the *c-fos* promoter by ET-1, consistent with a role for c-Src in ET-1 nuclear signaling.[69] One caveat in interpreting the results with SrcK- and Csk is that both proteins probably also inhibit closely related members of the Src family of PTKs such as Yes and Fyn. Thus it is difficult to formally rule out involvement of other Src-related kinases in ET-1 nuclear signaling. However, we have been unable to demonstrate activation by ET-1 of other Src family kinases expressed in mesangial cells (i.e., c-Yes). Taken together, the results with SrcK- and Csk support a role for c-Src in induction of the *c-fos* immediate early gene by ET-1. The results also suggest a wider role for nonreceptor PTKs in nuclear signaling by G protein-coupled receptors.

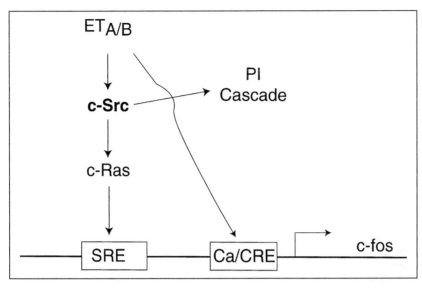

Fig. 12.1. Functions of c-Src PTK in nuclear signal transduction. ET$_A$ and ET$_B$ receptors activate c-Src by Ca^{2+}-dependent mechanisms. Src then transmits Ras-dependent signals to the *c-fos* serum response element (SRE) in the promoter. Src also amplifies the ET-induced PI cascade by mechanisms involving tyrosine phosphorylation of the G$_q$/G$_{11}$ α subunits that couple to ET receptors. Also note that c-Src, while necessary, is not sufficient for full activation of the *c-fos* promoter. Full activation requires a Src-independent pathway to the *c-fos* Ca^{2+}/cAMP response element (Ca/CRE). See text for more detail.

Src PTKs in ET-1 Signal Transduction

Although our understanding of nuclear signaling by ET-1 is still nascent, we are now in a position to suggest some working models of how Src and other PTKs function in ET-1 signaling (Fig. 12.1). First, ET-1-activated Src apparently sends signals to Ras, which in turn transmits signals to a variety of Ras-dependent effectors of gene expression such as the Raf-1/MAPK cascade. Second, accumulating evidence suggests that Src potentiates activation of the phosphoinositide (PI) cascade by ET-1, which by virtue of elevating intracellular free [Ca^{2+}]$_i$ and activating protein kinase C induces new patterns of gene expression.

In addition to activating c-Src, ET-1 also activates p21 Ras in mesangial cells, and expression of a dominant negative Ras mutant (Asn17 c-Ha-Ras) blocks activation of the *c-fos* promoter by ET-1.[72] c-Src acts upstream of c-Ras in the same ET-1 signaling pathway as dominant negative Ras blocks Src activation of the *c-fos* promoter but dominant negative Src mutants fail to block activation by Ras (Fig. 12.1).[69,72] Similar to systems in which Src functions upstream of Ras, ET-1-activated Src probably stimulates Ras by promoting P-Tyr and Grb2 association of the Shc adapter protein.[73] The ability of ET-1 to activate Ras undoubtedly accounts for the potent stimulation of p42,44 MAPK by ET-1.[74] Activation of a Src-Ras signaling cassette by ET-1 probably contributes to induction of genes in addition to *c-fos*.

Compelling evidence also points to a role for Src in the unusually extended activation of the PI cascade by ET-1. The PI cascade and its two major effectors, $[Ca^{2+}]_i$ and protein kinase C, are critical for ET-1 signal transduction. The ET-1-stimulated PI cascade is greatly amplified in fibroblasts transfected with v-Src.[75] The ability of Src to increase P-Tyr and activity of G_q/G_{11} G protein α subunits, which mediate ET-1 stimulation of phospholipase Cβ (see reference 1 for review), accounts for much of the potentiation of the PI cascade by Src and ET-1.[76] In fact, the ability of ET-1-Src to activate the same G protein α subunit that couples to ET-1 receptors provides a general mechanism for amplifying ET-1 signal transduction. Several other proteins relevant to nuclear signaling directly associate with c-Src and are tyrosine phosphorylated: phosphatidyl inositol-3-kinase, ras GAP, phospholipase Cγ, and Raf-1.[77,78] It remains to be determined if these effectors participate in ET-1-Src signaling pathways.

Conclusions

ET peptides regulate networks of gene expression that control surprisingly diverse phenotypic responses such as differentiation and cell growth. ET signaling controls gene expression in development and regulates differentiation of diverse cells and tissues, particularly cells derived from the neural crest. Induction of genes that control cell cycle by ET (i.e., mitogenic signaling) is important in compensatory remodeling of the vasculature in response to injury. Although relatively little is known about the signals by which ETs regulate expression of genes such as *c-fos*, the finding that ET-1 activates the c-Src nonreceptor PTK has shed some light on these signals. ET-1 activates c-Src by a Ca^{2+} influx-dependent pathway. ET-1-activated Src is proposed to have at least two functions in cell signaling (Fig. 12.1). First, Src activates a Ras-based pathway involving MAPK that stimulates the promoter of responsive genes such as *c-fos*. Second, Src phosphorylates and activates several effectors (i.e., G_q/G_{11} G protein α subunits) that potentiate activation of the PI cascade by ET-1.

Future studies to elucidate mechanisms of nuclear signaling by ETs might focus on the following questions. What PTKs in addition to c-Src and FAK are activated by ET-1 and how are they activated? What are the relevant P-Tyr target proteins in ET-1 signaling and what role, if any, do they play in induction of gene expression? In the case of *c-fos*, it is clear that Src activity alone is not sufficient for activation of the *c-fos* promoter by ET-1.[53,69] Not surprisingly, other signals such as protein kinase C and perhaps phosphatidyl inositol-3-kinase are required.[42,43,53,79] Future studies will no doubt address how these effectors interact with PTKs to regulate gene expression and will identify other signaling pathways that contribute to nuclear signaling by ETs.

Acknowledgments

The author thanks William Herman, Yuan Wang, Allison Rooney, Jennifer Jones, Patrick Rhoten, and Michael Dunn. Thanks also go to Aravinda Chakravarti, George Dubyak and Hsing-Jien Kung for many helpful suggestions. This work was supported by a grant from the National Institutes of Health, DK-46939.

References

1. Simonson MS. Endothelins: Multifunctional renal peptides. Physiol Rev 1993; 73:375-411.
2. Levin ER. Endothelins. New Engl J Med 1995; 333:356-363.
3. Yanagisawa M. The endothelin system: A new target for therapeutic intervention. Circulation 1994; 89:1320-1322.
4. Kohan DE. Endothelins in the normal and diseased kidney. J Kidney Dis 1997; 29:2-26.
5. Rubanyi GM, Polokoff MA. Endothelins: Molecular biology, biochemistry, pharmacology, physiology, and pathophysiology. Pharmacol Rev 1994; 46:325-415.
6. Simonson MS, Wann S, Mene' P et al. Endothelin stimulates phospholipase C, Na^+/H^+ exchange, c-fos expression, and mitogenesis in rat mesangial cells. J Clin Invest 1989; 83:708-712.
7. Komuro I, Kurihara H, Sugiyama T et al. Endothelin stimulates c-fos and c-myc expression and proliferation of vascular smooth muscle cells. FEBS Lett 1988; 238:249-252.
8. Badr KF, Murray JJ, Breyer MD et al. Mesangial cell, glomerular, and renal vascular responses to endothelin in the kidney. J Clin Invest 1989; 83:336-342.
9. Battistini B, Chailler P, D'Orleans-Juste P, Briere N, Sirois P. Growth regulatory properties of endothelins. Peptides 1993; 14:385-399.
10. Sakai S, Miyauchi T, Kobayashi M et al. Inhibition of myocardial endothelin pathway improves long-term survival in heart failure. Nature 1996; 384:353-355.
11. Benigni A, Zoja C, Corna D et al. A specific endothelin subtype A receptor antagonist protects against functional and structural injury in a rat model of renal disease progression. Kidney Int 1993; 44:440-444.
12. Douglas SA, Louden C, Vickery-Clark LM et al. A role for endogenous endothelin-1 in neointimal formation after rat carotid artery balloon angioplasty. Circ Res 1994; 75:190-197.
13. Lerman A, Edwards BS, Hallett JW et al. Circulating and tissue endothelin immunoreactivity in advanced atherosclerosis. New Engl J Med 1991; 325:997-1001.
14. Giad A, Yanagisawa M, Langleben D et al. Expression of endothelin-1 in the lungs of patients with pulmonary hypertension. New Engl J Med 1993; 328:1732-1739.
15. Ferrer P, Valentine M, Jenkins-West T et al. Orally active endothelin receptor antagonist BMS-182874 suppresses neointimal development in balloon-injured rat carotid arteries. J Cardiovasc Pharmacol 1995; 26:908-915.
16. Benigni A, Zoja C, Corna D et al. Blocking both type A and B endothelin receptors in the kidney attenuates renal injury and prolongs survival in rats with remnant kidney. Am J Kid Dis 1996; 27:416-423.
17. Hocher B, Liefeldt L, Thone-Reineke C et al. Characterization of the renal phenotype of transgenic rats expressing the human endothelin-2 gene. Hypertension 1996; 28:196-201.
18. Hocher B, Thone-Reineke C, Rohmeiss P et al. Endothelin-1 transgenic mice develop glomerulosclerosis, interstitial fibrosis, and renal cysts in an age and gender dependent manner. J Am Soc Nephrol 1996; 12:1633.
19. Fukada Y, Hirata Y, Yoshimi H et al. Endothelin is a potent secretagogue for atrial natriuretic peptide in cultured rat atrial myocytes. Biochem Biophys Res Comm 1988; 155:167-171.
20. Hu JR, Berninger UG, Lang RE. Endothelin stimulates atrial natriuretic peptide (ANP) release from rat atria. Eur J Pharmacol 1988; 158:177-180.
21. Sandok EK, Lerman A, Stingo AJ, Perrella MA, Gloviczki P, Burnett JC. Endothelin in a model of acute ischemic renal dysfunction: Modulating action of atrial natriuretic factor. J Am Soc Nephrol 1992; 3:196-202.

22. Neuser D, Knorr A, Stasch JP, Kazda S. Mitogenic activity of endothelin-1 and -3 on vascular smooth muscle cells is inhibited by atrial natriuretic peptides. Artery 1990; 17:311-324.

23. Simonson MS. Anti-AP-1 activity of all-trans retinoic acid in glomerular mesangial cells. Am J Physiol 1994; 267:F805-F815.

24. Zhou M, Sucov HM, Evans RM et al. Retinoid-dependent pathways suppress myocardial cell hypertrophy. Proc Natl Acad Sci 1995; 92:7391-7395.

25. Wu J, Garami M, Cheng T et al. 1,25 $(OH)_2$ Vitamin D_3 and retinoic acid antagonize endothelin-stimulated hypertrophy of neonatal rat cardiac myocytes. J Clin Invest 1996; 97:1577-1588.

26. Van Biesen T, Luttrell LM, Hawes BE et al. Mitogenic signaling via G protein-coupled receptors. Endocrine Rev 1996; 17:698-714.

27. Kurihara Y, Kurihara H, Suzuki H et al. Elevated blood pressure and craniofacial abnormalities in mice deficient in endothelin-1. Nature 1994; 368:703-710.

28. Kurihara Y, Kurihara H, Oda H et al. Aortic arch malformations and ventricular septal defect in mice deficient in endothelin-1. J Clin Invest 1995; 96:293-300.

29. Hosoda K, Hammer RE, Richardson JA et al. Targeted and natural (Piebald-Lethal) mutations of endothelin-B receptor gene produce megacolon associated with spotted coat color in mice. Cell 1994; 79:1267-1276.

30. Baynash AG, Hosoda K, Giaid A et al. Interaction of endothelin-3 with endothelin-B receptor is essential for development of epidermal melanocytes and enteric neurons. Cell 1994; 79:1277-1285.

31. Puffenberger EG, Hosoda K, Washington SS et al. A missense mutation of the endothelin-B receptor gene in multigenic Hirschsprung's disease. Cell 1994; 79:1257-1266.

32. Chakravarti A. Endothelin receptor-mediated signaling in Hirschsprung disease. Human Mol Genetics 1996; 5:303-307.

33. Lahav R, Ziller C, Dupin E et al. Endothelin 3 promotes neural crest cell proliferation and mediates a vast increase in melanocyte number in culture. Proc Natl Acad Sci 1996; 93:3892-3897.

34. Folkman J, D'Amore PA. Blood vessel formation: What is its molecular basis? Cell 1996; 87:1153-1155.

35. Chakravarthy U, Gardiner TA, Anderson P et al. The effect of endothelin 1 on the retinal microvascular pericyte. Microvasc Res 1992; 43:241-254.

36. Yamagishi S, Hsu C-C, Kobayashi K et al. Endothelin-1 mediates endothelial cell-dependent proliferation of vascular pericytes. Biochim Biophys Res Comm 1993; 191:840-846.

37. Guidry C, Hook M. Endothelins produced by endothelial cells promote collagen gel contraction by fibroblasts. J Cell Biol 1991; 115:873-880.

38. Villaschi S, Nicosia RF. Paracrine interactions between fibroblasts and endothelial cells in a serum-free coculture model. Lab Invest 1994; 71:291-299.

39. Rhoten RLP, Comair YG, Shedid D et al. Specific repression of the preproendothelin-1 gene in intracranial arteriovenous malformations. J Neurosurg 1997; 86:101-108.

40. Spetzler RF, Wilson CB, Weinstein P et al. Normal perfusion pressure breakthrough theory. Clin Neurosurg 1978; 25:651-672.

41. Spetzler RF, Martin NA. A proposed grading system for arteriovenous malformations. J Neurosurg 1986; 65:476-483.

42. Muldoon LL, Pribnow D, Roland KD et al. Endothelin-1 stimulates DNA synthesis and anchorage-independent growth of rat-1 fibroblasts throught of protein kinase C-dependent mechanism. Cell Reg 1990; 1:379-390.

43. Kusuhara M, Yamaguchi K, Kuranami M et al. Stimulation of anchorage-independent cell growth by endothelin in NRK 49F cells. Cancer Res 1992; 52:3011-3014.
44. Kusuhara M, Yamaguchi K, Nagasaki K et al. Production of endothelin in human cancer cell lines. Cancer Res 1990; 50:3257-3261.
45. Kar S, Yousem SA, Carr BI. Endothelin-1 expression by human hepatocellular carcinoma. Biochim Biophys Res Comm 1995; 216:514-519.
46. Nelson JB, Hedican SP, George DJ et al. Identification of endothelin-1 in the pathophysiology of metastatic adenocarcinoma of the prostate. Nature Med 1995; 1:944-949.
47. Helset E, Sildnes T, Konopski ZS. Endothelin-1 stimulates monocytes in vitro to release chemotactic activity identified as interleukin-8 and monocyte chemotactic protein-1. Mediators Inflamm 1994; 3:155-160.
48. Hahn AW, Regenass S, Kern F et al. Expression of soluble and insoluble fibronectin in rat aorta: Effects of angiotensin II and endothelin-1. Biochim Biophys Res Comm 1993; 192:189-197.
49. Cantley LC, Auger KR, Carpenter C et al. Oncogenes and signal transduction. Cell 1991; 64:281-302.
50. Huckle WR, Dy RC, Earp HS. Calcium-dependent increase in tyrosine kinase activity stimulated by angiotensin II. Proc Natl Acad Sci 1992; 89:8837-8841.
51. Nasmith PE, Mills GB, Grinstein S. Guanine nucleotides induce tyrosine phosphorylation and activation of the respiratory burst in neutrophils. Biochem J 1989; 257:893-897.
52. Force T, Kyriakis JM, Avruch J et al. Endothelin, vasopressin, and angiotensin II enhance tyrosine phosphorylation by protein kinase C-dependent and -independent pathways in glomerular mesangial cells. J Biol Chem 1991; 266:6650-6656.
53. Simonson MS, Herman WH. Protein kinase C and protein tyrosine kinase activity contribute to mitogenic signaling by endothelin-1: Cross-talk between G protein-coupled receptors and pp60^{c-src}. J Biol Chem 1993; 268:9347-9357.
54. Zachary I, Gil J, Lehmann W et al. Bombesin, vasopressin, and endothelin rapidly stimulate tyrosine phosphorylation in intact Swiss 3T3 cells. Proc Natl Acad Sci USA 1991; 88:4577-4581.
55. Zachary I, Sinnett-Smith J, Rozengurt E. Bombesin, vasopressin, and endothelin stimulation of tyrosine phosphorylation in Swiss 3T3 cells. J Biol Chem 1992; 267:19031-19034.
56. Weber H, Webb ML, Serafino R et al. Endothelin-1 and angiotensin-II stimulate delayed mitogenesis in cultured rat aortic smooth muscle cells: Evidence for common signaling mechanisms. Mol Endocrinol 1994; 8:148-158.
57. Simonson MS, Wang Y, Herman WH. Ca^{2+} channels mediate protein tyrosine kinase activation by endothelin-1. Am J Physiol 1996; 270:F790-F797.
58. Richardson A, Parsons JT. Signal transduction through integrins: a central role for focal adhesion kinase. BioEssays 1995; 17:229-236.
59. Simonson MS, Dunn MJ. Endothelin-1 stimulates contraction of rat glomerular mesangial cells and potentiates β-adrenergic-mediated cyclic adenosine monophosphate accumulation. J Clin Invest 1990; 85:790-797.
60. Zachary Is, Sinnett-Smith J, Turner CE et al. Bombesin, vasopressin, and endohthelin rapidly stimulate tyrosine phosphorylation of the focal adhesion-associated protein paxillin in Swiss 3T3 cells. J Biol Chem 1993; 268:22060-22065.
61. Zachary I, Sinnett-Smith J, Rozengurt E. Stimulation of tyrosine kinase activity in anti-phosphotyrosine immune complexes of Swiss 3T3 cell lysates occurs rapidly after addition of bombesin, vasopressin, and endothelin to intact cells. J Biol Chem 1991; 266:24126-24133.

62. Force T, Bonventre JV. Endothelin activates Src Tyrosine kinase in glomerular mesangial cells. J Am Soc Nephrol 1992; 3:491.

63. Zhao Y, Sudol M, Hanafusa H et al. Increased tyrosine kinase activity of c-Src during calcium-induced keratinocyte differentiation. Proc Natl Acad Sci 1992; 89:8298-8302.

64. Haneda M, Kikkawa R, Koya D et al. Endothelin-1 stimulates tyrosine phosphorylation of p125 focal adhesion kinase in mesangial cells. J Am Soc Nephrol 1995; 6:1504-1510.

65. Richardson A, Parsons JT. A mechanism for regulation of the adhesion-associated protein tyrosine kinase pp125FAK. Nature 1996; 380:538-540.

66. Twamley-Stein GM, Pepperkok R, Ansorge W et al. The Src family tyrosine kinases are required for platelet-derived growth factor-mediated signal transduction in NIH 3T3 cells. Proc Natl Acad Sci USA 1993; 90:7696-7700.

67. Roche S, Koegl M, Barone MV et al. DNA synthesis induced by some but not all growth factors requires Src family protein tyrosine kinases. Molec Cell Biol 1995; 15:1102-1109.

68. Herskowitz I. Functional inactivation of genes by dominant negative mutations. Nature 1987; 329:219-222.

69. Simonson MS, Wang Y, Herman WH. Nuclear signaling by endothelin-1 requires Src protein tyrosine kinases. J Biol Chem 1996; 271:77-82.

70. Sabe H, Knudsen B, Okada M et al. Molecular cloning and expression of chicken C-terminal Src kinase: lack of stable association with c-Src protein. Proc Natl Acad Sci U S A 1992; 89:2190-2194.

71. Imamoto A, Soriano P. Disruption of the csk gene, encoding a negative regulator of Src family tyrosine kinases, leads to neural tube defects and embryonic lethality in mice. Cell 1993; 73:1117-1124.

72. Herman WH, Simonson MS. Nuclear signaling by endothelin-1: A Ras pathway for activation of the c-fos serum response element. J Biol Chem 1995; 270:11654-11661.

73. Cazaubon SM, Ramos-Morales F, Fischer S et al. Endothelin induces tyrosine phosphorylation and GRB2 association of Shc in astrocytes. J Biol Chem 1994; 269:24805-24809.

74. Wang Y, Simonson MS, Pouyssegur J et al. Endothelin rapidly stimulates mitogen-activated protein kinase activity in rat mesangial cells. Biochem J 1992; 287:589-594.

75. Mattingly RR, Wasilenko WJ, Woodring PJ et al. Selective amplification of endothelin-stimulated inositol 1,4,5-trisphosphate and calcium signaling by v-src transformation of Rat-1 fibroblasts. J Biol Chem 1992; 267:7470-7477.

76. Liu W, Mattingly RR, Garrison JC. Transformation of Rat-1 fibroblasts with the v-src oncogene increases the tyrosine phosphorylation state and activity of the a subunit of Gq/G11. Proc Natl Acad Sci 1996; 93:8258-8263.

77. Brown MT, Cooper JA. Regulation, substrates, and functions of src. Biochim Biophys Acta 1996; 1287:121-149.

78. Courtneidge SA, Fumagalli S, Koegl M et al. The Src family of protein tyrosine kinases: regulation and functions. Development 1993; 57-64.

79. Sugawara F, Ninomiya H, Okamoto Y et al. Endothelin-1-induced mitogenic responses of Chinese hamster ovary cells expressing human endothelin A: The role of a wortmannin-sensitive signaling pathway. Mol Pharmacol 1996; 49:447-457.

CHAPTER 13

Mechanisms of Endothelin-Induced Mitogenesis and Activation of Stress Response Protein Kinases

Thomas L. Force

Introduction

The study of signal transduction mechanisms activated by the endothelin (ET) family of vasoactive peptides is in its infancy compared to the study of growth factor activated pathways. The most proximal mechanisms activated by ETs are well known- activation of phospholipase Cβ, and subsequently protein kinase C, and activation of plasma membrane Ca^{2+} channels. What is only starting to become clear is how these proximal signaling pathways, initially thought to be strikingly different from the proximal mechanisms activated by growth factors, share many similarities and both culminate in the activation of the same protein serine/threonine kinase cascade that has dominated the research on growth factor signaling over the past several years. This cascade, the c-Raf-1/ERK (for extracellular-signal regulated kinase, also known as MAP or mitogen-activated protein kinase) cascade) is critical to the mitogenic response to growth factors. To understand ET-induced mitogenesis it is critical to understand how G-protein-coupled receptors activate this kinase cascade since it appears that this cascade transduces signals from cell surface receptors to the nucleus, thus altering the transcription of genes which lead to the mitogenic response. In this chapter I will first explore the role of the ETs as growth factors, and then will focus on the c-Raf-1/ERK cascade, the mechanisms of its activation by ET, including putative roles of nonreceptor tyrosine kinases, and its role in the mitogenic response. I will also discuss possible roles for other pathways in the mitogenic response to ETs, including phospho-inositide-3 kinases. Finally, I will discuss the role of protein kinase cascades which are activated by cellular stresses, and which culminate in the activation of other members of the MAP kinase superfamily, the stress-activated protein kinases (SAPKs) and p38. These kinases do not appear to transduce mitogenic signals, but rather growth inhibitory or apoptotic signals. Much of the work that will be reviewed herein derives from studies in which endothelin-1 (ET-1) was used as agonist

Endothelin Receptors and Signaling Mechanisms, edited by David M. Pollock and Robert F. Highsmith. © 1998 Springer-Verlag and R.G. Landes Company.

in cells which signal primarily via the ET_A receptor. Although relatively little is known about mitogenic signaling and virtually nothing is known about stress signaling triggered by ETs acting via the ET_B receptor, I will briefly discuss potential roles of this receptor.

Endothelin as a Mitogen

ET-1 stimulates growth in a number of different cell lines including vascular and tracheal smooth muscle cells,[1-3] glomerular mesangial cells (the contractile cell of the renal glomerulus),[4-6] and a number of fibroblast cell lines.[7,8] Although ET-1 alone in some cells is weakly mitogenic,[2,8-10] the vast majority of experiments demonstrate that ET-1 is a comitogen, requiring the presence of low concentrations of serum (<0.5%), insulin (<5µg/ml), or other growth factors for maximal mitogenic effect (11, 12 and extensively reviewed in 13). When low concentrations of comitogens are present, the magnitude of the mitogenic effect of ET-1 is comparable to that of EGF and PDGF. ET-1 also produces a synergistic mitogenic effect in combination with submaximal concentrations of growth factors including EGF and PDGF, or with serum.[1,9,10,12]

The mitogenic effect of the endothelins in vascular smooth muscle cells and glomerular mesangial cells is mediated primarily via the ET_A receptor.[6,14,15] This conclusion is based on several lines of evidence: ET-3 is a much less potent mitogen than ET-1 (EC_{50} roughly two orders of magnitude greater for ET-3 vs. ET-1); the ET_B-selective agonists, sarafotoxin 6c and [Ala[1],Ala[3],Ala[11],Ala[15]]ET-1(6-21), do not stimulate DNA synthesis; and the ET_A-selective antagonist BQ123 completely inhibits DNA synthesis in response to ET-1.[6,15] This is not surprising since the ET_A receptor is the predominant one expressed in these cells in culture. Furthermore, the ET_A receptor appears to be the predominant one expressed in human arterial and venous vascular smooth muscle cells in situ.[16]

Activation of the ET_B receptor by endothelins can also trigger mitogenesis, however. In Chinese hamster ovary (CHO) cells stably expressing a transfected ET_B receptor, [^{3}H]thymidine uptake, and activation of the c-Raf-1 protein kinase cascade (see below), was activated by ET-1, ET-3, or sarafotoxin 6c.[17] Furthermore, in primary cultures of rat astrocytes, which predominantly express the ET_B receptor, enhanced DNA synthesis in response to ET-1 appears to be activated via this receptor subtype.[18] A mitogenic response to ET_B receptor occupation may be much more cell-type specific than the proliferative response induced by activation of the ET_A receptor since, for example, ET-1 binding to the ET_B receptor inhibits growth in human hepatic myofibroblastic Ito cells.[19] Importantly for this discussion, where examined, the mitogenic response to ET_B receptor activation appears to employ identical signaling mechanisms to those utilized by the ET_A receptor.

Reported half-maximal and maximal concentrations of ET-1 required to induce DNA synthesis in various cell lines have varied widely in the literature. Much of this variation is due to the use of different comitogens, or different concentrations of comitogens, but even where similar experimental protocols were used, differences remain. These differences are probably accounted for, in part, by differences in the number of receptors expressed by the cells. Not surprisingly, Kanse et al have reported a correlation between ET-1-induced mitogenesis and ET_A receptor number on human vascular smooth muscle cells.[16] The phenotypic state of the smooth muscle cell lines used (differentiated, contractile state vs. undifferentiated, proliferating state) also may lead to inconsistent results.[20] Simply put, we and others have observed a gradual reduction in responsiveness of glomerular mesangial cells and A10 vascular smooth muscle cells to ET-1, whether measured as stimulation of DNA synthesis, increases in cytosolic free $[Ca^{2+}]$, or enhanced tyrosine phosphorylation, as passage number increases. Although the reasons for this are not clear, we have found it to be a constant source of variability in experiments designed to examine the mitogenic response to ET-1.

Because of the above considerations, it is impossible to compare relative potencies of ET-1 with the other mitogenic vasoactive peptides (angiotensin II (AngII), and arginine vasopressin (AVP)) unless receptor number is known and cell type is taken into account. That said, however, qualitatively, mitogenic responses to the three peptides are similar.

Autocrine effects of ET-1 (and AngII) on mitogenesis must also be considered.[3] Following exposure of primary glomerular mesangial cells to ET-1, there is increased expression of PDGF-A and PDGF-B genes and increased secretion of PDGF dimers.[5] PDGF secretion was detected 12 hours after ET-1 and was maintained for 36 hours. Similar effects have been observed in vascular smooth muscle cells in response to AngII.[21,22] This "autocrine growth model" has been postulated to account for the delayed mitogenic response to ET-1 (and AngII) compared to PDGF (48 hours vs. 24 hours) observed in rat aortic smooth muscle cells.[3] In support of this model, AngII-induced mitogenesis in these cells is reversed by suramin, a nonspecific growth factor antagonist.[3] While this model may apply to some cell types, it is unlikely to account for all of the mitogenic effects of ET-1 since cycloheximide, an inhibitor of protein synthesis, fails to block ET-1-induced induction of c-*fos*, indicating synthesis of PDGF, or other growth factors, is not necessary for ET-1-induced induction of this immediate-early gene which correlates with the mitogenic response in many types of cells.[6,14]

Signal Transduction Pathways as Potential Modulators of the Mitogenic Response

Many signal transduction pathways have been proposed as mediators of the mitogenic response to ET-1 (reviewed in refs. 13, 24). When ET-1 binds to the ET_A receptor, phospholipase Cβ (PLCβ) is activated, resulting in increases in inositol trisphosphate (IP_3) and diacylglycerol. Although isolated reports have appeared suggesting that the vasoactive peptides may also activate the growth factor-activated PLCγ,[25] this does not appear to be the case in the majority of cells (reviewed in 26). IP_3 releases Ca^{2+} from intracellular stores when IP_3 binds to a specific receptor in the endoplasmic reticulum. Diacylglycerol, together with the increase in Ca^{2+}, activates certain isoforms of protein kinase C (PKC). The ET-1 receptor also couples to membrane Ca^{2+} currents. Each of these proximal events, release of Ca^{2+} from internal stores, activation of PKC, and influx of Ca^{2+}, has been postulated to mediate some or all of the mitogenic response to ET-1. In addition, activation of one or more tyrosine kinases has been proposed to be a mediator of ET-1-induced mitogenesis. We will consider the evidence for each below.

Increases in Cytosolic Free $[Ca^{2+}]$

Depletion of intracellular Ca^{2+} stores can arrest cells at G0, and Ca^{2+} transients can initiate the G0 to G1 transition and gene transcription in many types of cells (26, 27 and references therein). The mechanisms for this are unclear but may involve, in part, activation of the serine/threonine phosphatase, calcineurin, or activation of Ca^{2+}/calmodulin kinase with subsequent activation of the cAMP response element binding protein (CREB) (reviewed in refs. 27, 28).

Elevation of cytosolic free $[Ca^{2+}]$, whether from release of intracellular stores or influx via a receptor operated channel, a voltage dependent channel, or capacitative entry via a Ca^{2+} release-activated current (ICRAC), is one of the earliest responses of the cell to ET-1. Several groups have sought to prove a connection between the $[Ca^{2+}]$ transient and the mitogenic response. Pretreatment of cells with either of two blockers of the dihydropyridine-sensitive voltage-dependent Ca^{2+} channel, nifedipine or nicardipine, significantly reduced DNA synthesis by vascular smooth muscle cells in response to ET-1.[10,11,29] Similar results were reported for HeLa cells pretreated with the Ca^{2+} chelator EGTA.[30] Although nonspecific effects of these agents must be considered, these data suggest Ca^{2+} influx may be an important component of the mitogenic response to ET-1. As noted, however, mechanisms of this effect of Ca^{2+} influx are purely speculative and generation of a Ca^{2+} transient alone is not sufficient to induce mitogenesis.

PKC activation. Exposure of smooth muscle cells or fibroblasts to ET-1 induces a prolonged (>20 min) increase in diacylglycerol levels and, not surprisingly, activates PKC as determined by translocation to the membrane of kinase activity or phosphorylation of a PKC substrate.[31,32] Phorbol ester responsive isoforms of PKC are, in general, susceptible to "downregulation," the process of markedly reducing levels of certain isoforms of PKC by prolonged exposure (24 hours or more) to high concentrations (>100nM) of phorbol ester.[33] Most studies have suggested that downregulation of PKC attenuates the mitogenic response,[2,6-8,17] protein synthesis,[34] and induction of c-*fos*[14] in response to ET-1. In support of a requirement for a functioning PKC pathway, various inhibitors of PKCs, including H7 and sangivamycin, have also been reported to inhibit [³H]thymidine incorporation or protein synthesis.[6,34] Although phorbol ester sensitive PKC isoforms appear to modulate ET-1-induced mitogenesis, it is not clear how the PKCs contribute to the mitogenic effect. Activated PKC isoforms do potently activate ERK-1and -2, and this could be the mechanism of their pro-mitogenic effect, but the literature is divided on whether PKCs are necessary for ERK-1/-2 activation by ET-1. Some studies have clearly demonstrated that the phorbol ester sensitive PKCs are not required for ET-1-induced activation of the ERKs (see below). Thus the most logical site of action of the PKCs which might modulate the mitogenic response to ET-1, the c-Raf-1/ERK cascade, does not appear to be involved, at least in some cells. These uncertainties over the site of action of PKCs will be addressed in more detail below.

Clearly, activation of PKCs by phorbol esters or the cell permeant diacylglycerol analog, 1-oleoyl 2-acetyl-*sn*-glycerol, is not sufficient, by itself, to induce mitogenesis in the same cells (e.g., mesangial cells) in which PKCs have been reported to be necessary for ET-1-induced mitogenesis.[6] Another signaling pathway must be involved. Candidates for the additional factors necessary to transduce the mitogenic signal include phorbol ester-insensitive PKCs or, more likely, one or more nonreceptor tyrosine kinases.

Tyrosine kinase activation. The other signal transduction pathway that appears critical to ET-1-induced mitogenesis is activation of a tyrosine kinase. It has been clear for some time that ET-1 activates one or more tyrosine kinases as evidenced by the phosphorylation on tyrosine residues of several cytosolic proteins ranging in molecular weight from 45 kDa to 225 kDa.[33,35-37] Furthermore, activation of the tyrosine kinase pathway is independent of PKC downregulation and occurs despite pretreatment of cells with various PKC inhibitors, making it a candidate pathway to transduce the PKC-independent mitogenic signal.[33] The most convincing evidence to date that a tyrosine kinase(s) is critical to the mitogenic response to ET-1 is based on the use of two chemically unrelated tyrosine kinase inhibitors, genistein and herbimycin A. Pretreatment of glomerular mesangial cells with either of these agents virtually eliminated the mitogenic response to ET-1 (100 nM) whereas daidzein, an inactive compound related to genistein, was ineffective.[6] Furthermore, an ET-1-induced increase in expression of the immediate-early gene, c-*fos*, is only minimally decreased by PKC downregulation but is markedly decreased by the tyrosine kinase inhibitors.

These data, taken together, indicate that one or more isoforms of phorbol ester-sensitive PKCs and one or more nonreceptor tyrosine kinases are critical in activating mitogenesis in response to ET-1. For the PKCs, it is likely that the effect is mediated, at least in part, by the c-Raf-1 protein kinase cascade that culminates in the activation of the ERKs, which in turn appear to transduce the signal to the nucleus and activate transcription by phosphorylating certain transcription factors. In addition to PKC-induced activation of this cascade, there clearly are other mechanisms whereby ET-1 activates this cascade, and increasing evidence suggests activation of one or more Src tyrosine kinase family members may be critical (see below). Other tyrosine kinases, including the focal adhesion kinase (FAK), Pyk2, a nonreceptor tyrosine kinase which is activated by an increase in cytosolic free $[Ca^{2+}]$,[38] and possibly members of a third family, the Janus kinases or Jaks, could also play roles in ET-1-induced mitogenesis.[39,40] In the sections below, I will discuss each of these pathways, what is known about their activation, and how they might play a role in the transduction of the endothelin-induced mitogenic signal to the nucleus.

Signal Transduction Mechanisms of Mitogenesis

The c-Raf-1/ERK Cascade

The mitogen-activated protein kinases (MAP kinases), p44 and p42, also known as extracellular signal regulated kinases (ERKs)-1 and -2, were initially identified as two protein kinases that became phosphorylated on tyrosine in response to insulin and other growth factors.[41] The ERKs, in turn, phosphorylated and activated another serine/threonine kinase, or RSK, the ribosomal S6 protein kinase.[42] This, plus their activation by growth factors, strongly suggested the ERKs were involved in the mitogenic response, but it was not clear how they were activated or what role they played in transducing the mitogenic signal.

It had been known for some time that expression of the viral oncogene v-*raf* led to a grossly transformed phenotype in NIH3T3 fibroblasts.[43-45] v-*raf* encodes a constitutively active protein serine/threonine kinase. The protooncogene product, c-Raf-1, was, like the ERKs, potently activated by growth factors. The crucial observations placing these two kinases on a protein kinase cascade and identifying c-Raf-1 as "upstream" of the ERKs greatly advanced our understanding of how the mitogenic signal might be transduced from the growth factor receptor at the cell surface to the nucleus.[46-48]

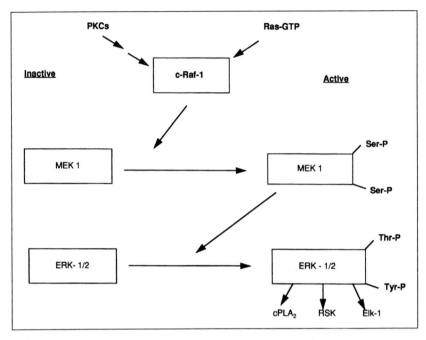

Fig. 13.1. The c-Raf-1/ERK protein kinase cascade. c-Raf-1, which is activated by agonists which cause GTP loading of Ras, or agonists which activate protein kinase Cs (PKCs), phosphorylates two serine (Ser-P) residues on MEK-1, activating it. Active MEK-1 then phosphorylates a threonine (Thr-P) and a tyrosine (Tyr-P) residue on ERK-1 and ERK-2, activating them. Various targets of the ERKs, including cytosolic phospholipase A_2, the 90 kDa ribosomal S6 kinase (RSK), and the transcription factor, Elk-1 (a Ternary Complex Factor), are also shown.

A wealth of data in multiple types of cells now suggest that activation of the c-Raf-1/ERK kinase cascade (Fig. 13.1) is both necessary and sufficient for the mitogenic response to a variety of stimuli.[49-52] Dominant interfering mutants of various components of the cascade inhibit the mitogenic response to growth factors and constitutively active components activate mitogenesis or transformation, even in the absence of growth factors. It is not surprising, therefore, that ET-1 activates this cascade in fibroblasts, glomerular mesangial cells, and vascular smooth muscle cells.[53-56] Clearly, to understand how ET-1 signals mitogenesis it is critical to understand this cascade and how ET-1 activates it. Although much of the pathway was elucidated in growth factor-stimulated cells, the identical pathway, at least from c-Raf-1 to the ERKs, is utilized in response to ET-1.

Using PDGF as an example, the cascade, as currently defined, is activated by simultaneous binding of a PDGF dimer to two receptors, forming a receptor dimer (Fig. 13.2).[57] The dimerized receptors, with intrinsic tyrosine kinase activity, then "autophosphorylate" or, more correctly, cross-phosphorylate several tyrosine residues within the receptor.[58] One phosphotyrosine residue within the kinase domain enhances catalytic activity of the receptor's tyrosine kinase. Several other phosphotyrosine residues outside the kinase domain act as docking sites for sig-

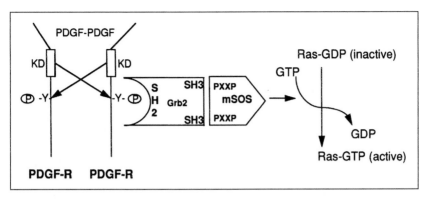

Fig. 13.2. Activation of Ras by growth factor receptors. Following binding of a PDGF dimer to two PDGF receptors (PDGF-R), the two receptors are brought into proximity and the tyrosine kinase domain (KD) of one receptor phosphorylates several tyrosine (Y) residues on the other receptor. One of these tyrosine phosphorylated residues (Y-P) binds the SH2 domain of Grb2, bringing Grb2 to the cell membrane. Since Grb2 is constitutively bound to mSOS via the former's two SH3 domains and two proline-rich sequences in the latter, mSOS is also brought to the membrane where it catalyzes the exchange of GDP for GTP by Ras.

nal transducing molecules. These signal transducing molecules interact with the various phosphotyrosine residues on the receptor via Src homology 2 (SH2) domains, first identified in the nonreceptor tyrosine kinase, Src. These SH2 domains are composed of approximately 100 amino acids and interact with phosphotyrosine residues but not unphosphorylated tyrosine residues. Which phosphotyrosine a particular SH2 domain interacts with is largely determined by the three amino acids immediately carboxy-terminal to the phosphotyrosine.[59]

SH2 domain-containing proteins bound to the receptor then propagate the signal by one of two mechanisms—by changing the subcellular localization of another signal transducing protein to which the SH2 domain-containing protein is also bound, or by inducing an allosteric change in a protein which alters its catalytic activity. The former mechanism is utilized in the growth factor-induced activation of Ras, and is mediated by the SH2-containing protein Grb2 (growth factor receptor bound protein 2) (Fig. 13.2). Grb2 exists in a prebound complex in the cytoplasm with mSOS (the guanine nucleotide exchange factor (GEF) homologous to Drosophila son of sevenless). Grb2 and mSOS bind via interaction of the former's SH3 domain with a short proline-rich sequence in mSOS (reviewed in 60). When, for example, the PDGF receptor autophosphorylates, Grb2 binds to a specific phosphotyrosine residue via its SH2 domain, bringing mSOS to the membrane where its substrate, Ras, is located. mSOS then catalyzes the exchange of GDP for GTP on Ras, and Ras is activated. Catalytic activity of mSOS does not appear to be regulated, and thus activation of Ras appears to occur solely as a result of the change in the localization of mSOS.

GTP-loaded Ras can then bind several signaling molecules, rasGAP (the GTPase-activating protein which enhances the intrinsic GTPase activity of Ras, thus inactivating it), and a phosphoinositide 3 kinase (PI3K, see below). Most importantly for this discussion, Ras interacts with c-Raf-1.[61] c-Raf-1 binds to Ras via a region in its N-terminus, residues 51-131, contained in the CR1 region (conserved region 1, a region conserved in mammalian, *Drosophila*, and *C. elegans* Raf homologs).[61] If CR1 is overexpressed in cells, it functions as a dominant negative inhibitor of signal transduction induced by activated Ras or growth factors, presumably by competing with full-length c-Raf-1 for Ras-GTP.[62]

It now seems clear that the primary function of activated Ras is to bring c-Raf-1 to the cell membrane since targeting c-Raf-1 to the membrane by attaching a membrane localization signal (CAAX box plus a polybasic domain) is sufficient to activate c-Raf-1, even in the presence of dominant negative Ras.[63,64] Additional components are required to activate c-Raf-1, however, since c-Raf-1 is not fully activated in this setting unless growth factor is also added. Furthermore, coincubation of purified GTP-loaded Ras with c-Raf-1 fails to activate the latter. Thus it appears that once c-Raf-1 arrives at the membrane, it interacts with an additional, mitogen-dependent factor that is responsible for full activation of c-Raf-1. Activation presumably occurs when the inhibitory amino-terminal regulatory domain of c-Raf-1, which normally masks the kinase domain, swings out of the way of the kinase domain, freeing it to interact with its target.[65] The importance of this regulatory domain is best seen with truncation mutants of c-Raf-1 lacking the regulatory domain, such as BXB-Raf, which are highly transforming.[46,65]

There are several candidate cofactors/activators of c-Raf-1 which have been identified in yeast two-hybrid screens or by coimmunoprecipitation or affinity chromatography. The most likely group are the 14-3-3 family of proteins which have been implicated in remarkably diverse roles including activation of enzymes involved in neurotransmitter release and of Ca^{2+}-activated phospholipase A2, and inhibition of PKC. When 14-3-3 is injected into Xenopus oocytes, c-Raf-1 becomes activated. Furthermore, activation of c-Raf-1 in yeast expressing Ras and c-Raf-1 is dependent upon yeast 14-3-3 proteins. The β isoform of 14-3-3, identified in a yeast two-hybrid screen with c-Raf-1 as the "bait," associates with the N-terminal regulatory domain of c-Raf-1.[66] 14-3-3 molecules exist as dimers, and each 14-3-3 molecule can bind two c-Raf-1 molecules (67, 68 and reviewed in 69). The 14-3-3 protein appears to allow c-Raf-1 molecules to interact more effectively when they are brought to the cell membrane by GTP-bound Ras. The interaction appears to be critical for activation of c-Raf-1 kinase activity, and may bring about cross-phosphorylation of one c-Raf-1 molecule by the other.[69]

The G-Protein Connection

The ERKs are clearly activated by ET-1[17,53,70] and in most types of cells, with exceptions (see below), this appears to proceed via activation of c-Raf-1.[55] It had not been clear until recently, however, how agonists with heterotrimeric G-protein-coupled receptors activated this cascade. It now seems that the ET-1 receptor may activate the c-Raf-1/ERK cascade via at least two mechanisms: one utilizing βγ subunits acting via a Ras-dependent pathway and the other, the αq subunit acting via a Ras-independent pathway (Fig. 13.3).

The βγ-Mediated Pathway

The pathway from heterotrimeric G-protein-coupled receptors to c-Raf-1 activation via Ras has largely been defined in studies employing Gi-coupled receptors, but evidence suggests a similar pathway is utilized by the ET-1 receptor. Gi-coupled receptors, such as the m2 muscarinic receptor, the α2-adrenergic receptor, the

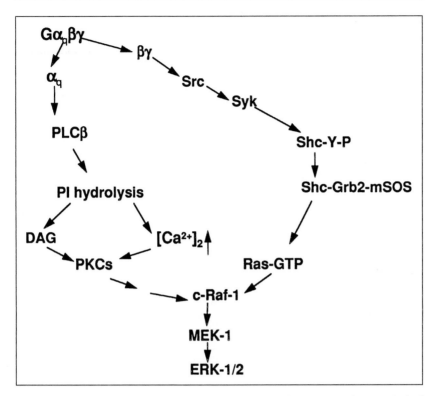

Fig. 13.3. Putative mechanisms of the activation of the c-Raf-1/ERK cascade by Gq-linked receptors. Following receptor occupancy, GTP-loaded α_q dissociates from $\beta\gamma$ subunits. Possibly depending on the type of cell, α_q, $\beta\gamma$, or both may transduce the signal to c-Raf-1 activation. For α_q, activation probably proceeds via phospholipase Cβ to produce diacylglycerol (DAG) and an increase in intracellular calcium ($[Ca^{2+}]_i$, activating one or more PKC isoforms. PKC may then activate c-Raf-1 via Ras-dependent or Ras-independent pathways, the latter possibly involving direct activation by phosphorylation. Free $\beta\gamma$ subunits may activate Ras via a cascade of events involving one or more tyrosine kinases and the adapter protein Shc which brings Grb2-mSOS to the cell membrane.

thrombin receptor, and the receptor for lysophosphatidic acid, activate the c-Raf-1/ERK cascade via Ras since expression of dominant negative Ras blocks activation of the cascade.[71-77] The activation of Ras appears to be mediated via $\beta\gamma$ subunits since expression of a $\beta\gamma$ subunit binding protein derived from the β adrenergic receptor kinase (βARK) blocks activation of Ras and of the c-Raf-1/ERK cascade.[75] In addition, expression of $\beta\gamma$ subunits, but not constitutively active α_i subunits, activates the c-Raf-1/ERK cascade.[78] Since expression of dominant negative Ras prevents ERK activation by $\beta\gamma$ subunits, it appears that Ras is indeed downstream of $\beta\gamma$.[75,76,79,80]

Lefkowitz and co-workers have explored the mechanism of activation of Ras by $\beta\gamma$ and have shown that expression of $\beta\gamma$ subunits leads to a sustained 2-fold increase in tyrosine phosphorylation of the adapter protein Shc.[76] Following exposure of cells to epidermal growth factor (EGF), Shc associates, via its SH2 domain, with a phosphotyrosine residue on the EGF receptor. Shc becomes phosphorylated upon a tyrosine residue, which leads to the association of the Grb2-mSOS complex with Shc. This activates Ras by bringing the Ras guanine nucleotide exchange factor (mSOS) to the membrane. Since LPA and $\alpha 2$ adrenergic receptor agonists not only induce tyrosine phosphorylation of Shc, but also cause formation of a Shc-Grb2-mSOS complex, both EGF and G-protein-linked receptors appear to utilize similar mechanisms to activate Ras. To confirm that these effects were mediated by $\beta \gamma$ subunits, Lefkowitz and co-workers demonstrated that the tyrosine phosphorylation of Shc, induced by occupation of the Gi-linked LPA or $\alpha 2$ adrenergic receptors, was prevented by expression of the $\beta \gamma$ subunit binding protein derived from βARK.[76]

Simonson and co-workers have clearly demonstrated that ET-1-induced activation of c-Raf-1 in glomerular mesangial cells proceeds via Ras and a dominant inhibitory mutant of Ras blocks ET-1 induction of c-*fos*.[54] This is consistent with experiments which demonstrate that cotransfection of the Gq-linked m1 muscarinic receptor with a dominant inhibitory mutant of *ras* or with Rap1a (which antagonizes effects of Ras by competing for similar substrates) blocked activation of the c-Raf-1/ERK cascade in response to carbachol.[79] Since in astrocytes, ET-1 induces tyrosine phosphorylation of Shc and formation of the Shc/Grb2/mSOS complex, it is likely that, at least in these cells and in mesangial cells, the ET receptor employs a similar $\beta \gamma$-dependent mechanism for activation of Ras to that utilized by the Gi-linked receptors.

What mediates Shc tyrosine phosphorylation? Growth factors employ the intrinsic tyrosine kinase activities of their receptors to activate mechanisms leading to Ras activation, but G protein-coupled receptors have no such activity. However, it had been known for some time that inhibitors of protein tyrosine kinases blocked activation of ERKs by G-protein linked receptors in some cells.[72,75-77] Recently it has been shown that G-protein coupled receptors employ one or more nonreceptor protein tyrosine kinases. There are several families of these kinases, but G protein-coupled receptors appear to employ three of them, the Src family, the Syk/ZAP-70 family, and, possibly, the Pyk2 family, to signal activation of the c-Raf-1/ERK cascade. Wan and co-workers generated B cell lines deficient in Lyn (a Src family member) or Syk by homologous deletion.[77] They found that in cells deficient in either Lyn or Syk, activation of MEK and the ERKs in response to occupation of the Gq-coupled m1 muscarinic acetycholine receptor was blocked. The Gi-linked m2 muscarinic receptor required only Syk for activation of MEK and the ERKs. Since m1-mediated activation of Syk was markedly reduced in Lyn(-) cells, it appears that Lyn may be upstream of Syk in the signaling cascade. Furthermore, one of the SH2 domains of Syk is critical to ERK activation, suggesting this domain may be required for association with the upstream activating Src family member. Although Lyn is predominantly expressed in cells of hematopoietic lineage, three other family members, Src, Lck, and Fyn were able to compensate for the absence of Lyn and allowed ERK activation by the m1 receptor.

These data suggest that Gq-coupled activation of c-Raf-1 is mediated by activation of Ras which proceeds via serial activation of a Src family member and Syk (Fig. 13.3). Shc may then associate with either Src or Syk via interaction of the SH2 domain of Shc with phosphotyrosine residues (on Syk or the Src family member), or via interaction of proline-rich regions of Shc with the SH3 domain(s) of Src or Syk. Either of these kinases may then directly phosphorylate Shc, initiating association of Grb2/mSOS with Shc. Since Src is localized to the cell membrane, this multiprotein complex will presumably form at the membrane. This achieves the membrane localization of mSOS which is critical to activation of Ras.

For this mechanism to be employed by the ET_A receptor, it is critical that Src be activated in response to ET-1. Src was the first identified tyrosine kinase which was activated in response to ET-1.[6,81] Src is activated only about 2- to 3-fold over control by ET-1 in A10 or BC3H1 smooth muscle cells or glomerular mesangial cells. However, this is equivalent to the degree of activation observed after the exposure of NIH3T3 cells to the potent mitogen, PDGF or after exposure of macrophages to CSF-1.[82-84] Furthermore, the observations by Simonson and co-workers that induction of c-*fos* in response to ET-1 requires Src, and that Ras is downstream of Src in this ET-1 signaling pathway, are entirely consistent with the mechanism of c-Raf-1/ERK cascade activation proposed by Wan et al from their work with muscarinic receptors.[54,77] More will be said about the role of Src in the mitogenic response to ET-1 (below) and in signaling the nucleus (see chapter 12).

Other cells may employ an additional nonreceptor tyrosine kinase to recruit SOS to the membrane and activate Ras. Pyk2 is a tyrosine kinase that is activated by a rise in the cytosolic free $[Ca^{2+}]$, induced by membrane depolarization or by several ligands with G protein-coupled receptors, including bradykinin.[38] Pyk2 undergoes rapid tyrosine phosphorylation when the cytosolic free $[Ca^{2+}]$ increases. Then, either Shc binds to the kinase and is tyrosine phosphorylated, bringing Grb2 and SOS to the complex, or Grb2 (and therefore SOS) may bind directly to Pyk2 via the former's SH2 domain. Although unproven, this mechanism may be an extremely important one in the response to ET-1 of cells of the central nervous system and the kidney, where Pyk2 is expressed.

Recently, Gutkind and co-workers identified another candidate in the signaling cascade from G-protein-linked receptors to Ras.[85] PI3-kinases which are activated by growth factors consist of heterodimers of a p110 catalytic domain and a p85 adapter protein which mediates binding to growth factor receptors and to IRS-1 via its SH2 domain. PI3Kγ does not appear to interact with p85 subunits, but can be activated in vitro by either α or βγ subunits of G proteins. Based on experiments employing overexpression of PI3Kγ and of various dominant inhibitory or constitutively active mutants of components of the cascade described above, it appears that free βγ subunits recruit PI3Kγ to the membrane where its substrates are located. PI3Kγ then activates Src or a Src-like kinase, which then triggers tyrosine phosphorylation of Shc, its association with Grb2/mSOS, activation of Ras and the c-Raf-1/ERK cascade. These studies of the role of PI3Kγ were performed with the Gi-linked m2 muscarinic receptor, and must be repeated with the ET receptor before any conclusions can be drawn, but they may explain the observation that a wortmannin-sensitive signaling pathway modulates the mitogenic response to ET-1 in CHO cells expressing the ET_A receptor.[86] Although intriguing, the observations were made in CHO cells which expressed an average of 250,000 receptors/cell, more than 10 times the number expressed on fibroblasts or smooth muscle cells. Thus interactions of the receptor with signaling molecules might occur in this setting that have little physiological relevance.

The αq-Mediated Pathway

Other evidence suggests the ET-1 receptor can also activate the c-Raf-1/ERK cascade via a Ras-independent pathway, and that this pathway might employ αq subunits in some types of cells. For example, in Rat-1 fibroblasts, ET-1 stimulation does not increase GTP loading of Ras but does activate c-Raf-1 and the ERKs, suggesting a Ras-independent pathway to c-Raf-1 activation. That this pathway is activated by αq and not βγ is suggested by several lines of evidence.[71,72] First, expression of a constitutively active mutant of αq in COS cells (but not PC12 pheochromocytoma cells) activates the c-Raf-1/ERK cascade.[80,87] Most importantly, Lefkowitz and co-workers found that activation of the c-Raf-1/ERK cascade by the Gq-coupled α1B adrenergic receptor and the m1 muscarinic receptor were not blocked by expression of the βARK peptide, by dominant negative Ras, or by protein tyrosine kinase inhibitors, but were inhibited by dominant negative c-Raf-1 or by PKC depletion with prolonged phorbol ester treatment.[75] Thus it appears that Gq-coupled receptors can also utilize αq to activate phosphoinositide hydrolysis via a Ras-independent mechanism, leading to PKC activation and, subsequently, activation of c-Raf-1. The major questions remaining to be answered are how do PKCs activate c-Raf-1, and how important is this pathway in ET-1-induced activation of the cascade.

Activation of the c-Raf-1/ERK Cascade by Protein Kinase C Isoforms

Phorbol esters are potent activators of c-Raf-1 and the ERKs.[88,89] Realizing the limitations of equating phorbol ester-induced activation with PKC-induced activation (see below), these data imply a role for PKC isoforms in c-Raf-1/ERK activation.[90] In some cells, phorbol ester-induced activation of the c-Raf-1/ERK cascade is Ras-dependent, indicating the PKCs are acting at sites proximal to Ras, such as one of the nonreceptor tyrosine kinases.[89,91] This is not the mechanism of αq-mediated activation of the c-Raf-1/ERK cascade, however, and, therefore, there must be an alternative target of the PKCs which leads to activation of the cascade. This alternative mechanism was suggested by studies in which expression of a constitutively active mutant of PKCδ, or overexpression of wild type PKCε activated the ERK cascade, and the activation was Ras-independent, but c-Raf-1-dependent, suggesting PKCs might activate c-Raf-1 directly.[92,93] PKC isoforms have been proposed to be direct activators of c-Raf-1. c-Raf-1 can be phosphorylated by PKCs in vitro at Ser 499, a residue within the kinase domain, and this phosphorylation is associated with activation of c-Raf-1.[94,95] In addition, coexpression of PKCα and c-Raf-1 in SF9 insect cells activates c-Raf-1, albeit modestly.[95] This has not been a consistent observation, however.[96] Furthermore, mutation of Ser 499, the residue phosphorylated by PKCα, does not prevent activation of c-Raf-1 in insect cells when coexpressed with Ras and Src, or activation of c-Raf-1 in response to a number of stimuli.[95,97] It appears that PKC isoforms may increase c-Raf-1 autokinase (or autophosphorylating) activity, but not its kinase activity directed at its physiologic substrate, MEK1.[96] Thus in those cell types in which c-Raf-1 and the ERKs are activated by phorbol esters, it is likely that the PKCs are not acting directly upon c-Raf-1, and how the PKCs activate the cascade remains unknown.

Role of PKCs in ET-1-Induced Activation of the c-Raf-1/ERK Cascade

Although the mechanisms are not known, it is clear that activated PKCs are capable of activating the ERK cascade. This does not mean, however, that ET-1-induced activation is dependent upon PKCs. It has proven difficult to determine the role of PKCs in the ET-1-induced activation of the ERK cascade. This is, in part, due to the fact that phorbol esters, which are traditionally employed as activators of PKCs or to deplete cells of PKCs (if cells are exposed for prolonged periods), have intracellular targets other than the PKCs. These include Vav, a guanine nucleotide exchange protein which may be involved in the activation of Ras, and n-chimerin, a GTPase activating protein which inactivates another small G protein, Rac.[92]

With this caveat in mind, where examined, ET-1-induced activation of the c-Raf-1/ERK cascade is partially dependent upon PKCs based on experiments utilizing downregulation of PKCs or putative specific inhibitors of PKCs. In an early report, downregulation of PKCs in glomerular mesangial cells partially inhibited ET-1-induced activation of the ERK cascade.[70] Furthermore, ET-1-induced activation of c-Raf-1 in ventricular myocytes also appears to be partially dependent upon classical PKC isoforms.[98,99] The interpretation of these studies is markedly complicated by the fact that prolonged phorbol ester exposure results in up to a four-fold increase in basal c-Raf-1 activity.[78,99] This makes calculation of fold-stimulation of c-Raf-1 activity following ET-1 difficult to interpret. Furthermore, this chronic stimulation of the c-Raf-1/ERK cascade can lead to counter-regulatory mechanisms, such as induction of ERK-inactivating phosphatases (see below), which could depress ERK activation in response to ET-1 or any agonist. This would lead to the erroneous conclusion that ET-1-induced ERK activation requires PKCs.

Experiments employing other Gq-linked receptors are not of much help in defining a role for the PKCs since they are divided on whether PKCs are necessary for c-Raf-1/ERK activation, even when identical receptors are employed. For example, in COS cells transfected with the m1 muscarinic (Gq-linked) receptor and NIH3T3 cells stably expressing the receptor, carbachol-induced c-Raf-1 (and ERK-2) activation was only minimally inhibited by PKC downregulation by prolonged phorbol ester treatment or by a putatively specific PKC inhibitor, GF 109203X.[78,100] Furthermore, DNA synthesis and focus formation were unaffected by PKC inhibition. In contrast, activation of the ERKs by the m1 muscarinic receptor (and the α1B adrenergic receptor) in COS or CHO cells, or by the arginine vasopressin receptor in vascular smooth muscle cells, was inhibited by PKC depletion.[75,101]

In summary, the importance of the ET-1-activated, αq-mediated pathway to activation of the c-Raf-1/ERK cascade, which appears to signal via activation of classical PKC isoforms, remains unclear at this time. This is largely due to uncertainties over the role of PKCs, and this uncertainty is due to methodologic problems confounding interpretation of data and to contradictory results. In those cells in which c-Raf-1 activation is partially dependent upon PKCs, the activation is likely not to be via direct phosphorylation of c-Raf-1.

The role of the novel isozymes of PKC, PKCζ and PKCλ, in ET signaling are unclear. These isoforms are not downregulated by prolonged exposure of cells to phorbol esters, making it difficult to study their role in ligand-induced activation of the ERK cascade. Although a constitutively active mutant of PKCζ is capable of activating the ERK (and the SAPK, see below) cascade, and a dominant negative mutant inhibits serum-induced activation of the ERKs, it remains to be demonstrated that PKCζ plays any role in the response to ET-1.

Other routes to c-Raf-1 activation. Phosphatidylcholine-specific phospholipase C (PC-PLC)(reviewed in 102), presumably acting via its primary product, diacylglycerol, has also been proposed to be a c-Raf-1 activator.[103] Expression of PC-PLC activates c-Raf-1 and the activation is not prevented by downregulation of PKCs. We have directly assayed species of diacylglycerol for their ability to activate c-Raf-1, and found the activation to be minimal (1.5 fold vs. greater than 100-fold for PKC).[104] Thus diacyglycerol does not appear to play any direct role in the activation of c-Raf-1 and a role for PC-PLC in the activation of c-Raf-1 remains speculative.

Tyrosine phosphorylation of c-Raf-1 at residues 340 and/or 341 may play a role in the activation of c-Raf-1 by oncogenic Src.[105] Activation of nononcogenic Src family members has been suspected to play a direct role in activation of c-Raf-1 in response to T cell receptor activation and IL-2 stimulation (reviewed in ref. 65). Since c-Raf-1 is not phosphorylated to any extent on tyrosine in response to ET-1 or in response to growth factors, this mechanism probably plays no direct role in the activation of c-Raf-1 by ET-1.[53,105] As noted above, however, Src may play a critical role in the phosphorylation of Shc and subsequent activation of Ras.

It is difficult to summarize the often conflicting data concerning the mechanisms of activation of c-Raf-1 by ET-1. ET-1 clearly activates c-Raf-1 in most cells, and this is the predominant mechanism of activation of the ERKs. Where examined, the mechanism of activation of c-Raf-1 appears to be via activation of Ras. Ras activation, in turn, appears to be modulated by the formation of a complex of proteins including Shc, Grb2, and mSOS. Complex formation appears to be triggered by the tyrosine phosphorylation of Shc, catalyzed by one or more nonreceptor tyrosine kinases. In other cells, a PKC-dependent pathway, probably activated by αq, may activate the c-Raf-1/ERK cascade via a Ras-independent pathway, but in those cells, the mechanism of activation of c-Raf-1 remains unclear.

Inhibition of c-Raf-1 by Activators of Protein Kinase A

Agents which increase cellular levels of cAMP have been known for some time to interfere with the actions of growth factors and, more recently, the actions of ET-1 in ventricular myocytes.[99] This effect appears, at least in part, to be via inhibition of activation of the c-Raf-1/ERK cascade by inhibiting the coupling of GTP loaded Ras to c-Raf-1.[106,107] The mechanism involves cAMP-induced activation of protein kinase A which then phosphorylates c-Raf-1 on Ser43 within the regulatory domain of c-Raf-1. This phosphorylation reduces the affinity with which c-Raf-1 binds to Ras.[106] Other mechanisms may play a role in cAMP-induced inactivation of c-Raf-1, however, since B-Raf, which does not have a site analogous to Ser43, is also inhibited by activators of PKA. These data raise the possibility that ET-1-induced inhibition of adenylyl cyclase could play a role in enhancing c-Raf-1 activation. Although this mechanism is unlikely to be the primary mechanism of activation of c-Raf-1, simultaneous exposure of cells to ET-1 and an activator of protein kinase A can be expected to reduce the mitogenic effect of ET-1.[55]

Downstream of c-Raf-1. Components of the signaling cascade downstream of c-Raf-1 have been identified and their mechanisms of activation determined (Fig. 13.1). Once activated, c-Raf-1 initiates activation of a cascade of serine/threo-

Table 13.1 MEK1/2 Kinases

Kinase	Expressed in	Comments
c-Raf-1	Ubiquitous	Activated by mitogens inc. ET-1
A-Raf	Urogenital tissues, heart	Activated by ET-1 in myocytes
B-Raf	CNS, testis	Activated by mitogens
Tp12/Cot	Spleen, thymus. liver, lung	T cell activation
Mos	Oocytes	Meiotic cell cycle
MEKKs	Ubiquitous	MEKK3>>MEKK1,2

The first five protein kinases are protooncogene products. It is not known if there are oncogenic variants of MEKK3.

nine kinases which culminate in signaling the nucleus. The major physiologic substrates of c-Raf-1 are MAP/ERK Kinase-1 (MEK-1, also known as MAP kinase or MKK1) and MEK-2, although the former appears to be the preferred substrate.[108] MEK-1 is phosphorylated on two serine residues within kinase subdomain VIII by c-Raf-1.[109] Expression of constitutively active forms of MEK-1, made by mutating these phosphorylation sites to acidic amino acids which function as phosphorylated residues, is sufficient to activate mitogenesis in fibroblasts.[50,51] These data strongly support the concept that activation of the ERKs is sufficient to activate mitogenesis and that the key downstream kinases are the ERKs, not components of a divergent pathway activated by Ras or c-Raf-1.

MEK kinase-1 (MEKK-1) can also phosphorylate and weakly activate MEK-1.[110] This, plus its similarity to a yeast kinase, Sterile 11 (Ste11, so named because yeast with mutations in the gene do not mate normally in response to pheromone), which is downstream of a heterotrimeric G protein in yeast (reviewed in 111), led some to postulate it might be the missing connection in the G-protein pathway to the ERKs. Activation of MEK-1 by MEKK-1 appears to be an artifact of overexpression of MEKK-1, however, and is probably not physiologically relevant.[112] Rather, MEKK-1 appears to be on the pathway culminating in activation of the SAPKs (see below).[112]

This does not mean that there are no other physiologically relevant activators of MEK1 than c-Raf-1 (Table 13.1). The protooncogene product, Mos, is a MEK1 kinase that regulates the meiotic cell cycle in vertebrate oocytes and when expressed ectopically in somatic cells, causes transformation.[113,114] Similarly, the Tpl2/Cot protooncogene product also functions as a MEK1 (and SEK1, see below) kinase, but its primary role thus far appears to be in the activation of T cells.[115] In airway smooth muscle cells, ET-1 strongly activates the ERKs but only weakly activates c-Raf-1, suggesting that ET-1 employs an alternative route to MEK1 and ERK activation in these cells.[55] The identity of this MEK kinase is not known at this point. The cloning of additional MEK kinases, one of which, MEKK3, preferentially activates the MEK1/ERK pathway, suggests this pathway or ones utilizing other MEKKs, may account for observations like those in airway smooth muscle cells.[116] Identification of the components of these alternative pathways of activation of the ERKs by ET-1 is a critical step in understanding the mitogenic response to ET-1.

MEK-1, and the closely related MEK-2, are dual specificity kinases, phosphorylating tyrosine as well as serine/threonine residues. MEK-1 phosphorylates a TEY motif in subdomain VIII of ERK-1 and ERK-2.[117] At present, the only known physiologic substrates of MEK-1 are ERK-1 and ERK-2.

There seems to be no question that the ERKs are major effectors of the mitogenic signal, whether generated by a growth factor or an agonist with a G-protein coupled receptor. They are "proline-directed" kinases, having as their consensus phosphorylation sequence motif, serine or threonine followed by a proline residue. In contrast to the remarkable substrate specificity of c-Raf-1 and MEK-1, which probably serves to isolate this pathway from other mammalian MAP kinase pathways (see below), the ERKs have many substrates, including cPLA2 and other protein kinases including the 90 kDa (but not the 70 kDa) ribosomal S6 kinase, RSK. The critical substrates of the ERKs which might transduce the mitogenic signal to the nucleus are the transcription factors. Although these pathways are covered in much more detail (see chapter 12), a brief discussion of nuclear targets of the ERKs is warranted in the current context.

Regulation of Transcription by the ERKs

Translocation to the nucleus. Although the ERKs have been characterized as cytosolic proteins, following mitogenic stimulation, a portion of the cytosolic pool translocates to the nucleus.[118,119] Pouyssegur and co-workers have studied the time-course of ERK activation in response to growth factors and agonists with heterotrimeric G protein-coupled receptors, and determined how this correlates with the mitogenic response.[118,120] From their data it is clear that irrespective of the type of agonist, if the agonist leads to prolonged activation of the ERKs (greater than 3 hours), the ERKs translocate to the nucleus. Furthermore, they found that prolonged activation and nuclear translocation correlated highly with induction of mitogenesis. The intensity and percentage of nucleii immunolabeled with anti-ERK antibody correlated well with the relative ability of various mitogens (serum, FGF, thrombin, phorbol esters, thrombin receptor peptide) to stimulate DNA synthesis. This work was done with thrombin, which is linked to a Gi-coupled receptor, as agonist, and it is not clear that sustained activation of the ERKs is necessary for ET-1-induced mitogenesis. In Swiss 3T3 fibroblasts, ET-1 (10 nM) caused only very transient activation of the ERKs whereas basic FGF (5 ng/ml) produced sustained activation. Despite these differences, both agonists were highly mitogenic.[121]

Simonson and Dunn have found that exposure of glomerular mesangial cells to ET-1 induces translocation of both ERK-1 and ERK-2 into the nucleus (discussed in 118). This is a critical observation because it confirms that following ET-1, as with the growth factors, the activated ERKs have access to those potential substrates which determine the genetic response to ET-1, transcription factors.

After translocation of the ERKs to the nucleus, they phosphorylate and activate two transcription factors, Elk-1, which is critical to induction of genes such as c-*fos* that are regulated by a promoter element called the serum response element, and c-Myc. These activated transcription factors play critical roles in the induction of immediate early genes and in the mitogenic response. The RSK protein kinase family, which are substrates of the ERKs, also translocate to the nucleus when cells are exposed to activators of the c-Raf-1/ERK cascade.[122] One family member, RSK2, may also play an important role in immediate early gene induction and mitogenesis since it appears to phosphorylate the cAMP response element binding protein

(CREB) at Ser133, a critical residue for activation of CREB and for expression of c-*fos* in response to growth factors.[123] The existence of two Ets domain-containing transcription factors in Drosophila that are also ERK substrates suggests many other mammalian transcription factors will be identified which are regulated by this pathway. Identifying these substrates, how they are regulated by the ERKs, and what role they play in ET-1-induced mitogenesis will be a fruitful and vitally important area of research.

Interaction with the cell cycle. Inevitably, the ET-1-activated signal transduction systems must interact with the cell cycle machinery to cause cells to enter S phase. Sustained, as opposed to transient, activation of the ERKs appears to be required for many cells to pass the G1 restriction point and to enter S phase, in which cellular DNA is replicated.[49,52] Although ERK activation is clearly linked to the cell cycle, it has not been clear where the ERKs might interact with the cell cycle machinery. Expression of the D-type cyclins, which are the regulatory (activating) subunits for the cyclin-dependent kinase 4 and 6 (cdk4 and cdk6) catalytic subunits, appear to control the early stages of the transition toward S phase. Levels of D type cyclins rise in response to growth factor stimulation (reviewed in 124). A critical link between signal transduction and the cell cycle has recently been suggested by the finding that expression of dominant inhibitory mutants of MEK-1 or ERK-1, or expression of the MAP kinase phosphatase, MKP-1, which dephosphorylates and inactivates the ERKs (see below), inhibited growth factor-dependent expression of cyclin D1. Expression of a constitutively active mutant of MEK-1 increased cyclin D1 expression.[125] While the mechanisms of this effect are not clear, these data represent the first clear link between activation of the ERK cascade and cell cycle events leading to entry into S phase.

Turning Off the ERKs

It is obviously critical for the cell to have mechanisms of inactivating the ERKs. The importance of these mechanisms may be exemplified by the failure to isolate constitutively active ERKs by genetic methods or by site-directed mutagenesis. Cobb postulates that the cell simply cannot tolerate unrestrained ERK activity.[89] ERKs are activated by phosphorylation, catalyzed by MEK-1, on a threonine and a tyrosine residue in the motif Thr-Glu-Tyr within the kinase domain. Not surprisingly, incubation of activated ERKs with either serine/threonine or tyrosine phosphatases in vitro inactivates the kinase. It has now been convincingly demonstrated that dual specificity phosphatases (dephosphorylating Ser/Thr and Tyr residues), the MKPs (MAP kinase phosphatases), likely inactivate the ERKs by a similar mechanism in vivo.[126,127] Regulation of MKP-1 activity appears to be predominantly at the level of transcription and marked increases in MKP-1 mRNA are seen within the first 60 min after exposure of cells to mitogens. Most importantly, overexpression of MKP-1 inhibits Ras- and serum-induced DNA synthesis, suggesting MKP-1 is a major negative modulator of the mitogenic response to agonists, such as the growth factors, which act via Ras.[126] Induction of MKP-1 has been demonstrated in response to angiotensin II in vascular smooth muscle cells,[128] and it is likely that ET-1 will have a similar effect.

The c-Raf-1/ERK cascade may be negatively regulated at other sites as well. I have discussed the negative modulation of Ras/c-Raf-1 by activators of protein kinase A. In addition, MEK-1, the immediate upstream activator of the ERKs, appears to be negatively regulated by phosphorylation on two threonine residues.[129]

Phosphorylation at these sites overrides the activating phosphorylations catalyzed by c-Raf-1. These threonine residues lie in a sequence which is predicted to be a substrate for the kinase, p34*cdc2*, a key regulator of the cell division cycle during M phase.[129] While it is not clear that p34*cdc2* catalyzes these phosphorylations in vivo, it would make teleologic sense since this would prevent what Rossomando et al have termed "an untimely activation" of the ERK pathway during M phase.[129]

Finally, a recently described family of genes encoding RGS proteins (for regulators of G-protein signaling), has been shown to negatively regulate the activation of the ERK cascade by various G-protein coupled receptors.[130] Although the mechanisms of action are unclear, the site of action appears to be at the level of the G protein or possibly upstream of that (affecting coupling to the receptor), and may function as a desensitization mechanism.

Tyrosine Kinase Signaling

The other major signal transduction system activated by ET-1 which is likely to modulate the mitogenic response is the activation of tyrosine kinases. Nonreceptor tyrosine kinases known or suspected to be activated by ET-1 include Src and possibly related family members, the focal adhesion kinase, FAK, the Janus kinases (Jaks), and, conceivably Pyk2 (discussed above), any or all of which may play important roles in the response to ET-1.

ET-1 enhances tyrosine phosphorylation of a number of cellular proteins via activation of one or more tyrosine kinases.[33,35-37] Furthermore, activation of the tyrosine kinase(s) appears indispensable for ET-1-induced mitogenesis in glomerular mesangial cells expressing the ET_A receptor.[6]

It is unclear what role, if any, FAK activation plays in transducing a mitogenic signal. Clearly, based on the findings of several laboratories using a variety of inhibitors, one or more tyrosine kinases are required for both focal adhesion formation and DNA synthesis.[6,131-133] Many questions remain including which tyrosine kinases are critical, are they located in the focal adhesion, and if so, how do events occurring in the focal adhesion get transmitted to the nucleus. How does the nucleus know the cell is anchored so that DNA synthesis may proceed? FAK has not been demonstrated to translocate to the nucleus and the only likely substrates of FAK identified thus far are cytoskeleton-associated proteins.[133,134] Sorting out the role of the focal adhesion in the mitogenic response and what role FAK may play in the formation of the focal adhesion remains a major challenge to understanding mitogenic signaling by the endothelins.

We will focus on two pathways: activation of the Src family, and activation of the Jak/STAT pathway, since it is these two which are candidate modulators of the mitogenic response.

The Role of Src

As noted above, Src is activated by ET-1, and tyrosine kinase inhibitors reduce ET-1-stimulated ³H-thymidine incorporation. Although this is not necessarily due to inhibition of Src or its family members, Src does appear to play a vital role in the mitogenic response to growth factors. In NIH3T3 cells, a kinase-inactive Src mutant functioned as a dominant inhibitor of PDGF-activated signaling and inhibited entry of PDGF-stimulated cells into S phase.[135] A similar requirement for Src has been demonstrated for CSF-1- and EGF-induced mitogenesis.[135-137] Although Src "knock-outs" might seem to be an ideal way in which to address the role of Src

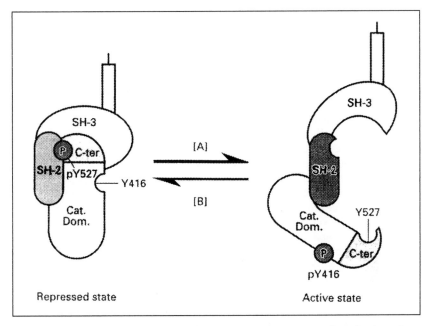

Fig. 13.4. Mechanisms of activation and inactivation of Src. Tyr 527 (Y527), when phosphorylated by the tyrosine kinase, Csk, or another unidentified tyrosine kinase, interacts with the SH2 domain of Src, in effect, masking the kinase domain (repressed state). Following ET-1, an unidentified tyrosine phosphatase dephosphorylates Tyr 527, exposing the kinase domain and allowing phosphorylation of the autophosphorylation site, Tyr 416, and interaction of both the SH2 domain and the kinase domain, and possibly the SH3 domain, with Src targets (active state). Adapted from ref. 215.

in mitogenesis, Src is a member of a large family of nonreceptor tyrosine kinases, including Lyn, Fyn, Lck, and Yes, and it appears that there may be significant overlap in function of these kinases.[138]

In cells transformed by the oncogene, v-src, which encodes a constitutively active Src mutant, Shc is found complexed with Grb2 and mSOS.[139] As discussed above, a similar cast of characters appears to transduce early steps in the mitogenic response to activation of G protein-coupled receptors. Briefly, one or more Src family members are believed to couple βγ to Ras activation by directly or indirectly (via activation of another tyrosine kinase of the Syk/ZAP-70 family) phosphorylating Shc which triggers complex formation with Grb2 and mSOS, bringing the guanine nucleotide exchange factor to Ras.[77]

Src kinase activity is tightly negatively regulated by phosphorylation of Tyr 527, at least in part catalyzed by the carboxy-terminal Src kinase (Csk) and possibly by Src itself (Fig. 13.4).[140-144] Kinase activity is repressed by binding of the SH2 domain of Src to phosphorylated Tyr 527, in effect masking the kinase domain (reviewed in 145). The oncogenic variant, v-Src, lacks Tyr 527 and is constitutively active as are those in which Tyr 527 has been mutated. Src is activated by dephosphorylation of Tyr 527 by an as yet unidentified tyrosine phosphatase. CD45 probably is responsible for activation of Src following T cell receptor activation and another transmembrane protein tyrosine phosphatase (RPTPα) may be involved in neuronal differentiation, but it is most unlikely that these transmembrane tyrosine phosphatases play any role in the response to ET-1.[146] Following dephosphorylation of Tyr 527, Tyr 416, a residue within the kinase domain, is autophosphorylated, further increasing Src kinase activity (Fig. 13.4). Not surprisingly, ET-1-induced Src activation in smooth muscle cells is accompanied by transient dephosphorylation of Tyr 527, and activation of Src is inhibited by the tyrosine phosphatase inhibitor, sodium orthovanadate.[81]

When Src is activated, its SH2 domain releases from Tyr 527 and the SH2 domain is then free to interact with phosphotyrosine residues on various receptors and signaling molecules.[82,84,147-149] In addition, the SH3 domain of Src may be critical for binding to Shc and other proteins which have proline-rich peptide motifs containing the sequence PXXP.[150] Finally, signaling proteins containing SH2 domains may complex to Src, presumably via interaction of their SH2 domains with phosphotyrosine residues on Src (reviewed in 60, 151). These multiple SH2-phosphotyrosine and SH3-proline-rich region interactions allow the assembly of multiprotein complexes containing signaling proteins which are not only candidate modulators of the mitogenic signal but also of cytoskeletal reorganization. The importance of these complexes in mitogenic signaling is best illustrated by the observations that overexpression of any of three so-called docking proteins, Crk, Shc, or Nck, which contain SH2 and SH3 domains but no obvious enzymatic activity, induces cellular transformation.[152-155]

Identification of candidate Src substrates or Src associated proteins, other than Shc, which might transduce mitogenic signals has been difficult. The p85 subunit of the PI-3-kinase which has been identified as complexed with v-Src, is an unlikely candidate since there is no evidence that the p85/p110 PI-3-kinase is activated in response to ET-1 (see below).[149] Similarly, we have found that the SH2 domain-containing protein tyrosine phosphatase, Syp, which is highly tyrosine phosphorylated in v-*src* transformed cells, does not appear to be regulated in response to ET-1 since Syp is not tyrosine phosphorylated (which appears to be its mechanism of activation) in response to ET-1 and does not form a complex with Src.[156,157] Although there are many other proteins which interact with Src in v-*src* transformed cells, the role of these proteins in the response to ET-1, especially the mitogenic response, are unknown.[148] In summary, while it is likely that Src activation plays a role in ET-1-induced mitogenesis, as it does in the response to growth factors, probably by activating complex formation of Shc/Grb2/mSOS, neither the phosphatase responsible for activating Src in response to ET-1, nor substrates other than Shc which might be responsible for transducing the mitogenic signals, are known.

Role of the JAK/STAT Pathway in ET-induced Mitogenesis

The Jak/STAT pathway was first discovered as a signal transduction pathway involved in gene expression induced by interferon (IFN)-α and IFN-γ (reviewed in 39, 40, 158). Members of the Jak family, for Janus kinase, include Jak1, Jak2, and Tyk2, and are notable for a tyrosine kinase domain and a second kinase-like domain. This second kinase-like domain was the impetus for naming them after the Roman god of gates and doorways, Janus, who is usually depicted with two faces.[60] Based on somatic cell genetic complementation experiments, the Jak family was shown to be critical in IFN responsiveness. A mutant cell line defective in IFN-α and -β signaling was rescued by transfection with genomic DNA encoding Tyk2 (reviewed in 39, 40, 158). More recently, these kinases have been found to play a role in signal transduction from virtually all cytokine receptors, including the mitogenic cytokines, erythropoietin, several interleukins, and GM-CSF and G-CSF as well as growth hormone and prolactin. Most importantly for this discussion, the vasoactive peptide, AngII, rapidly induces tyrosine phosphorylation of Jak2 and Tyk2 (but not Jak1) in rat aortic smooth muscle cells and enhances kinase activity of Jak2.[159]

Signal transduction through this pathway typically is initiated when a cytokine binds to its cognate receptor and induces dimerization of receptors. Jaks, which associate with a juxtamembrane domain of the cytoplasmic portion of the receptors, are brought into proximity, allowing the Jaks to transphosphorylate a tyrosine residue within the catalytic domain which enhances the kinase's activity. In the case of the AngII (AT$_1$) receptor, Jak 2 appears to directly associate with the AT$_1$ receptor based on coimmunoprecipitation. It is not known which portion of the AT$_1$ receptor is the Jak family target.

The activated kinases also phosphorylate residues within the cytokine receptors, possibly providing binding sites for other signaling molecules (e.g., Shc, p85 subunit of PI3K, and the tyrosine phosphatase, Syp), and for the second arm of this signaling pathway, the STATs (signal transducers and activators of transcription), which contain an SH2 domain.[158]

The STAT family includes at least six members at present. STATs, which in resting cells are largely cytoplasmic, associate with the various receptors after ligand stimulation. Interaction with and activation of STATs by the various cytokine receptors are specific, that is each receptor associates with a particular group of STATs. The specificity does not appear to be determined by which Jak family member associates but rather by the relative affinities of the STAT SH2 domain for the phosphotyrosine residue in the receptor. After the STATs associate with the receptor, they are phosphorylated by a Jak family member, at a conserved tyrosine residue in the C-terminus. Following this, the STATs dimerize. Once they dimerize, they are DNA-binding competent. The dimerization appears to be driven by interaction of the SH2 domain of one STAT with the phosphotyrosine residue of the other STAT, and vice versa.

Following dimerization, the STATs move into the nucleus via an unknown mechanism. All STATs, except for STAT2 which binds DNA weakly, if at all, bind very similar DNA sequences: TTCCNGGAA.[40,160] Any specificity of STAT binding to promoter sequences is determined by the preference of different STATs for the central nucleotide. For example, STAT1 prefers G in this position, and STAT5 prefer A.[6] Thus which genes are transcribed in response to a particular stimulus is determined in large part by which STATs are activated, and, in turn, which STATs are activated is determined by STAT SH2 domain interactions with specific receptors.

In rat aortic smooth muscle cells, AngII leads to tyrosine phosphorylation of STAT1α and β, and STAT2 with a slightly delayed time course compared to IFNs. STAT3 is phosphorylated minimally and only very late following AngII.[159] STAT1 translocation into the nucleus has been demonstrated following AngII, albeit with a markedly delayed time course compared to IFNs (30 min vs. 5 min), suggesting the possibility of an autocrine or paracrine mechanism.[159] Recently, the observations that phosphorylation of Ser 727 of STAT1 modulates transcriptional activating activity of STAT1, that Ser 727 lies within a consensus ERK phosphorylation motif, and that ERK-2 coimmunoprecipitates with STAT1 and are critical for STAT regulated gene expression, has suggested another mechanism by which G protein coupled receptors could utilize the STAT1 pathway.[161] That the ERKs are capable of phosphorylating STAT1 is also suggested by the observation that in NIH3T3 cells expressing a consitutively active MEK1, STAT1 is hyperphosphorylated.

The major functions of the Jak/STAT pathways are to transduce signals from cytokine receptors, and what role, if any, these pathways play in growth regulation in response to vasoactive peptides is unclear. Based on findings in "knock-out mice," STAT1, the only STAT to date which has been postulated to play a role in G protein coupled receptor signaling (and in growth factor and growth hormone signaling), does not appear to play an important role in the mitogenic response. Specifically, the only nonredundant role of STAT1 is to transduce signals from the IFN receptors and thereby regulate a set of genes controlling innate immunity (reviewed in 160). The knock-out mice responded normally to growth hormone and EGF administration, suggesting STAT1 played no nonredundant role in signaling from these receptors. The creation of these knock-out mice and of dominant inhibitory mutants of various components of the pathway, should allow a thorough evaluation of the role of STAT1 in G protein receptor signaling.

Phosphoinositide 3 Kinase

The precise role of the lipid products of phosphoinositide 3 kinase (PI3K), phosphatidylinositol (3,4,5)-trisphosphate and phosphatidylinositol (3,4)-bisphosphate, in cell signaling are unknown but they have been implicated in processes as diverse as membrane ruffling, the respiratory burst (superoxide generation) in neutrophils, and the regulation of cell growth (reviewed in 162). Not surprisingly, the role, if any, of PI3K enzymes and their products in endothelin signal transduction is not clear. The growth factor receptor-associated and Ras-modulated PI3K,[163] which consists of an 85 kDa regulatory subunit and a 110 kDa catalytic subunit, probably plays little, if any role in ET-1 signal transduction, based on the observations that there is no tyrosine phosphorylation of p85 or any increase in PI3K activity in anti-phosphotyrosine or anti p85 immunoprecipitates from ET-1 stimulated mesangial cells (T. Force, unpublished observations). Recently, however, a novel PI3K activity in myeloid cells was found to be activated in a pertussis toxin-sensitive manner by fMetLeuPhe, which signals via a G protein-coupled receptor.[164] This PI3K activity was stimulated by Gβγ but not by heterotrimeric G proteins and was immunologically and chromatographically distinct from the p85/p110 growth factor-activated PI3K.[164] Furthermore, wortmannin, which potently inhibits the p85/p110 PI3K activity, only modestly inhibited the Gβγ-activated PI3K. Hawkins and co-workers noted the similarities between growth factor-induced activation of PLCγ vs. G protein-linked receptor activation of PLCβ, and activation of the p85/p110 PI3K vs. the novel PI3K.[164]

More recently, experiments utilizing transfection of PI3Kγ and βγ subunits strongly suggest a role for 3 phosphorylated phosphoinositides as signaling molecules in the pathway from Gi-coupled receptors to a Src-like tyrosine kinase, Shc, and activation of Ras and the c-Raf-1/ERK cascade. It is possible that ET-1 may activate this novel PI3K, but as yet, ET-1-induced production of 3 phosphorylated phosphinositides has not been demonstrated. Furthermore, we have found no accumulation of phosphatidylinositol (3,4,5)-trisphosphate in glomerular mesangial

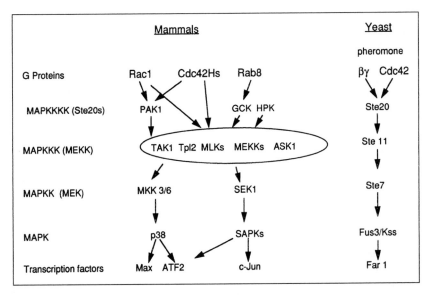

Fig. 13.5. The stress-activated protein kinase cascades. Two mammalian stress-activated MAP kinase cascades, culminating in the activation of the SAPKs or p38, are aligned with components of the yeast pheromone response pathway which culminates in activation of two MAP kinases, Fus3 and Kss1. The small GTP-binding proteins, Rac, Cdc42Hs, and Rab8 (which has been shown to interact with GCK in a yeast two hybrid screen), are also shown.

cells exposed to ET-1 but clear increases in cells exposed to PDGF (T. Force, unpublished observations). Thus the role of 3-phosphorylated phosphoinositides in mitogenic signaling in response to ET-1 remains unknown.

The Stress-activated MAP Kinase Signaling Cascades

Recently there has been an explosion of interest in other members of the MAP kinase superfamily of protein serine/threonine kinases. The remarkable conservation of the yeast pheromone pathway in mammals and the multiple MAP kinase pathways in yeast suggested the existence of other MAP kinases in humans. To date, seven mammalian MAP kinases have been cloned, and an eighth has been tentatively identified. ERK-5/BMK-1 may be involved in signaling in response to oxidant stress, and ERK-6 appears to be involved in skeletal muscle differentiation.[165-166] Fos regulating kinase (Frk) has been tentatively identified as a MAP kinase based on its substrate preference, though it has not been cloned.[167] This section will focus on the stress-activated protein kinases (SAPKs; also known as the c-Jun amino-terminal kinases or JNKs) and p38/RK, the mammalian homolog of HOG-1, a yeast kinase involved in the response to hyperosmolar stress (Fig. 13.5).[168-172] p38 and the SAPKs, as their name suggests, are activated by cellular stresses including inflammatory cytokines (TNFα and IL-1β), heat shock, osmolar stress, UV and ionizing radiation, and reperfusion of ischemic kidney or reversible ATP depletion induced by chemical anoxia in renal tubular epithelial cells.[168,173,174]

Very little has been published on the role of these other MAP kinase family members in ET signal transduction. Exposure of airway smooth muscle cells to ET activates the SAPKs and, importantly, the degree of activation is equivalent to that induced by TNFα, a potent activator of the SAPKs. Furthermore, the fold-activation of the SAPKs by ET was greater than the fold activation of the ERKs by ET.[55] More recently, the SAPKs were reported to be activated by ET-1 in glomerular mesangial cells and the magnitude of activation was equivalent to that of the ERKs.[56] These data raise the possibility that the SAPKs may be a major signaling arm of the ET receptor, and suggest these kinases might play a role in the ET-induced hypertrophic adaptation of myocytes and vascular smooth muscle cells. Supporting this, another agent which induces the hypertrophic phenotype and signals via Gq, Angiotensin II, also activates the SAPKs to a much greater degree than it activates the ERKs, and, as with ET-1 in glomerular mesangial cells, the activation is PKC-independent but calcium-dependent.[175] Finally, occupation of either the Gq-linked m1 muscarinic receptor or the Gi-linked m2 receptor in Rat 1a cells activates the SAPKs.[176] The m2 receptor also activated the ERK cascade, but the m1 receptor did not. SAPK activation was calcium-dependent for both receptors. The calcium transient was postulated to provide a mechanism for differential activation of the SAPK vs. the ERK cascades since the authors found that if the carbachol-induced calcium transient was inhibited by BAPTA, SAPK activation by the m2 receptor was prevented but activation of the ERK cascade was not. These data suggest differential activation by ET-1 of the SAPKs vs. the ERKs in different cell types may relate to the magnitude of the calcium transient. To date, it is not known whether ET activates p38, although most activators of one cascade also activate the other.

Like the ERKs, these kinases are downstream components of protein kinase cascades that have been conserved throughout millions of years of evolution from the simplest eukaryotes. The cascades (Fig. 13.5) consist of a MEK kinase (MEKK), which phosphorylates and activates a MEK, which, in turn, phosphorylates and activates a MAP kinase (the SAPKs or p38). Not only is this three-tiered module conserved, but within each tier, there are marked homologies between yeast and human kinases (reviewed in 174, 177).

Not surprisingly, given the wide array of activators of the SAPKs and p38, there is a remarkable diversity of upstream activators of the cascades (Fig. 13.5). MEKK1 and MEKK2 preferentially activate the SAPKs via activation of SEK-1 (also known as MKK4 or SAPKKK1), whereas MEKK3 preferentially activates the ERKs via MEK1 or MEK2. If the MEKKs are overexpressed, however, they become more promiscuous, a fact that has led to some confusion over what components of the cascades are physiologically relevant. No MEKKs have been identified which are in the cascade culminating in the activation of p38, but the identification of four other kinases, ASK1 (apoptosis signal regulating kinase), TAK1 (TGFβ activated kinase), Tpl2, and mixed lineage kinase-3 (MLK-3), two of which (ASK1 and MLK-3) activate the p38 cascade in transfection experiments, suggest that other MEKK-like kinases will be identified which are upstream components of the p38 cascade.[115,178-180]

In the yeast pheromone response pathway, epistasis analyses suggested a fourth tier of kinases, typified by Ste20, which is upstream of the yeast MEKK, Ste11, might play a role in the activation of the SAPKs and p38 in mammalian cells (Fig. 13.5). This was confirmed when two human Ste20-like kinases, germinal center kinase

(GC kinase, which is ubiquitously expressed but is highly so in the germinal centers of lymph nodes) and Pak1 (for p21 activated kinase) were found to be capable of activating the SAPKs and p38.[181,182]

Pak1, or a related kinase, is a candidate to mediate ET-induced activation of the SAPKs. This hypothesis is based on the activation of Pak1 (and p38) by the f-MetLeuPhe receptor in neutrophils which is coupled to a pertussis toxin sensitive G-protein.[182] Pak1 is activated by binding to members of the Rho family of small GTP binding proteins, Rac1 and Cdc42Hs, which regulate actin cytoskeletal reorganization, producing membrane ruffling and lamellopodia. When complexed to GTP, Rac and Cdc42Hs activate Pak1 by binding to a highly conserved region in the regulatory domain of Pak1. The binding then stimulates the autophosphorylation and activation of the kinase. Not surprisingly, constitutively active variants of Rac1 and Cdc42Hs markedly activate the SAPK and p38 cascades, and the activation appears to proceed via Pak1.[183-185]

It is not clear, however, whether ET-induced activation of the SAPKs proceeds via Rac/Cdc42Hs and Pak1 since to date, activation of the small GTP-binding proteins by ET has not been demonstrated. This may be due, in part, to the lack of an adequate antibody, analogous to the Ras antibody, which would allow one to directly examine GTP loading of Rac or Cdc42Hs. In lieu of this, translocation of the small G proteins to the membrane fraction has been employed as a marker of activation.[186] In Swiss 3T3 fibroblasts, ET-1 induced the translocation of RhoA (which does not activate the SAPKs), but not Rac or Cdc42Hs, to a membrane fraction. These data suggest that ET-1 may utilize a Rac/Cdc42Hs and Pak1-independent pathway to SAPK activation. Given the rapidly expanding list of Ste20-like kinases from the GC kinase subfamily, it is possible that one of these transduces the upstream signal to SAPK activation.

Although the intervening steps are still unclear, several groups have demonstrated that G-protein-coupled receptors, in addition to the ET, AngII, and fMetLeuPhe receptors, activate the SAPKs and/or p38. Gi-coupled receptors, including the α1-adrenergic receptor and the m2 muscarinic receptor, activate the SAPKs, and the thrombin receptor activates p38.[187-190] In NIH3T3 cells, the Gq-linked m1 muscarinic receptor also activates the SAPKs and the activation is independent of PKCs.[190] The proximal mechanisms of activation of the cascade appear to employ βγ subunits since overexpression of βγ markedly activated the SAPKs. In contrast, αs, αi2, αq, and α13 did not.[190] Other groups have presented evidence that expression of GTPase deficient variants of αq, as well as α12 and α13 induces a persistent, albeit modest (two to three-fold) activation of the SAPKs, but this could be due to a secondary stress response to the constitutively active α subunits.[87,191]

Activation of the SAPKs by βγ subunits is dependent upon both Ras and Rac since expression of a dominant interfering mutant of either of the small G proteins blocks activation by βγ.[190] The mechanism of activation is unclear but in yeast, βγ activates Cdc42 which then binds to and activates Ste20, the yeast homolog of Pak1. In mammals, the most likely mechanism is that βγ recruits a guanine nucleotide exchange factor(s) for Ras and Rac, and GTP-loaded Rac then activates Pak1 and the rest of the cascade. Thus we have a skeleton of the pathway utilized by the ET receptor and other G protein-linked receptors to activate the SAPKs, but even more so than with the ERK cascade, many uncertainties remain. Identifying these components is another of the major challenges for those studying ET signal transduction.

The Biology of SAPK and p38 Activation

Substrates of the SAPKs and p38

Like the ERKs, following phosphorylation and activation, the SAPKs and p38 translocate to the nucleus where they phosphorylate and activate several targets. Some clues to the biology of SAPK/p38 activation can be found by examining those targets. The SAPKs are the predominant stress-activated c-Jun transcriptional activation domain kinases, and phosphorylation within this domain markedly increases transcriptional activating activity of c-Jun.[168,173,192] The transcription factor ATF-2 (a member of the CREB family of transcription factors) is also phosphorylated within its transactivation domain by both the SAPKs and p38.[193-198] Both kinases increase transactivating activity of ATF-2, and the SAPKs also appear to enhance the DNA binding activity of ATF-2 via phosphorylation of the DNA binding/dimerization domain of the transcription factor.[193-195,197-199] Since c-Jun and ATF-2 appear to control induction of c-*jun* in response to certain cellular stresses, it is likely that the SAPKs transduce the signal to the nucleus which leads to induction of c-*jun*. Another immediate early gene which is induced following cellular stress is c-*fos*. Elk-1 and SAP-1a, which are Ternary Complex Factors (TCFs) and, along with Serum Response Factor are critical to induction of c-*fos* in response to stress, are also substrates for p38 and the SAPKs, and p38 has been shown to play a critical role in the induction of c-*fos* in response to UV irradiation.[200] p38 also phosphorylates Max, a basic helix-loop-helix protein that binds the transcription factor, c-Myc, and is essential for its DNA binding and transcriptional activating activity.[201] The role of this phosphorylation is not known at present. Finally, the p38 pathway may also control activation of CREB and the related ATF-1. One of the substrates of p38 is MAPKAP kinase-2 (for mitogen-activated protein kinase activated protein kinase-2), a kinase which initially was thought to be a substrate of MAPKs (ERKs). MAPKAP kinase-2 phosphorylates CREB at Ser133 in vitro, and this phosphorylation increases transcriptional activating activity of CREB.[202] CREB is phosphorylated on Ser133 following cellular stress (sodium arsenite), and it is likely that this is mediated via activation of p38 and its target, MAPKAP kinase-2, since the specific p38 inhibitor, SB203580, markedly inhibits CREB phosphorylation and activation by arsenite.

p38 has other substrates with roles in signal transduction and the response to stress. p38 plays an important role in the aggregation of platelets in response to a variety of agonists since pretreatment of platelets with SB203580 inhibits aggregation.[189] This has been postulated to be related to phosphorylation by p38 of a critical residue (Ser 505) of cytosolic phospholipase A2 (cPLA2), a residue that is also phosphorylated by the SAPKs. Phosphorylation of this residue by the ERKs has been reported to increase the activity of cPLA2.[203] Although it remains unclear if p38 activates cPLA2 in platelets, given the potent activation of cPLA2 by ET, the role of p38 and the SAPKs in this activation needs to be explored.[187]

MAPKAP kinase-2, which, as noted above, is activated by p38 and may phosphorylate CREB and ATF-1, also phosphorylates the small heat shock protein, Hsp25/HSP27, whose phosphorylation is a prominent part of the response to cytokines, cellular stress (ATP depletion, heat shock, etc.), and some growth factors. While the physiological significance of the phosphorylation is unknown and the function of Hsp25/HSP27 is not clear, it likely helps the cell respond to these stresses.

Biological Effects of SAPK/p38 Activation

As noted, insights into the biological function of the SAPKs and p38 can be gained by identifying downstream targets. For example, ET-1 induces c-*jun* and c-*fos*, and at least for c-*jun*, this presumably involves phosphorylation by the SAPKs of c-Jun and ATF-2 (see above and chapter 12). However, since c-*jun* is induced in response to a wide variety of stimuli, it is difficult to draw any conclusions about the biological responses initiated by the SAPKs and p38 once they are activated. These must be examined directly.

Based on a series of studies, it is now clear that, in susceptible cell types, activation of the SAPK and/or p38 cascades can trigger apoptosis or programmed cell death.[204-207] A wide array of stimuli, including withdrawal of NGF from PC12 cells, oxidant stress, UV irradiation, DNA damaging agents such as cis-platinum and ionizing radiation, heat shock, and TNFα, induce apoptosis in susceptible cells. Since expression of dominant inhibitory mutants of components of the SAPK or p38 pathways can block apoptosis, and expression of constitutively active components can induce apoptosis, it is likely that these pathways transduce critical signals in the response to certain apoptotic stimuli. In PC12 cells, expression of constitutively active MEK1, the immediate upstream activator of the ERKs, prevents apoptosis induced by NGF withdrawal, suggesting that the decision to initiate apoptosis or not may depend on the balance between anti-apoptotic signals transduced by the ERK cascade and pro-apoptotic signals transduced by the SAPK/p38 cascades. c-Raf-1 may be involved as well since it may phosphorylate and inactivate the pro-apoptotic Bcl-2 family member, Bad.[208,209] Clearly, however, other protein kinase cascades may modulate the apoptotic response. For example, in Swiss 3T3 cells, expression of a constitutively active MEKK1 (see Fig. 13.5) markedly activates the SAPKs (but not p38) and initiates apoptosis, but coexpression of a dominant inhibitor of SAPK activation fails to block MEKK1-induced apoptosis.

ET has not been reported to initiate apoptosis. It is conceivable, since ET-induced activation of the SAPKs is at least as great as activation of the ERKs, that some type of cell will be found which undergoes apoptosis in response to ET. Although one report of apoptosis in response to Ang II has appeared, it seems unlikely that apoptosis will be found to be an important, physiologically relevant response to ET in normal cells.[210]

It seems much more likely that ET-induced SAPK and/or p38 activation will play a prominent role in the induction of the hypertrophic phenotype in susceptible cells. I have already noted that several hypertrophic stimuli besides ET, including Ang II and α-adrenergic agonists, markedly activate the SAPKs and the activation is more pronounced than the ERKs.[211-213] Transfection of neonatal rat ventricular myocytes with constitutively active MEKK1, the upstream activator of the SAPKs, induced an increase in myocyte size and transcriptional changes characteristic of the hypertrophic response (increased promoter activity of the "marker" genes, atrial natriuretic factor, β-myosin heavy chain, and skeletal muscle α-actin).[212] In contrast, activation of the ERK cascade produces similar changes in promoter activity, but no increase in myocyte size. Although another marker of the hypertrophic response, an increase in organization of the myofilaments, was not seen after transfection of MEKK1, indicating the SAPK cascade is not solely responsible for full induction of the hypertrophic response, it is likely to be an important component. Completing the link from the ET receptor through a cytosolic signaling cascade to the hypertrophic phenotype will be a critical link to understanding the biology of the response of cells to ET.

Finally, in some cells, activation of the p38 cascade and, in some cells, the SAPK cascade, does not induce mitogenesis, apoptosis, or hypertrophy, but causes cell cycle arrest at the G1/S transition.[214] Although much work has focused on the role of ETs as mitogens or comitogens, since activation of the SAPKs is a prominent component of the response to ET-1, it will be equally important to determine whether ETs might also force cells out of the cell cycle, toward a postmitotic, differentiated phenotype. Indeed, the ET-1 induced marked activation of the SAPKs, in parallel with the activation of the ERKs, is distinctly different than the response to most growth factors which predominantly, or exclusively, activate the ERKs. The balance between ERK activation and SAPK activation may help determine whether the cell decides to respond to ET by undergoing cell cycle arrest/differentiation or mitogenesis. Determining how the cell makes this decision is clearly one of the most important challenges in the field.

Conclusions

ET-1 has been known for many years to be a mitogen in various types of cells, yet a picture of the pathways transducing the mitogenic signal to the nucleus has only recently begun to develop. Although sometimes based on limited data, a recurring theme has emerged: despite the marked differences in proximal signal transduction mechanisms of the G-protein coupled receptors and the growth factors, both groups of mitogens share several distal pathways. Most important among these appear to be the c-Raf-1/ERK cascade and activation of nonreceptor tyrosine kinases, both of which have been repeatedly demonstrated to be critical for growth factor-induced mitogenesis and presumably are as well for ET. A key difference between the growth factors and ET-1 are that the latter also importantly activates stress response kinases. Major questions remain including what role do the stress-activated MAP kinase cascades play in the biologic response to ET? Does this response arm account for some of the specificity in the transcriptional and biological responses of cells to growth factors vs. ET? Only by addressing these questions can we understand the mechanisms regulating the pleiotropic biological responses of cells to this peptide.

References

1. Bobik A, Grooms A, Millar JA et al. Growth factor activity of endothelin on vascular smooth muscle. Am J Physiol 1990; 258:C408-C415.
2. Malarkey K, Chilvers ET, Lawson MF et al. Stimulation by endothelin-1 of mitogen-activated protein kinases and DNA synthesis in bovine tracheal smooth muscle cells. Br J Pharmacol 1995; 116:2267-2273.
3. Weber H, Webb ML, Serafino R et al. Endothelin-1 and Angiotensin-II stimulate delayed mitogensis in cultured rat aortic smooth muscle cells: evidence for common signaling mechanisms. Mol Endocrinol 1994; 8:148-156.
4. Simonson MS, Wann S, Mene P et al. ET stimulates phospholipase C, Na$^+$/H$^+$ exchange, c-fos expression, and mitogenesis in rat mesangial cells. J Clin Invest 1989; 83:708-712.
5. Jaffer FE, Knauss TC, Poptic E et al. Endothelin stimulates PDGF secretion in cultured human mesangial cells. Kidney Int 1990; 38:1193-1198.
6. Simonson MS, Herman WH. Protein kinase C and protein tyrosine kinase activity contribute to mitogenic signaling by endothelin-1. J Biol Chem 1993; 13:9347-9357.

7. Muldoon L, Pribnow D, Rodland KD et al. ET-1 stimulates DNA synthesis and anchorage-independent growth of rat-1 fibroblasts through a protein kinase C-dependent mechanism. Cell Reg 1990; 1:379-390.

8. Takuwa N, Takuw Y, Yanagisawa M et al. A novel vasocative peptide stimulates mitogenesis through inositol lipid turnover in Swiss 3T3 fibroblasts. J Biol Chem 1989; 264:7856-7861.

9. Weissberg PL, Witchell C, Davenport AP et al. The endothelin peptides ET-1, ET-2, ET-3 and sarafotoxin S6b are comitogenic with PDGF for VSMC. Atherosclerosis 1990; 85:257-262.

10. Hirata Y, Takagi Y, Fukuda Y et al. ET is a potent mitogen for rat vascular smooth muscle cells. Atherosclerosis 1989; 78:225-228.

11. Komuro I, Kurihara H, Sugiyama T et al. ET stimulates c-fos and c-myc expression and proliferation of vascular smooth muscle cells. FEBS Lett 1988; 238:249-252.

12. Imokawa G, Yada Y, Kimura M. Signaling mechanisms of endothelin-induced mitogenesis and melanogenesis in human melanocytes. Biochem J 1996; 314:305-312.

13. Battistini B, Chailler P, D'Orleans-Juste P et al. Growth regulatory properties of endothelins. Peptides 1993; 14:385-399.

14. Simonson MS, Jones J, Dunn M. Differential regulation of fos and jun gene expression and AP-1 cis-element activity by endothelin isopeptides. J Biol Chem 1992; 267:8643-8649.

15. Ohlstein EH, Arleth A, Bryan H et al. The selective endothelin receptor antagonist BQ-123 antagonizes endothelin-1-mediated mitogenesis. Eur J Pharmacol 1992; 225:347-350.

16. Kanse S, Wijelath E, Kanthou C et al. The proliferative responsiveness of human vascular smooth muscle cells to endothelin correlates with endothelin receptor density. Lab Invest 1995; 72:376-382.

17. Wang Y, Rose PM, Webb ML et al. Endothelin stimulates mitogen-activated protein kinase cascade through either ET_A or ET_B. Am. J Physiol 1994; 267:C1130-1135.

18. Lazarini F, Strosberb AD Couraud PO et al. Coupling of ET-B endothelin receptor to mitogen-activated protein kinase stimulation and DNA synthesis in primary cultures of rat astrocytes. J Neurochem 1996; 66:459-465.

19. Mallat A, Fouassier L, Preaus A et al. Growth inhibitory properties of endothelin-1 human hepatic myofibroblastic Ito cells. J Clin Invest 1995; 96:42-49.

20. Serradeil-Le Gal C, Herbert JM, Garcia C et al. Importance of the phenotypic state of vascular smooth muscle cells on the binding and mitogenic activity of endothelin. Peptides 1991; 12:575-579.

21. Naftilan AJ, Re P., Dzau VJ. Induction of platelet derived growth factor A-chain and c-myc gene expression by angiotensin II in cultured rat vascular smooth muscle cells. J Clin Invest 1989; 83:1419-1424.

22. Gibbons GH, Pratt RE, Dzau VJ. Vascular smooth muscle cell hypertrophy vs. hyperplasia. J Clin Invest 1992; 90:456-461.

23. Weber H, Taylor DS, Molloy CJ. Angiotensin II induces delayed mitogenesis and cellular proliferation in rat aortic smooth muscle cells: correlation with the expression of specific endogenous growth factors and reversal by suramin. J Clin Invest 1994; 93:788-798.

24. Simonson MS, Dunn MJ. Cellular signaling by peptides of the endothelin gene family. FASEB J 1990; 4:2989-3000.

25. Marrero M, Paxon W, Duff J et al. Angiotensin II stimulates tyrosine phosphorylation of phospholipase C-g1 in vascular smooth muscle cells. J Biol Chem1994; 269:10935-10939.

26. Clapham DE. Calcium signalling. Cell 1995; 80:259-268.

27. Ghosh A, Greenberg ME. Calcium signalling in neurons: molecular mechanisms and cellular consequences. Science 1995; 268:239-247.

28. Means AR. Calcium, calmodulin, and cell cycle regulation. FEBS Lett 1994; 347:1-4.

29. Nakaki T, Nakayama M, Yamamoto S et al. ET-mediated stimulation of DNA synthesis in VSMC. Biochem Biophys Res Commun 1989; 158:880-883.

30. Shichiri M, Hirata Y, Nakajima T et al. ET-1 is an autocrine/paracrine growth factor for human cancer cell lines. J Clin Invest 1991; 87:1867-1871.

31. Muldoon, L, Rodland KD, Forsythe ML et al. Stimulation of phosphatidylinositol hydrolysis, diacylglycerol release, and gene expression in response to endothelin, a potent new agonist for fibroblasts and smooth muscle cells. J Biol Chem 1989; 264:8529-8536.

32. Griendling k, Tsuda T, Alexander RW. Endothelin stimulates diacylglycerol accumulation and activates protein kinase C in cultured vascular smooth muscle cells. J Biol Chem 1989; 264:8237-8240.

33. Force T, Kyriakis JM, Avruch J et al. Endothelin, vasopressin, and angiotensin II enhance tyrosine phosphorylation by protein kinase C-dependent and -independent pathways in glomerular mesangial cells. J Biol Chem 1991; 266:6650-6656.

34. Chua BHL, Krebs CJ, Chua CC et al. ET stimulates protein synthesis in SMC. Am J Physiol 1992; 262:E412-E416.

35. Huckle WR, Prokop CA, Dy RC et al. Angiotensin II stimulates protein-tyrosine phosphorylation in a calcium-dependent manner. Mol Cell Biol 1990; 10:6290-6298.

36. Zachary I, Sinnett-Smith J, Rozengurt E. Stimulation of tyrosine kinase activity in anti-phosphotyrosine immune complexes of Swiss 3T3 cell lysates occurs rapidly after addition of bombesin, vasopressin, and endothelin to intact cells. J Biol Chem 1991; 266:24126-24133.

37. Zachary I, Sinnett-Smith J, Rozengurt E. Bombesin, vasopressin, and endothelin stimulation of tyrosine phosphorylation in Swiss 3T3 cells. J Biol Chem 1992; 267:19031-19034.

38. Lev S, Moreno H, Martinez R et al. Protein tyrosine kinase PYK2 involved in Calcium-induced regulation of ion channel and MAP kinase functions. Nature 1995; 376:737-745.

39. Darnell JE, Kerr IM, Stark GR. Jak-STAT pathways and transcriptional activation in response to IFNs and other extracellular signaling proteins. Science 1994; 264:1415-1421.

40. Ihle JN, Witthuhn BA, Quelle FW et al. Signaling by the cytokine receptor superfamily: JAKs and STATs. TIBS 1994; 19:222-227.

41. Ray BL, Sturgill TW. Insulin-stimulated microtubule-associated protein kinase is phosphorylated on tyrosine and threonine in vivo. Proc Natl Acad Sci 1988; 85:3753-3757.

42. Sturgill TW, Ray LB, Erikson E et al. Insulin-stimulated MAP-2 kinase phosphorylates and activates ribosomal protein S6 kinase II. Nature 1988; 334:715-718.

43. Rapp UR, Goldsborough MD, Mark GE et al. Structure and biological activity of v-raf, a unique oncogene transduced by a retrovirus. Proc Natl Acad Sci 1983; 80:4218-4222.

44. Rapp UR, Heidecker G, Huleihel M et al. Raf family serine/threonine protein kinases in mitogen signal transduction. Cold Spring Harbor Symp on Quant Biol 1988; 53:173-184.

45. Rapp UR. Role of Raf-1 serine/threonine protein kinase in growth factor signal transduction. Oncogene 1991; 6:495-500.

46. Kyriakis JM, App H, Shang X et al. Raf-1 activates MAP kinase-kinase. Nature 1992; 358:417-421.

47. Dent P, Haser W, Haystead TAJ et al. Activation of mitogen-activated protein kinase kinase by v-Raf in NIH 3T3 cells and in vitro. Science 1992; 257:1404-1407.

48. Howe LR, Leevers SJ, Gomez N et al. Activation of the MAP kinase pathway by the protein kinase raf. Cell 1992; 71:335-342.

49. Pages G, Lenormand P, L'Allemain G et al. Mitogen-activated protein kinases p42mapk and p44mapk are required for fibroblast proliferation. Proc Natl Acad Sci 1993; 90:8319-8323.

50. Cowley S, Paterson H, Kemp P et al. Activation of MAP kinase kinase is necessary and sufficient for PC12 differentiation and for transformation of NIH 3T3 cells. Cell 1994; 77:841-852.

51. Mansour S, Matten W, Hermann A et al. Transformation of mammalian cells by contitutively active MAP kinase kinase. Science 1994; 265:966-970.

52. Marshall C. Specificity of receptor tyrosine kinase signalling: transient versus sustained extracellular signal-regulated kinase activation. Cell 1995; 80:179-185.

53. Kyriakis JM, Force TL, Rapp UR et al. Mitogen regulation of c-Raf-1 protein kinase activity toward mitogen-activated protein kinase kinase. J Biol Chem 1993; 268:16009-16019.

54. Herman WH, Simonson MS. Nuclear signaling by endothelin-1. J Biol Chem 1995; 270:11654-11661.

55. Shapiro PS, Evans JN, Davis RJ et al. The seven-transmembrane-spanning receptors for endothelin and thrombin cause proliferation of airway smooth muscle cells and activation of the extracellular regulated kinase and c-Jun NH_2-terminal kinase groups of mitogen-activated protein kinases. J Biol Chem 1996; 271: 5750-5754.

56. Araki S, Haneda M, Togawa M et al. Endothelin-1 activates c-Jun NH2-terminal kinase in mesangial cells. Kidney Int 1997; 51:631-639.

57. Heldin C-H. Dimerization of cell surface receptors in signal transduction. Cell 1995; 80:213-223.

58. Ullrich A, Schlessinger J. Signal transduction by receptors with tyrosine kinase activity. Cell 1990; 61:203-212.

59. Songyang Z, Shoelson SE, Chauduri M et al. SH2 domains recognize specific phosphopeptide sequences. Cell 1993; 72:767-778.

60. Cohen GB, Ren R, Baltimore D. Modular binding domains in signal transduction proteins. Cell 1995; 80:237-248.

61. Zhang X-F, Settleman J, Kyriakis J et al. Normal and oncogenic p21ras proteins bind to the amino-terminal regulatory domain of c-Raf-1. Nature 1993; 364:308-313.

62. Bruder JT, Heidecker G, Rapp UR. Serum-, TPA-, and Ras-induced expression from Ap-1/Ets-driven promoters requires Raf-1 kinase. Genes & Devlopment 1992; 6:545-556.

63. Leevers SJ, Paterson HF, Marshall CJ. Requirement for Ras in Raf activation is overcome by targeting Raf to the plasma membrane. Nature 1994; 369:411-414.

64. Stokoe D, Macdonald SG, Cadwallader K et al. Activation of Raf as a result of recruitment to the plasma membrane. Science 1994; 264:1463-1467.

65. Daum G, Eisenmann-Tappe I, Fries H-W et al. 1994. The ins and outs of Raf kinases. TIBS. 19:474-480.

66. Li S, Janosch P, Tanji M et al. Regulation of Raf-1 kinase activity by the 14-3-3 family of proteins. EMBO J 1995; 14:685-696.

67. Luo Z, Tzivion G, Belshaw PJ et al. Oligomerization activates c-Raf-1 through Ras dependent mechanism. Nature 1996; 383:181-185.

68. Farrar MA, Ila-Alberola J, Perlmutter RM. Activation of the Raf-1-kinase cascade by coumermycin-induced dimerization. Nature 1996; 383:178-181.

69. Marshall CJ. Raf gets it together. Nature 1996; 383:127-128.

70. Wang Y, Simonson MS, Pouyssegur J et al. Endothelin rapidly stimulates mitogen-activated protein kinase activity in rat mesangial cells. Biochem J 1992; 287:589-594.

71. Winitz S, Russell M, Qian N-X et al. Involvement of Ras and Raf in the Gi-coupled acetylcholine muscarinic m2 receptor activation of mitogen-activated protein (MAP) kinase kinase and MAP kinase. J Biol Chem 1993; 268:19196-19199.

72. Van Corven EJ, Hordijk PL, Medema RH et al. Pertussis toxin-sensitive activation of p21ras by G protein-coupled receptor agonists in fibroblasts. Proc Natl Acad Sci 1993; 90:1257-1261.

73. Ablas J, van Corven EJ, Hordijk PL et al. Gi-mediated activation of the p21ras - mitogen-activated protein kinase pathway by a2-adrenergic receptors expressed in fibroblasts. J Biol Chem 1993; 268:22235-22238.

74. Howe LR, and Marshall CJ. Lysophosphatidic acid stimulates mitogen-activated protein kinase activation via a G-protein-coupled pathway requiring p21Ras and p74Raf-1. J Biol Chem 1993; 268:20717-20720.

75. Hawes BE, van Biesen T, Koch W et al. Distinct pathways of Gi-and Gq-mediated mitogen-activated protein kinase activation. J Biol Chem 1995; 270:17148-17153.

76. van Blesen T, Hawes BE, Luttrell DK et al. Receptor-tyrosine-kinase-and Gβγ-mediated MAP kinase activation by a common signalling pathway. Nature 1995; 376:781-784.

77. Wan,Y, Kurosaki T, Huang X. Tyrosine kinases in activation of the MAP kinase cascade by G-protein-coupled receptors. Nature 1996; 380:541-544.

78. Crespo P, Xu N, Daniotti JL et al. Signaling through transforming G protein-coupled receptors in NIH 3T3 cells involves c-Raf activation. J Biol Chem 1994; 269:21103-21109.

79. Crespo P, Xu N, Simonds WF et al. Ras-dependent activation of MAP kinase pathway mediated by G-protein βγ subunits. Nature 1994; 369:418-420.

80. Faure M, Yasenetskaya-Voyno TA, Bourne HE. cAMP and subunits of heterotrimeric G proteins stimulate the mitogen-activated protein kinase pathway in COS-7 cells. J Biol Chem 1994; 269:7851-7854.

81. Force T, Bonventre JV. Endothelin activates Src tyrosine kinase in glomerular mesangial cells. J Am Soc Nephrol 1992; 3:491.

82. Kypta RM, Goldberg Y, Ulug ET et al. Association between the PDGF receptor and members of the src family of tyrosine kinases. Cell 1990; 62:481-492.

83. Gould K, Hunter T. Platelet-derived growth factor induces multisite phosphorylation of pp60c-src and increases its protein-tyrosine kinase activity. Mol Cell Biol 1988; 8:3345-3356.

84. Courtneidge SA, Dhand R, Pilat D et al. Activation of Src family kinases by colony stimulating factor-1, and their association with its receptor. EMBO J 1993; 12:943-950.

85. Lopez-Ilasaca M, Crespo P, Pellici Giuseppe P et al. Linkage of G protein-coupled receptors to the MAPK signaling pathway through PI 3-kinase. Science 1997; 275:394-397.

86. Sugawara F, Ninomiya H, Okamoto Y et al. Endothelin-1-induced mitogen responses of Chinese hamster ovary cells expressing human endothelinA: The role of a wortmannin-sensitive signaling pathway. Mol Pharm 1996; 49:447-457.

87. Heasley LE, Storey B, Fanger G et al. GTPase-deficient Gα16 and Gαq induce PC12 cell differentiation and persistent activation of c-Jun NH2-terminal kinases. Mol Cell Biol 1996; 16:648-656.

88. L'Allemain G, Sturgill TW, Webber MJ. Defective regulation of mitogen-activated protein kinase activity in a 3T3 cell variant mitogenically nonresponsive to tetradecanoyl phorbol acetate. Mol Cell Biol 1991;1002-1008.

89. Cobb MH, Goldsmith EJ. How MAP kinases are regulated. J Biol Chem 1995; 270:14843-14846.

90. Leevers SJ, Marshall CJ. Activation of extracelluar signal-regulated kinase, ERK-2, by p21ras oncoprotein. EMBO J 1992; 11:569-574.

91. Robbins DJ, Cheng M, Zhen E et al. Evidence for a Ras-dependent extracellular signal-regulated protein kinase (ERK) cascade. Biochemistry 1992; 89:6924-6928.

92. Ueda Y, Hirai S, Osada S et al. Protein kinase C delta activates the MEK-ERK pathway in a manner independent of Ras and dependent on Raf. J Biol Chem 1996; 271:23512-23519.

93. Cacace AM, Ueffing M, Philipp A et al. PKC epsilon functions as an oncogene by enhancing activation of the Raf kinase. Oncogene 1996; 13:2517-2526.

94. Sozeri O, Vollmer K, Kiyanage M et al. Activation of the c-Raf protein kinase by protein kinase C phosphorylation. Oncogene 1992; 7:2259-2262.

95. Kolch W, Heidecker G, Kochs G et al. Protein kinase Ca activates RAF-1 by direct phosphorylation. Nature 1993; 364:249-252.

96. MacDonald SG, Crews CM, Wu L et al. Reconstitution of the Raf-1-MEK-ERK signal transduction pathway in vitro. Mol Cell Biol 1993; 13:6615-6620.

97. Whitehurst CE, Owaki H, Bruder JT et al. The MEK kinase activity of the catalytic domain of RAF-1 is regulated independently of Ras binding in T cells. J Biol Chem 1995; 270:5594-5599.

98. Bogoyevitch M, Clerk A, Sugden PH. Activation of the mitogen-activated protein kinase cascade by pertussis toxin-sensitive and -insensitive pathways in cultured ventricular cardiomyocytes. J Biochem 1995; 309:437-443.

99. Bogoyevitch MA, Marshall CJ, Sugden PH. Hypertrophic agonists stimulate the activities of the protein kinases c-Raf and A-Raf in cultured ventricular myocytes. J Biol Chem 1995; 270:26303-26310.

100. Qian N-X, Winitz S, Johnson GL. Epitope-tagged Gq alpha subunits: expression of GTPase-deficient alpha subunits persistently stimulates phosphatidylinositol-specific phospholipase C but not mitogen-activated protein kinase activity regulated by the M1 muscarinic acetylcholine receptor. Proc Natl Acad Sci 1993; 90:4077-4081.

101. Kribben A, Wieder ED, Li X et al. AVP-induced activation of MAP kinase in vascular smooth muscle cells is mediated through protein kinase C. Am J Physiol 1993; 265:C939-C945.

102. Exton JH. Phosphatidylcholine breakdown and signal transduction. Biochim Biophys Acta 1994; 1212:26-42.

103. Cai H, Erhardt P, Troppmair J et al. Hydrolysis of phosphatidylcholine couples Ras to activation of Raf protein kinase during mitogenic signal transduction. Mol Cell Biol 1993; 13:7645-7651.

104. Force T, Bonventre JV, Heidecker G et al. Enzymatic characteristics of the c-Raf-1 protein kinase. Proc Natl Acad Sci 1994; 91:1270-1274.

105. Marais R, Light Y, Paterson HF et al. Ras recruits Raf-1 to the plasma membrane for activation by tyrosine phosphorylation. EMBO J 1995; 14:3136-3145.

106. Wu J, Dent P, Jelinek T et al. Inhibition of the EGF-activated MAP kinase signaling pathway by adenosine 3',5' - monophosphate. Science 1993; 262:1065-1069.

107. Cook SJ, McCormick F. Inhibition by cAMP of Ras-dependent activation of Raf. Science 1993; 262:10691072.

108. Jelinek T, Catling AD, Reuter CW et al. RAS and RAF-1 form a signalling complex with MEK-1 but not MEK-2. Mol Cell Biol 1994; 14:8212-8218.

109. Zheng C-F, Guan K-L. Activation of MEK family kinases requires phosphorylation of two conserved Ser/Thr residues. EMBO J 1994; 13:1123-1131.

110. Lange-Carter CA, Pleiman CM, Gardner AM et al. A divergence in the MAP kinase regulatory network defined by MEK kinase and Raf. Science 1993; 260:315-319.

111. Herskowitz I. MAP kinase pathways in yeast: for mating and more. Cell 1995; 80:187-197.

112. Yan M, Dai T, Deak JC et al. Activation of stress-activated protein kinase by MEKK1 phosphorylation of its activator SEK1. Nature 1994; 372:798-800.

113. Okazaki K, Sagata N. The Mos/MAP kinase pathway stabilizes c-Fos by phosphorylation and augments its transforming activity in NIH 3T3 cells. EMBO J 1995; 14:5048-5059.

114. Posada J, Yew N, Ahn NG et al. Mos stimulates MAP kinase in Xenopus oocytes and activates a MAP kinase kinase in vitro. Mol Cell Biol 1993; 13:2546-2553.

115. Salmeron A, Ahmed TB, Carlile GW et al. Activation of MEK-1 and SEK-1 by Tpl-2 protooncoprotein, a novel MAP kinase kinase kinase. EMBO J 1996; 15:817-826.

116. Blank JL, Gerwin P, Elliott EM et al. Molecular cloning of mitogen activated pro tein/ERK kinase kinases (MEKK) 2 and 3. J Biol Chem 1996; 271:5361-5368.

117. Payne DM, Rossomando AJ, Martino P et al. Identification of the regulatory phosphorylation sites in pp42/mitogen-activated protein kinase (MAP kinase). EMBO J 1991; 10:885-892.

118. Lenormand P. Sardet C, Pages G et al. Growth factors induce nuclear translocation of MAP kinases (p42 and p44) but not of their activator MAP kinase kinase (p45) in fibroblasts. J Cell Biol 1993; 122:1079-1088.

119. Gonzalez F, Seth A, Raden D et al. Serum-induced translocation of mitogen-activated protein kinase to the cell surface ruffling membrane and the nucleus. J Cell Biol 1993; 122:1089-1101.

120. Meloche S, Pages G, Pouyssegur J. Biphasic and synergistic activation of p44MAPK (ERK1) by growth factors: correlation between late phase activation and mitogenicity. Mol Endocrinol 1992; 6:845-854.

121. Sakurai T. Abe Y, Takuwa N et al. Activin A stimulates mitogenesis in Swiss 3T3 fibroblasts without activation of mitogen-activated protein kinases. J Biol Chem 1994; 269:14118-14122.

122. Chen R-H, Sarnecki C, Blenis J. Nuclear localization and regulation of erk-and rsk-encoded protein kinases. Mol Cell Biol 1992; 12:915-927.

123. Xing J, Ginty DD, Greenberg ME. Coupling of the RAS-MAPK pathways to gene activation by RSK2, a growth factor-regulated CREB kinase. Science 1996; 273:959-960.

124. Grana X, Reddy EP. Cell cycle control in mammalian cells. Oncogene 1995; 11:211-219.

125. Lavoie JN, L'Allemain G, Brunet A et al. Cyclin D1 expression is regulated positively by the p42/44 MAPK and negatively by the p38/HOG MAPK pathway. J Biol Chem 1996; 271:20608-20616.

126. Sun H, Tonks NK, Bar-Sagi D. Inhibition of Ras-induced DNA synthesis by expression of the phosphatase MKP-1. Science 1994; 266:285-288.

127. Sun H, Charles C, Lau L et al. MKP-1 (3CH134), an immediate early gene product, is a dual specificity phosphatase that dephophorylates MAP kinase in vivo. Cell 1993; 75:487-493.

128. Duff JL, Monia BP, Berk BC. Mitogen-activated protein (MAP) kinase is regulated by the MAP kinase phosphatase (MKP-1) in vascular smooth muscle cells. Effect of actinomycin D and antisense olionucleotides. J Biol Chem 1995; 270:7161-7166.

129. Rossomando AJ, Dent P, Sturgill TW et al. Mitogen-activated protein kinase kinase 1 (MKK1) is negatively regulated by threonine phosphorylation. Mol Cell Biol 1994; 14:1594-1602.

130. Druey KM, Blumer KJ, Kang VH et al. Inhibition of G-protein-mediated MAP kinase activation by a new mammalian gene family. Nature 1996; 379:742-746.

131. Ridley AJ, A Hall. Signal transduction pathways regulating Rho-mediated stress fibre formation: requirement for a tyrosine kinase. EMBO J 1994; 13:2600-2610.

132. Seckl M, Rozengurt E. Tyrphostin inhibits bombesin stimulation of tyrosine phosphorlation, c-fos expression, and DNA synthesis in Swiss 3T3 cells. J Biol Chem 1993; 268:9548-9554.

133. Burridge K, Turner CE, Romer LH. Tyrosine phosphorylation of paxillin and pp125FAK accompanies cell adhesion to extracellular matrix: a role in cytoskeletal assembly. J Cell Biol 1992; 119:893-903.

134. Clark EA, Brugge JS. Integrins and signal transduction pathways: the road taken. Science 1995; 268:233-239.

135. Twamley-Stein GM, Pepperkok R, Ansorge W et al. The Src family tyrosine kinases are required for platelet-derived growth factor-mediated signal transduction in NIH3T3 cells. Proc Natl Acad Sci 1993; 90:7696-7700.

136. Wilson LK, Luttrell DK, Parsons JT et al. Src tyrosine kinase, myristylation, and modulatory domains are required for enhanced mitogenic responsiveness to epidermal growth factor seen in cells overexpressing c-src. Mol Cell Biol 1989; 9:1536-1544.

137. Roche S, Koegl M, Barone MV et al. DNA synthesis induced by some but not all growth factors requires Src family protein tyrosine kinases. Mol Cell Biol 1995; 15:1102-1109.

138. Soriano P, Montgomery C, Geske R et al. Targeted disruption of the c-src proto-oncogene leads to osteopetrosis in mice. Cell 1991; 64:693-702.

139. McGlade J, Cheng A, Pelicci G et al. Shc proteins are phosphorylated and regulated by the v-Src and v-Fps protein-tyrosine kinases. Proc Natl Acad Sci 1992; 89:8869-8873.

140. Shenoy S, Chackalaparampil I, Bagrodia S et al. Role of p34cdc2-mediated phosphorylations in two-step activation of pp60c-src during mitosis. Proc Natl Acad Sci 1992; 89:7237-7241.

141. Chackalaparamil I, Bagrodia S, Shalloway D. Tyrosine dephosphorylation of pp60c-src is stimulated by a serine/threonine phosphatase inhibitor. Oncogene 1994; 9:1947-1955.

142. Sabe H, Hata A, Okada M et al. Analysis of the binding of the Src homology 2 domain of Csk to tyrosine-phosphorylated proteins in the suppression and mitotic activation of c-Src. Proc Natl Acad Sci 1994; 91:3984-3988.

143. Nada S, Okada M, MacAuley A et al. Cloning of a complementary DNA for a protein -tyrosine kinase that specifically phosphorylates a negative regulatory site of p60c-src. Nature 1991; 351:69-72.

144. Chow LML, Fournel M, Davidson D et al. Negative regulation of T-cell receptor signalling by tyrosine protein kinase p50csk. Nature 1993; 365:156-160.

145. Cooper JA. Howell B. The when and how of Src regulation. Cell 1993; 73:1051-1054.

146. den Hertog J, Pals C, Peppelenbosch MP et al. Receptor protein tyrosine phosphatase activates pp60-c-src and is involved in neuronal differentiation. EMBO J 1993; 12:3789-3798.

147. Bibbins KB, Boeuf H, Varmus HE. Binding of the Src SH2 domain to phosphopeptides is determined by residues in both the SH2 domain and the phosphopeptides. Mol Cell Biol 1993; 13:7278-7287.

148. Koch CA, Moran MF, Anderson D et al. Multiple SH2-mediated interactions in v-src-transformed cells. Mol Cell Biol 1992; 12:1366-1374.

149. Fukui Y, Hanafusa H. Requirement of phosphatidylinositol-3 kinase modification for its association with p6osrc. Mol Cell Biol 1991; 11:1972-1979.

150. Weng Z, Thomas SM, Rickles RJ et al. Identification of Src, Fyn, and Lyn SH3-binding proteins: implications for a function of SH3 domains. Mol Cell Biol 1994; 14:4509-4521.

151. Cantley LC, Auger KR, Carpenter C et al. Oncogenes and signal transduction. Cell 1991; 64:281-302.

152. Li W, Hu P, Skolnik EY et al. The SH2 and SH3 domain-containing Nck protein is oncogenic and a common target for phosphorylation by different surface receptors. Mol Cell Biol 1992; 12:5824-5833.

153. Chou MM, Fajardo JE, Hanafusa H. The SH2- and SH3-containing Nck protein transforms mammalian fibroblasts in the absence of elevated phosphotyrosine levels. Mol Cell Biol 1992; 12:5834-5842.

154. Meisenhelder J, Hunter T. The SH2/SH3 domain-containing protein Nck is recognized by certain anti-phospholipase C-gamma1 monoclonal antibodies and its phosphorylation on tyrosine is stimulated by platelet-derived growth factor and epidermal growth factor treatment. Mol Cell Biol 1992; 12:5843-5856.

155. Pelicci G, Lanfrancone L, Grignani F et al. A novel transforming protein (SHC) with an SH2 domain is implicated in mitogenic signal transduction. Cell 1992; 70:93-104.

156. Vogel W, Lammers R, Huang J et al. Activation of a phosphotyrosine phosphatase by tyrosine phosphorylation. Science 1993; 259:1611-1614.

157. Feng G-S, Hui C-C, Pawson T. SH2-containing phosphotyrosine phosphatase as a target of protein-tyrosine kinases. Science 1993; 259:1607-1611.

158. Taniguchi T. Cytokine signalling through nonreceptor protein tyrosine kinases. Science 1995; 268:251-255.

159. Marrero M, Schleffer B, Paxton W et al. 1995; Direct stimulation of JAK/STAT pathway by the angiotensin II AT1 receptor. Nature 375:247-250.

160. Ihle JN. STATs: signal transducers and activators of transcription. Cell 1996; 84:331-334.

161. David M, Petricoin E, Benjamin C et al. Requirement for MAP kinase (ERK2) activity in Interferon a- and Interferon β-stimulated gene expression through STAT proteins. Science 1995; 269:1721-1723.

162. Kapeller R, Cantley LC. Phosphatidylinositol 3-kinase. Bioessays 1994; 16:565-576.

163. Rodriquez-Viciana P, Waren PH, Dhand R et al. Phosphatidylinositol-3-OH kinase as a direct target of Ras. Nature 1994; 370:527-532.

164. Stephens L, Smrcka A, Cooke FT et al. A novel phosphoinositide 3 kinase activity in myeloid-derived cells is activated by G protein βg subunits. Cell 1994; 77:83-93.

165. Zhou G, Bao ZQ, Dixon JE. Components of a new human protein kinase signal transduction pathway. J Biol Chem 1995; 270:12665-12669.

166. Lechner C, Zahalka MA, Giot JF et al. ERK6, a mitogen-activated protein kinase involved in C2C12 myoblast differentiation. Proc Natl Acad Sci 1996; 93:4355-4359.

167. Deng T, Karin M. c-Fos transcriptional activity stimulated by H-Ras-activated protein kinase distinct from JNK and ERK. Nature 1994; 371:171-175.

168. Kyriakis JM, Banerjee P, Nikolakaki E et al. A MAP kinase subfamily activated by cellular stress and tumour necrosis factor. Nature 1994; 369:156-160.

169. Derijard B, Hibi M, Wu IH et al. JNK1: A protein kinase stimulated by UV light and Ha-Ras that binds and phosphorylates the c-Jun activation domain. Cell 1994; 76:1025-1037.

170. Han J, Lee JD, Bibbs L et al. A MAP kinase targeted by endotoxin and hyperosmolarity in mammalian cells. Science 1994; 265:808-811.

171. Freshney NW, Rawlinson L, Guesdon F et al. Interleukin-1 activates a novel protein kinase cascade that results in the phosphorylation of Hsp27. Cell 1994; 71:1039-1049.

172. Rouse J, Cohen P, Trigon S et al. A novel kinase cascade triggered by stress and heat shock that stimulates MAPKAP kinase-2 and phosphorylation of the small heat shock proteins. Cell 1994; 78:1027-1037.

173. Pombo CP, Bonventre JV, Avruch J et al. The stress-activated protein kinases (SAPKs) are major c-Jun amino-terminal kinases activated by ischemia and reperfusion. J Biol Chem 1994; 269:26546-26551.

174. Force T, Pombo CM, Avruch JA et al. Stress-activated protein kinases in cardiovascular disease. Circ Res 1996; 78:947-953.

175. Zohn IE, Yu H, Li X et al. Angiotensin II stimulates calcium-dependent activation of c-Jun N-terminal kinase. Mol Cell Biol 1995; 15:6160-6168.

176. Mitchell FM, Russell M, Johnson GL. Differential calcium dependence in the activation of c-Jun kinase and mitogen-activated protein kinase by muscarinic acetylcholine receptors in rat 1a cells. Biochem J 1995; 309:381-384.

177. Kyriakis JM, Avruch J. 1996; Sounding the alarm: protein kinase cascades activated by stress and inflammation. J Biol Chem 271:24313-24316.

178. Ichijo H, Nishida E, Irie K et al. Induction of apoptosis by ASK1, a Mammalian MAPKKK that activates SAPK/JNK and p38 signaling pathways. Science 1997; 275:90-94.

179. Yamaguchi K, Shirakabe K, Shibuya H et al. Indentification of a member of the MAPKKK family as a potential mediator of TGF-β signal transduction. Science 1995; 270:2008-2011.

180. Tibbles LA, Ing YL, Kiefer F et al. MLK-3 activates the SAPK/JNK and p38/RK pathways via SEK1 and MKK3/6. EMBO J 1996; 16:7026-7035.

181. Pombo C, Kehrl J, Sanchez I et al. Activation of the SAPK pathway by the human STE20 homolog germinal centre kinase. Nature 1995; 377:750-754.

182. Knaus U, Morris S, Dong HJ et al. Regulation of human leukocyte p21-activated kinases through G protein-coupled receptors. Science 1995; 269:221-223.

183. Minden A, Lin A, Claret F-X et al. Selective activation of the JNK signalling cascade and c-Jun transcriptional activity by the small GTPases Rac and Cdc42Hs. Cell 1995; 81:1147-1157.

184. Coso OA, Chiariello M, Yu JC et al. The small GTP-binding proteins Rac and Cdc42 regulate the activity of the JNK/SAPK signalling pathway. Cell 1995; 81:1137-1146.

185. Zhang S, Han J, Sells MA et al. Rho family GTPases regulate p38 MAP kinase through the downstream mediator Pak1. J Biol Chem 1995; 270:23934-23936.

186. Fleming IN, Cassondra M, Elliott M et al. Differential translocation of Rho family GTPases by lysophosphatidic acid, endothelin-1, and platelet-derived growth factor. J Biol Chem 1996; 271:33067-33073.

187. Kramer RM, Roberts EF, Um SL et al. p38 mitogen-activated protein kinase phosphorylates cytosolic phospholipase A2 in thrombin-stimulated platelets. J Biol Chem 1996; 271:27723-27729.

188. Kramer RM, Roberts EF, Strifler BA et al. Thrombin induces activation of p38 MAP kinase in human platelets. J Biol Chem 1995; 270:27395-27398.

189. Saklatvala J, Rawlinson L, Waller RJ et al. Role for p38 mitogen-activated protein kinase in platelet aggregation caused by collagen or a thromboxane analogue. J Biol Chem 1996; 271:6586-6589.

190. Coso OA, Teramoto H, Simonds WF et al. Signaling from G protein-coupled receptors to c-Jun kinase involves βg subunits of heterotrimeric G proteins acting on a Ras and Rac1-dependent pathway. J Biol Chem 1995; 271:3963-3966.

191. Prasad MVVSV, Dermott JM, Heasley LE et al. Activation of Jun kinase/stress-activated protein kinase by GTPase-deficient mutants of Gα12 and Gα13. J Biol Chem 1995; 270:18655-18659.

192. Pulverer B, Kyriakis JM, Avruch J et al. Phophorylation of c-jun mediated by MAP kinases. Nature 1991; 353:670-674.

193. Gupta S, Campbell D, Derijard B et al. Transcription factor ATF2 regulation by the JNK signal transduction pathway. Science 1995; 267:389-393.

194. Morooka H, Bonventre JV, Pombo CM et al. Ischemia and reperfusion enhance ATF-2 and c-Jun binding to cAMP response elements and to an AP-1 binding site from the c-jun promoter. J Biol Chem 1995; 270:30084-30092.

195. Abdel-Hafiz HA, Heasley LE, Kyriakis JM et al. Activating transcription factor-2 DNA-binding activity is stimulated by phosphorylation catalyzed by p42 and p54 microtubule-associated protein kinases. Mol Endo 1992; 6:2079-2089.

196. Binetruy B, Smeal T, Karin M. Ha-Ras augments c-jun activity and stimulates phosphorylation of its activation domain. Nature 1991; 351:122-127.

197. Livingstone C, Patel G, Jones N. ATF2 contains a phosphorylation dependent transcriptional activation domain. EMBO J 1995; 14:1785-1797.

198. van Dam H, Wilhelm D, Herr I et al. ATF-2 is preferentially activated by stress-activated protein kinases to mediate c-jun induction in response to genotoxic agents. EMBO J 1995; 14:1798-1811.

199. Abdel-Hafiz HA, Chen CY, Marcell T et al. Structural determinants outside of the leucine zipper influence the interactions of CREB and ATF-2: interaction of CREB with ATF-2 blocks E1a-ATF-2 complex formation. Oncogene 1993; 8:1161-1174.

200. Price M, Cruzalegui FH, Treisman R. The p38 and ERK MAP kinase pathways cooperate to activate ternary complex factors and c-fos transcription in response to UV light. EMBO J 1996; 15:6552-6563.

201. Zervos AS, Faccio L, Kyriakis JM et al. MXI2, a MAP kinase that recognizes and phosphorylates Max. Proc Natl Acad Sci 1995; 92:10531-10534.

202. Tan Y, Rouse J, Zhang A et al. FGF and stress regulate CREB and ATF-1 via a pathway involving p38 MAP kinase and MAPKAP kinase-2. Embo J 1996; 15:4629-4642.

203. Lin L-L, Wartmann M, Lin AY et al. cPLA2 is phosphorylated and activated by MAP kinase. Cell 1993; 72:269-278.

204. Johnson NL, Gardner AM, Diener KM et al. Signal transduction pathways regulated by mitogen-activated/extracellular response kinase kinase kinase induce cell death. J Biol Chem 1996; 271:3229-3237.

205. Xia Z, Dickens M, Raingeaud J et al. Opposing effects of ERK and JNK-p38 MAP kinases on apoptosis. Science 1995; 270:1326-1331.

206. Verheij M, Bose R, Hua Lin X et al. Requirement for ceramide-initiated SAPK/JNK signalling in stress-induced apoptosis. Nature 1996; 380:75-79.

207. Zanke BW, Boudreau K, Rubie E et al. The stress-activated protein kinase pathway mediates cell death following injury induced by cis-platinum, UV irradiation or heat. Current Biol 1996; 6:5:505-613.

208. Zha J, Harada H, Yang E et al. Serine phosphorylation of death agonist BAD in response to survival factor results in binding to 14-3-3 not BCL-X. Cell 1996; 87:619-628.

209. Wang H-G, Rapp UR, R. J.C. Bcl-2 targets the protein kinase Raf-1 to mitochondria. Cell 1996; 87:629-638.

210. Yamada T, Horiuchi M, Dzau VJ. Angiotensin II type 2 receptor mediates programmed cell death. Proc Natl Acad Sci 1996; 93:156-160.

211. Xu Q, Liu Y, Gorospe M et al. Acute hypertension activates mitogen-activated protein kinases in arterial wall. J Clin Invest 1996; 97:508-514.

212. Bogoyevitch MA, Gillespie-Brown J, Ketterman AJ et al. Stimulation of the stress-activated mitogen-activated protein kinase subfamilies in perfused heart. Circ Res 1996; 79:162-173.

213. Bogoyevitch MA, Ketterman AJ, Sugden PH. Cellular stresses differentially activate c-Jun N-terminal protein kinases and extracellular signal-regulated protein kinases in cultured ventricular myocytes. J Biol Chem 1995; 270:29710-29717.

214. Molnar A, Theodoras AM, Zon LI et al. Cdc42Hs, but not Rac1, inhibits werum-stimulated cell cycle pregression at G1/S through a mechanism requiring p38/RK. J Biol Chem 1997; 272:in press.

215. Malarkey K, Belham CM, Paul A et al. The regulation of tyrosine kinase signalling pathways by growth factor and G-protein-coupled receptors. Biochem J 1995; 309:361-375.

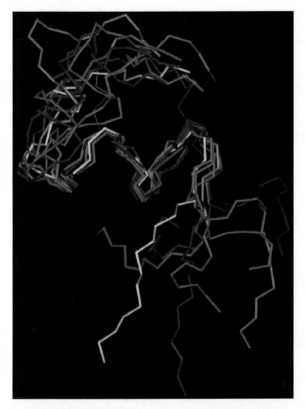

A

Fig. 6.4. NMR and X-ray crystal structure of ET-1. (A) NMR of the peptide backbone for eight low energy conformers of endothelin-1 superimposed for best fit of the helix (residues 9-15). Each structure is displayed in a different color with the disulfides shown in one structure as a thin red line. (B) X-ray structure of endothelin. The backbone trace is in magenta and the side chains are colored by functional group as follows: aromatic = orange, basic = blue, acidic = red, disulfides = yellow, aliphatic and hydrophobic = green, neutral and hydrophilic = light blue.

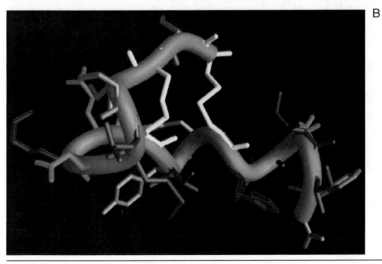

B

Endothelin Receptors and Signaling Mechanisms, edited by David M. Pollock and Robert F. Highsmith. © 1998 Springer-Verlag and R.G. Landes Company.

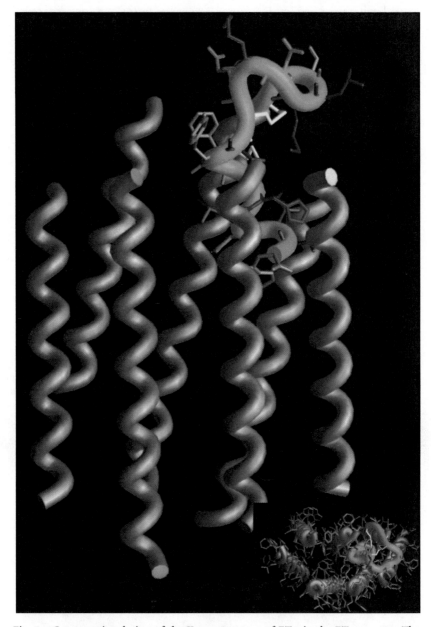

Fig. 6.5. Cross-sectional view of the X-ray structure of ET-1 in the ET$_A$ receptor. The backbone trace of the receptor is colored blue and the sidechains of the receptor were removed for clarity. The side chains and backbone trace of ET-1 are colored as in Figure 6.4. The inset is a view of the receptor from the extracellular surface. The C-terminal tail of endothelin inserts into the cavity formed by transmembrane domain 1, 2, 3, and 7.

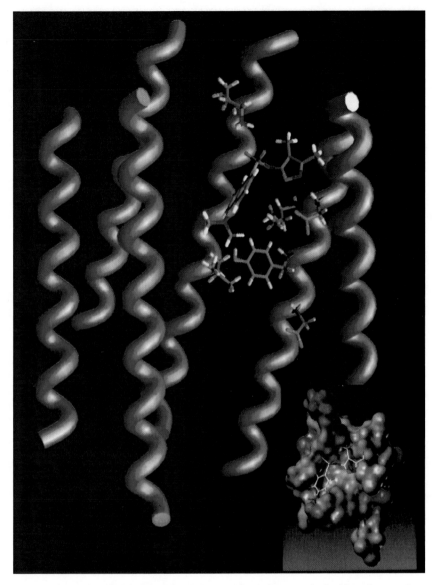

Fig. 6.6. Molecular model of the ET_A receptor with BMS-182874 docked into the putative binding cavity. The backbone trace of the transmembrane regions are colored blue with transmembrane region 7 removed for clarity. Selected side chains of the receptor that interact with BMS-182874 are displayed. BMS-182874 is colored as follows: C = green, N = blue, S = orange, O = red, and H = yellow. The inset shows the molecular surface of the binding pocket in the ET_A receptor colored by lipophilic potential (green & orange = lipophilic, red & blue = polar). The cavity has a complementary shape to the ligand and contains both hydrophilic and hydrophobic interactions.

Index

A

A-127722 24, 29-31, 33-36
A-182086 9
A-192621 9
A-Raf 155, 191
A7r5 118, 120, 136, 137
ABT-627 8
Acetylcholine 41, 57, 61
Adenylyl cyclase 123, 151, 153, 154, 190
ADP-ribose 136
Aldosterone 59, 60, 67, 124, 167
Amiloride 106, 152
Angioplasty 8, 57, 58, 95
Angiotensin converting enzyme 57
Angiotensin II (Ang II) 1, 24
Antisense oligodeoxynucleotides 54, 55
Atherosclerosis 57, 89, 92, 94, 96, 101, 147, 164
Atrial natriuretic peptide (ANP) 10, 164
AVP 105-108, 179

B

β adrenergic receptor kinase (βARK) 185
β-adrenergic stimulation 149, 154
Bay K 8644 117
Big ET-1 58, 91
BMS 182874 8
BQ-123 4, 6, 7, 17-19, 24, 28, 29-31, 33, 53, 54, 58, 66, 70, 71, 78
BQ-3020 25
BQ-788 7, 54
Bradykinin 1, 41, 57, 121, 122, 187

C

c-fos 163, 170-172, 179-181, 186, 187, 192, 193, 202, 206, 207
c-Raf-1 177, 178, 180-188, 190-194, 198, 203, 204
c-Src 167-172
c-Yes 169
Ca^{2+} 43, 45-51, 103-107, 115-125, 131, 132, 134-140, 147-150, 152-154, 169-172, 177, 179-181, 184, 185, 187
Ca^{2+} channels 105, 115-125

Calcium 28, 41, 43, 45, 46, 49, 60, 103, 115, 131, 132, 134-138, 185, 200
Calmidazolium 43, 45
cAMP response element binding protein (CREB) 180, 192
Carboxy-terminal Src kinase (Csk) 195
Cardiac contractility 148, 154
Cardiac output 57-59, 149
Caveolae 25, 68
Cell cycle 164, 165, 168, 169, 172, 191, 193, 204
CHO 8, 24, 35, 42-49, 54, 55, 69, 178, 187, 189
Cholera toxin (CTX) 134
Cl$^-$ channels 120, 121, 125
CMZ 43, 45, 46
Collecting duct 10, 11, 51, 101, 103, 105-109
Congestive heart failure (CHF) 57, 58, 61
COOH-terminal Src kinase (Csk) 170
cAMP 104
Cyclic AMP (cAMP) 123
Cyclic GMP 47, 48, 54
Cyclooxygenase 2, 68, 103, 105, 107, 136
Cyclosporine 60, 103
Cytokines 59, 91-95, 197, 199, 202

D

Dihydropyridine (DHP) 116
Diltiazem 116, 117

E

Endothelin agonists 3, 108
Endothelin converting enzyme (ECE) 1
Endotoxemia 59, 60, 91, 95
ENOS 41-46, 49, 53-56, 59, 61
Epidermal growth factor (EGF) 186
ET-1 1-4, 6-8, 10, 11, 17-19, 23-25, 28-36, 41-56, 58-61, 67-71, 75, 77-79, 81, 91, 94, 95, 101-110, 115-124, 131-139, 163-172, 177-182, 184, 186-196, 198-201, 203, 204
ET-2 1, 3, 10, 23, 68-70, 75, 131, 164
ET-3 1-3, 5, 10, 11, 17-19, 23-25, 42, 53, 54, 61, 68-71, 74, 75, 102, 104, 107, 108, 116, 121, 131, 133, 135, 165, 178

ET$_A$ receptors 3, 4, 6-8, 10, 11, 17, 19, 23,
      47-49, 51, 70, 78, 110, 118, 133, 134, 137,
      140, 150, 166
ET$_B$ receptors 3-8, 10, 11, 17-19, 21, 23, 24,
      28, 43, 44, 46, 53, 54, 69-71, 74, 75, 78,
      102, 103, 106-108, 110, 121, 133, 140, 150,
      151, 164, 171
ET$_C$ 5, 18, 19, 71, 132, 133
Extracellular signal regulated kinase
      (ERK) 177, 180-182, 184-194, 197-204

F

Fibronectin 167
Focal adhesion kinase (FAK) 168, 169,
      172, 181, 194
FR139317 4, 6, 7, 24, 29, 30

G

G-protein 2, 24, 68, 131-135, 137-139, 151,
      154, 155, 177, 184, 186, 187, 191, 192, 194,
      201, 204
Gene expression 61, 137, 163-165, 170-172,
      196, 198
Genistein (GEN) 44, 46, 170, 181
Glomerular filtration rate (GFR) 4, 108
Growth factor receptor bound protein 2
      (Grb2) 171, 183, 185-187, 190, 195, 196
GTP 24, 29, 48, 49, 68, 133, 139, 151, 182-185,
      188, 190, 199, 201
GTPase 133, 134, 183, 189, 201
Guanine nucleotide exchange factor
      (GEF) 183

H

Heart failure 1, 57, 58, 89, 93, 110, 133, 147,
      150, 154, 164
Hepato-renal syndrome 59, 60
Hirschsprung's disease 74, 163, 165
Human umbilical vein endothelial cells
      (HUVEC) 42, 44-46, 55, 92, 122
Hypertension 1, 2, 8, 41, 57, 59-61, 89, 95,
      96, 101, 133, 147, 154

I

ICRAC 180
IFN-α 196
IFN-γ 196
IL-2 92, 190
Imodipine 116
Indomethacin 108
Inducible NO synthase (iNOS) 59
Inositol 1,4,5-triphosphate (IP3) 131, 132,
      134, 135
Intracellular free calcium concentration
      131
Ionomycin 43, 45, 46, 169
IRL-1620 17, 19

J

Janus kinase 167, 169, 181, 194, 196

K

K$^+$ channel 120-125, 133, 152

L

L-arginine 43, 53, 61
Lipoxygenase 103, 136
Liver cirrhosis 59-61

M

Mastoparan 48, 49
MEK 155, 182, 186, 191-193, 200
MEK kinase-1 (MEKK-1) 191
Mesangial cells 116, 120, 122, 123, 133, 136,
      151, 164, 165, 167-171, 178-182, 186, 187,
      189, 192, 194, 198, 200
Migration 53-57
Mitogen activated protein kinase (MAP)
      132, 148, 153, 155, 165, 171, 172, 177, 181,
      191-193, 199, 200, 202, 204
mSOS 183, 185-187, 190, 195, 196
Myofibrillar proteins 148, 152, 153
Myofilament 139, 147, 149, 154

N

Na$^+$ channels 116
Na$^+$ H$^+$ antiporter 148, 152, 153
Na$^+$/H$^+$ exchange 102, 103
Na$^+$/K$^+$ ATPase 102
Neointimal proliferation 95
Nicardipine 116, 117, 136, 180
Nitric oxide (NO) 1, 41, 164
NO synthase 41, 46, 49, 52-54, 57, 59
Nuclear signaling 163, 167, 170-172

O

Osmotic water permeability 107

P

P38 177, 199-204
PD 012527 8
PD 142893 4, 70
PD 145065 6, 7, 33
PD 156707 8, 24, 29, 30, 33, 34, 36
Pertussis toxin (PTX) 48, 107, 133-135, 137, 151, 153, 154, 198, 201
Phorbol esters 132, 137, 139, 152-155, 180, 188, 189, 192
Phosphatidic acid (PA) 132, 137, 155
Phosphatidylinositol (Pl) 131, 134, 148, 151, 198
Phosphatidylinositol 4,5-bisphosphate (PIP2) 131, 134, 135, 138, 148, 155
Phospholamban 149
Phospholipase A2 (PLA) 103, 107, 132, 133, 136, 182, 184, 202
Phospholipase C (PLC) 107, 121, 131, 132, 134, 135, 137, 138, 151, 155, 190
Phospholipase D (PLD) 132, 137, 138, 148, 154, 155
Plasma ET-1 2, 59, 60, 103
Platelet activating factor 1
Platelet derived growth factor (PDGF) 92, 178, 179, 182, 183, 187, 194, 199
Podokinesis 55
Prolactin 67, 124, 197
Prostacyclin 1, 4, 67, 133, 140
Prostaglandin 3, 57, 67, 107, 132, 136
Prostaglandin E2 57, 107
Protein kinase A (PKA) 104, 105, 123, 135, 149, 152, 154, 190, 193, 202
Protein kinase C (PKC) 102-107, 116, 121, 132, 134-139, 148, 150-155, 165, 166, 169, 171, 172, 177-182, 184, 185, 188-190, 200, 201, 203

Protein tyrosine kinase (PTK) 46, 134, 137, 148, 155, 163, 165, 167-172, 186, 188
Proximal tubule 102-106
Pulmonary hypertension 1, 89, 95, 101, 133
Pyk2 181, 186, 187, 194

R

Raf-1 167, 171, 172, 177, 178, 180-194, 198, 203, 204
Ras 171, 172, 182-191, 193, 195, 198, 201
Rat microvascular endothelial cells (RMVEC) 54
Rat-1 fibroblast 166, 188
Receptor internalization 31, 33
Renal medulla 10, 51, 53, 61, 106
Renal plasma flow (RPF) 108
Reperfusion injury 89, 95, 147
RES701 19
Restenosis 8, 95, 96, 147, 164
Retinoic acid 165
Ro 46-8443 6, 7
Ro 47-0203 (Bosentan) 6, 8, 29, 30-34, 58
RPTPα 196
Ryanodine Ca2$^+$ 136

S

Sarafatoxin 69
SB 209670 8, 57, 78
SB 217242 8, 34
Septic shock 95, 96
Shc 171, 185-187, 190, 195-198
Shear stress 41, 44, 57
SIN-1 47, 48, 49
Sn1,2-diacylglycerol (DAG) 104, 105, 107, 131, 132, 134, 137, 138, 148, 151, 152, 155, 185
SNP 47, 48
Sodium nitroprusside 47, 53, 61
Soluble guanylyl cyclase 41
Src 167-172, 181, 183, 186-188, 190, 194-196, 198
STAT 194, 196-198
Stress-activated protein kinase (SAPK) 177, 190, 199-204
Suramin 179
Syk 186, 187, 195

T

Thrombin 44, 103, 185, 192, 201
Tissue plasminogen activator 1
Transcription factors 41, 155, 181, 192, 193, 202
Transforming growth factor β (TGFβ) 92, 106, 200
Troponin I (Tn-1) 149, 152
Tyrosine kinase 44, 46, 103, 137, 139, 148, 180-183, 186-188, 194-196, 198
Tyrosine phosphorylation (P-Tyr) 167-169, 171, 172

V

Vascular endothelial growth factor (VEGF) 55
Vasopressin 60, 92, 105, 118, 124, 179, 189
Verapamil 116, 117, 119, 136
Voltage-gated Ca2+ channels 115, 122

W

Wound healing 54, 55, 57, 166

Z

ZAP-70 186, 195